The Arbitration Act 1996

A COMMENTARY

The Arbitration Act 1996

A COMMENTARY

Second Edition

Bruce Harris, *FCIArb, FRSA*
Rowan Planterose, *MA (Cantab), LLB, FCIArb, Barrister*
Jonathan Tecks, *MA (Cantab), FCIArb, Barrister*

FOREWORD by
The Rt. Hon. The Lord Bingham of Cornhill
Lord Chief Justice of England

PUBLISHED IN CONJUNCTION WITH
THE CHARTERED INSTITUTE OF ARBITRATORS

Blackwell
Science

© 1996, 2000 by Bruce Harris,
Rowan Planterose and Jonathan Tecks

Blackwell Science Ltd
Editorial Offices:
Osney Mead, Oxford OX2 0EL
25 John Street, London WC1N 2BL
23 Ainslie Place, Edinburgh EH3 6AJ
350 Main Street, Malden
   MA 02148 5018, USA
54 University Street, Carlton
   Victoria 3053, Australia
10 rue Casimir Delavigne
   75006 Paris, France

Other Editorial Offices:

Blackwell Wissenschafts-Verlag GmbH
Kurfüstendamm 57
10707 Berlin, Germany

Blackwell Science KK
MG Kodenmacho Building
7–10 Kodenmacho Nihombashi
Chuo-ku, Tokyo 104, Japan

First edition published 1996
Reprinted 1997 (twice)
Second edition published 2000

Set in 10 on 12pt Palatino
by DP Photosetting, Aylesbury, Bucks
Printed and bound in Great Britain by
MPG Books Ltd, Bodmin, Cornwall

DISTRIBUTORS

Marston Book Services Ltd
PO Box 269
Abingdon
Oxon OX14 4YN
(Orders: Tel: 01235 465500
     Fax: 01235 465555)

USA
Blackwell Science, Inc.
Commerce Place
350 Main Street
Malden, MA 02148 5018
(*Orders:* Tel: 800 759 6102
       781 388 8250
    Fax: 781 388 8255)

Canada
Login Brothers Book Company
324 Saulteaux Crescent
Winnipeg, Manitoba R3J 3T2
(*Orders:* Tel: 204 837-2987
    Fax: 204 837-3116)

Australia
Blackwell Science Pty Ltd
54 University Street
Carlton, Victoria 3053
(*Orders:* Tel: 03 9347 0300
    Fax: 03 9347 5001)

A catalogue record for this title is available
from the British Library

ISBN 0-632-05063-2

Library of Congress
Cataloging-in-Publication Data
is available

For further information on Blackwell Science,
visit our website: www.blackwell-science.com

To:

Javier
Elizabeth
Frances, Peter and Caroline

Contents

Contents

Biographical Note

Bruce Harris is a full-time commercial and maritime arbitrator who has been involved in more than 7000 arbitrations and made more than 1300 awards. He has written, lectured and talked extensively on arbitration, and was President of the London Maritime Arbitrators Association in 1990/2 and Chairman of the Chartered Institute of Arbitrators in 1992/3. He was a member of the DAC.

Rowan Planterose practises from 12 Gray's Inn Square, Gray's Inn, London WC1R 5JP. He is a commercial and construction lawyer, and much of his time is devoted to arbitration, either as counsel or as arbitrator. He is a member of the Council of the Chartered Institute of Arbitrators and has sat on its Professional Development Committee for several years, directing or tutoring on many of its courses on arbitration, under the law prior to the new Act, under the Model Law and under the present Act.

Jonathan Tecks practises from 12 Gray's Inn Square, Gray's Inn, London WC1R 5JP. He is a commercial and construction lawyer who arbitrates in respect of contract claims and financial services. A Chartered Arbitrator, he directs and lectures on Institute training courses for arbitrators. He holds a Senior Lectureship in Law at the University of the West of England.

Foreword to Second Edition

by the Rt. Hon. The Lord Bingham of Cornhill
Lord Chief Justice of England
Former Master of the Rolls
Past President of the Chartered Institute of Arbitrators

If (as Samuel Johnson memorably observed) a second marriage may be a triumph of hope over experience, the second edition of a book such as this is a triumph of hope vindicated by experience – for it shows that those who have bought or used the first edition are expected to come back for a second helping, and that those who have not yet tasted the feast are expected to dine.

The first edition, written when the 1996 Act was on the point of coming into force, seized an historic opportunity. So does the second. The Civil Procedure Rules are changing the face of civil litigation. They will provide new challenges, but also new opportunities and new lessons for the practice of arbitration. To a greater extent than its predecessor, this edition takes note of what is happening in the courts. And, valuably, the new edition plots the progress of the 1996 Act over its first three years of active existence.

This second edition is as welcome as the first. It should enjoy the same success.

Foreword to First Edition

by the Rt. Hon. The Lord Bingham of Cornhill
Lord Chief Justice of England

The authors of this book predict that the Arbitration Act 1996 will lead to a revolution in the practice of arbitration in England, Wales and Northern Ireland. They are right. The Act is a new departure, a fresh and very hopeful attempt to lay down a clear, flexible, fair, efficient, modern, accessible and intelligible set of rules to govern the conduct of arbitration in this country.

But those who conceived and framed the 1996 Act have, in the manner of all wise revolutionaries, sought to retain what was best in our existing law of arbitration while making changes to bring the practice of arbitration closer to the older and simpler ideals by which it was once animated. So the emphasis is on the speedy and cost-effective resolution of disputes by an impartial (probably expert, often non-legal) tribunal, in accordance with procedures very largely determined by the parties, and with no, or at worst minimal, intervention by the courts.

The 1996 Act provides all those involved in the practical conduct of arbitration in this country with a new set of tools. The success of the Act will depend on those tools being skilfully used to fashion the product for which they were designed. This means, above all, that they should be knowledgeably used, with an understanding of their origin and of why they were designed as they were. It also requires an understanding of the possible risks, and the available safeguards against misuse.

This is an intensely practical and admirably user-friendly book, written to guide practitioners and arbitrators as they convert from the old practice of arbitration to the new. Like the Act itself, the book is written in clear and simple language; it supplies the reader with the references he or she is likely to need, and leaves out those likely to be unhelpful; it eschews academic debate, while sounding occasional notes of caution; it very largely dispenses with the citation of previous decided cases. In style and content this is a fitting companion-piece to a modern and workmanlike Act. It is much to be hoped that they will, together, raise the reputation of arbitration in this country to new and unprecedented heights.

Preface to Second Edition

In the Introduction to the first edition of this work we said that a revolution in the practice of arbitration was the likely consequence of the Arbitration Act 1996. But we also anticipated that the changes in practice that we saw as being required were likely to come about relatively slowly. Experience has not shown us to be wrong in that regard.

Shifts in the culture of arbitration in England, Wales and Northern Ireland are beginning to become apparent. Whilst they are presently manifesting somewhat sporadically, they do so with increasing frequency and force. Many arbitrators are beginning to take control of their cases to a greater extent than before, and to use the possibilities provided by the Act to advance their cases in accordance with the Act's spirit. Sometimes this tendency seems to be more apparent amongst lay arbitrators than it is amongst those from the legal profession. That observation is reflected in the fact that, even now, many practitioners seek to run cases as they would have done under the previous legislation.

But the climate is changing. The Woolf reforms (the Civil Procedure Rules, or 'CPR') have been introduced in the courts and can only help change attitudes. The various institutions responsible for appointing arbitrators and administering arbitrations have changed their rules in the light of the Act. There have been three years of experience of the Act, in which a not insubstantial number of cases on it have been decided by the courts. All these factors more than justify this second edition.

We have updated the text of the original Commentary so as: (i) to incorporate references to the significant decided cases (and a Table of Cases now appears immediately before the index); (ii) to deal with the new institutional rules etc.; and (iii) to include detailed references to the appropriate parts of the CPR. In this latter respect, we have included at Section H of the Materials a full note on the CPR and arbitration applications. Detailed descriptions of procedures to be followed in respect of particular arbitration applications appear at appropriate places throughout the commentary.

We have also taken the opportunity to revise some original text where it appeared out of date or where, on reflection, we have thought it could be improved or brought up-to-date. For example, we have used words and phrases found in the CPR rather than their predecessors. We have also discontinued references to foreign cases under the Model Law, and have dealt with the latest Unfair Terms in Consumer Contracts Regulations and the Contracts (Rights of Third Parties) Act 1999.

As a result more than four-fifths of the sections which follow have been altered to a greater or lesser extent. More or less substantial changes have been made to the commentary on ss. 1, 3, 6, 7, 9–14, 18, 20, 21, 24, 25, 27–39, 41–53, 55–59, 61, 63–74, 76–80, 84–92, 103 and 105. More modest changes have been effected to the commentary on ss. 2, 4, 5, 8, 10, 11, 13, 19, 23, 26, 40, 54, 60, 81, 93, 94, 96, 99, 101, 102, 104, 107 and 109.

There are also some modest changes to the Tables and to sections A, B and D of the Materials, as well as the additional note (mentioned above) on the CPR and arbitration applications in section H.

We have retained the original Preface unamended. We have also kept the original Introduction, although that has had to be subject to a number of changes because, when it was first written, it was looking forward from a time when the Act had not come into effect. We believe that both these parts of the work can be of continued assistance to both students and practitioners.

Bruce Harris
Rowan Planterose
Jonathan Tecks   *December 1999*

Preface to First Edition

For some years the provisions of a restated arbitration law have been taking shape, culminating in the Arbitration Act 1996. The Act introduces a radical new approach designed to keep arbitration in England, Wales and Northern Ireland in the forefront of choice for international parties, and new powers for arbitrators to permit them to redefine arbitration as a dispute resolution process with unique qualities.

It occurred to us, who are variously involved in the practice of arbitration and the teaching of its techniques and skills, that a commentary on the new law was called for, written with the participants in the arbitration process in mind. We have set out both to explain the philosophy of the new Act with its background in the UNCITRAL Model Law and international arbitration concepts, and to point out its practical effects and the changes it makes. We have also sought to provide materials that will help with implementing the Act.

Thus our section by section commentary moves from basic information about the 'status' (*mandatory* or *non-mandatory*) and the derivation of each section, through a brief summary of its effect, to a detailed consideration of the main points to be noted. We have outlined the various regimes established by the Act and, by cross-referring different sections, have tried to show the pattern of its operation. Where appropriate, we have included comparisons with Model Law provisions, and have commented on the application of institutional arbitration rules.

The 'Tables' section comprises a brief summary of basic information about the Act and the major changes introduced by it. There are also tables that allow ready reference to the sources of the Act in previous legislation and in the Model Law, and that indicate where those sources may now be found in the Act.

One of the particular features of the Act is the range of choices made available to arbitrators and parties at various stages and in respect of various aspects of the arbitration. Our 'Materials' section includes checklists dawn up to indicate those choices in the early stages of the arbitration; as to the powers of the tribunal and the court; in respect of preliminary meetings; and in relation to costs. We have also made drafting suggestions in relation to arbitration clauses, agreements on resignation and awards. Further drafting suggestions appear – coupled with commentary – in the main text concerning peremptory orders (41E); interest (49F); 'capping' orders (65D); and costs awards (61E).

It remains for us to thank all our partners, family members and friends who have so generously given us their support and understanding during the writing of this book.

Bruce Harris
Rowan Planterose
Jonathan Tecks   *September 1996*

References and Abbreviations

Mustill, Sir Michael J. & Boyd, Stewart C. (1989) *Commercial Arbitration*, 2nd edn. Butterworths, London.

Redfern, Alan & Hunter, Martin (1991) *The Law and Practice of International Commercial Arbitration*, 2nd edn. Sweet & Maxwell, London.

CIArb	Chartered Institute of Arbitrators
CIMAR	Construction Industry Model Arbitration Rules
CPR	Civil Procedure Rules
DAC	Departmental Advisory Committee
FCEC	Federation of Civil Engineering Contractors
IBA	International Bar Association
ICC	International Chamber of Commerce
ICE	Institution of Civil Engineers
LCIA	London Court of International Arbitrators
LMAA	London Maritime Arbitrators Association
RSC	Rules of the Supreme Court
UNCITRAL	United Nations Commission on International Trade Law

Law Report Abbreviations

AC	Appeal Cases
ADRLJ	The Arbitration and Dispute Resolution Law Journal
All ER	All England Law Reports
All ER (Comm.)	All England Law Reports Commercial Cases
BLR	Building Law Reports
CILL	Construction Industry Law Letter
CLC	Commercial Law Cases
CLD	Construction Law Digest
Con LR	Construction Law Reports
E&B	Ellis and Blackburn (reprinted in English Reports at volumes 118 to 120)

EG	Estates Gazette
HKLR	Hong Kong Law Reports
HL Cas	House of Lords Cases
KB	King's Bench Cases
Lloyd's Rep	Lloyd's Law Reports
LMLN	Lloyd's Maritime Law Newsletter
P	Probate
QB	Queen's Bench Cases
WLR	Weekly Law Reports
WWR	Western Weekly Reports (Canada)

Part 1

Introduction

When, over 18 years ago, the United Nations Commission on International Trade Law (UNCITRAL) decided to prepare a draft uniform law on arbitration procedure, its members could hardly have imagined that their proposal would lead to a revolution in the practice of arbitration in England. Yet such a revolution is the likely consequence of the Arbitration Act 1996, and that Act would almost certainly never have come into being but for the Model Law which UNCITRAL produced in 1985.

The Model Law

Whilst the United Kingdom took a full part in the discussions at UNCITRAL which led to the Model Law, it was finally decided that it should not be adopted (save in Scotland) when the government accepted the recommendation to that effect contained in the June 1989 Report (the Mustill Report) of the Departmental Advisory Committee (DAC) which was chaired at that time by Lord Justice Mustill, as he then was.

This recommendation was based on a number of considerations. Whilst the Model Law was seen as being possibly suitable for states with no developed law or practice of arbitration, for those with a reasonably modern law but not much practice and for those with outdated or inaccessible laws, it was not thought suitable for a country such as England, where the law of arbitration is up to date and where there is extensive current practice.

The Model Law was also seen to be incomplete in many respects; it was not obvious that changing what was generally regarded as a satisfactory system of law would either attract arbitrations to England or prevent them from being driven away; beneficial provisions of the Model Law largely existed already in English law, and other aspects of it were thought to be detrimental or at least debatable.

1

The Mustill recommendation

However, though rejecting adoption of the Model Law for England, the DAC considered that because English law on arbitration was widely dispersed (in statutes and, more particularly, in cases) and fragmentary, the situation could be improved by the adoption of a new Arbitration Act. Whilst suggesting that the idea of a full code setting out the whole of English arbitration law should be retained as a long-term objective, it recommended an intermediate solution:

> 'in the shape of a new Act with a subject-matter so selected as to make the essentials of at least the existing statutory arbitration law tolerably accessible, without calling for a lengthy period of planning and drafting, or prolonged parliamentary debate. It should in particular have the following features.
>
> (1) It should comprise a statement in statutory form of the more important principles of the English law of arbitration, statutory and (to the extent practicable) common law.
> (2) It should be limited to those principles whose existence and effect are uncontroversial.
> (3) It should be set out in logical order, and expressed in language which is sufficiently clear and free from technicalities to be readily comprehensible to the layman.
> (4) It should in general apply to domestic and international arbitrations alike, although there may have to be exceptions to take account of treaty obligations.
> (5) It should not be limited to the subject-matter of the Model Law.
> (6) It should embody such of our proposals for legislation as have by then been enacted . . .
> (7) Consideration should be given to ensuring that any such new statute should, so far as possible, have the same structure and language as the Model Law, so as to enhance its accessibility to those who are familiar with the Model Law.'

The Mustill Report continued by saying that the DAC was satisfied that these requirements could be met and 'within a time-scale which would answer the needs of keeping English arbitration law up to date and remaining in the vanguard of the various systems currently enjoying the preference of regular international users.'

The drafting exercise

No-one should hereafter be tempted to cite the seven year gap between this Report and the enactment of the 1996 Act in support of any argument to the effect that seven years is not 'a lengthy period'. Delay in proceeding in accordance with the DAC's recommendation first arose because of the reluctance of government to undertake the drafting exercise required. It was, indeed, only thanks to the efforts of a private group organised by Mr Arthur Marriott, an English solicitor, that funds were raised enabling a private drafting exercise to start. This group's contribution – and in particular that of Mr Marriott – should not go unacknowledged by anyone commenting on the new Act. Without them, England would almost certainly still be trying to survive with its previous legislation.

But even once under way, the drafting process did not go smoothly. The approach of the first former Parliamentary Counsel to be engaged was such that an alternative draughtsman had to be found. The result was a draft Bill whose title said it was 'to consolidate, with amendments, the Arbitration Act 1950, the Arbitration Act 1975, the Arbitration Act 1979 and related enactments'.

That draft and a consultation document were circulated for comments in February 1994. By April 1995, the DAC (by now under the chairmanship of Lord Justice Saville) reported upon the reactions to that draft. Although it was undoubtedly a highly skilful piece of work, the DAC said, it appeared that the draft did not carry into effect what most users in fact wanted. What was called for was much more along the lines of a restatement of the law, in clear and 'user-friendly' language, following, as far as possible, the structure and spirit of the Model Law, rather than simply a classic exercise in consolidation.

Thanks to the enlightened approach of Lord Bingham, now Lord Chief Justice but then Master of the Rolls, considerable time was made available to Lord Justice Saville which enabled him and Mr Toby Landau of the Bar to prepare their own, entirely fresh, provisional draft. That was then worked upon by Parliamentary Counsel, Mr Geoffrey Sellers, for the government had meanwhile taken on the burden of the exercise, thus relieving the private group of the need to find further sponsorship.

In the result, a new draft Bill was circulated for public consultation in July 1995. This had the title:

'An Act to restate and improve the law relating to arbitration pursuant to an arbitration agreement...'

which may be contrasted with the title to the February 1994 draft. This draft met with an extremely favourable response and formed the basis of what is now the Arbitration Act 1996. The Act governs arbitrations commenced on and after 31 January 1997.

Satisfying the Mustill Report requirements

There is no doubt that, in the main, the seven requirements identified by the Mustill Report as being the features that should characterise any new legislation have been met. It seems to us that no important principle of English arbitration law has been omitted; the order and language are logical and clear (there are even, most unusually, helpful cross-references); the Act covers both domestic and international arbitration; and full regard has been had to the Model Law, though not by way of a limitation. Where there may be a deviation from the Mustill Report features is in relation to the second of them, namely that new legislation should 'be limited to those principles whose existence and effect are uncontroversial'.

The Act in fact contains many new provisions that were not previously to be found either in legislation or in the common law. And some of its provisions may be thought to be controversial, although they did not give rise to any appreciable difficulty during either the consultation period or the parliamentary debates.

Uncontroversial?

Immunity

By s.29, an arbitrator is given immunity from liability for anything done or omitted in the discharge or purported discharge of his functions as arbitrator, absent bad faith. Similar (although more limited) immunity is given to arbitral institutions by s.74. For years before the passing of the Act there was debate, sometimes highly animated, as to whether arbitrators in fact had any such immunity, and as to whether indeed they should have it. Neither the legal nor the moral positions were thought to be clear. Yet, at a stroke, Parliament has – and without any difficulty – given arbitrators (and institutions) protection.

Interest

In the area of the awarding of interest there has been another dramatic change in the law. By s.49, arbitrators may henceforth award simple '*or compound*' interest. Even the courts (save in very limited and rarely-encountered circumstances) have no power to award compound interest! The possibility of some such power being created has been before Parliament on a number of occasions, but has never been taken up. And even when detailed provisions were enacted for the awarding of interest by both arbitrators and the courts in 1982, the statutory power given was expressly limited to the awarding of *simple* interest.

Arbitrators' powers

Subject to any contrary agreement by the parties, arbitrators henceforth have very wide-ranging powers over the way in which proceedings are to be conducted, including (and many may think most extremely) whether to hold an oral hearing or not, and if so for what purposes (s.34(2)(h)). When it comes to the costs of an arbitration, s.65 contains a wholly new power (which the parties may overrule by their agreement) enabling arbitrators to direct that the recoverable costs of an arbitration shall be limited to a specified amount. In other words they may 'cap' recoverable costs, leaving parties free to spend whatever they like but in the knowledge that any recovery they make at the end of the day will be limited to the sum fixed by the arbitrators. This is an entirely new concept, at least in England, and as with many other provisions in the Act, it remains to be seen how it will work out in practice.

Overriding principles

The object of arbitration, and how arbitrations must be run
Whatever new powers arbitrators have, those powers – and indeed the whole Act – are subject to a number of principles which, unusually, the Act sets out. Firstly, the object of arbitration is defined in s.1 as being 'to obtain the fair resolution of disputes by an impartial tribunal without unnecessary delay or expense'. The same section goes on to provide that the parties 'should be free to agree how their disputes are resolved, subject only to such safeguards as are necessary in the public interest'. S.33 imposes upon arbitrators a duty to act 'fairly and impartially as between the parties, giving each party a reasonable opportunity of putting his case and dealing with that of his opponent' and they are also to 'adopt procedures suitable to the circumstances of the particular case, avoiding unnecessary delay or expense, so as to provide a fair means for the resolution' of disputes. Lastly, s.40 imposes on parties a duty to 'do all things necessary for the proper and expeditious conduct' of the proceedings.

The principles embodied in these sections must now inform the approach taken by parties, arbitrators and indeed the courts to the conduct of arbitrations and to the application of the Act. Thus, to take but one example, an arbitrator wanting to exercise the power to 'cap' costs under s.65 must ask himself whether so doing might hinder the fair resolution of the case, whether he is acting impartially, whether he is helping the parties to avoid unnecessary expense and whether capping the costs might prevent either party from having a reasonable opportunity to put his case and deal with that of his opponent.

Party autonomy
It is fundamental to the whole approach of the Act that, so far as is consistent with the requirements of public policy, parties to arbitration agreements should have the maximum possible freedom to choose how their tribunals are to be structured, how their cases are to be run, what their awards are to contain, and so on. Therefore, most of the Act's provisions are non-mandatory, in the sense that parties can agree their own regimes. The mandatory provisions, which must apply whatever the parties may choose to agree, have been kept to a minimum.

Court intervention

It is consistent with the policy of party autonomy, and indeed with the Model Law, that court intervention should be available to the minimum extent possible, and then only to support arbitration, not to interfere with it otherwise. Thus s.1(c) provides that in matters governed by Part I of the Act (which contains all its main provisions) 'the court should not intervene except as provided by this Part'. It has been suggested that this provision is not directive, the argument being that 'should' means 'ought'. We disagree: in our view it is the subjunctive that is – correctly – used here.

New points

The text that follows identifies in some detail the principal new provisions in the Act.

Application and 'seat'

Ss.2 and 3 deal with the applicability of the Act and introduce the concept of 'the seat of the arbitration' for the first time into English law.

Institutions

S.3 recognises the part that arbitral institutions have to play in many arbitrations. Other references, some indirect, occur in, for instance, ss. 4, 14 ('a person other than a party to the proceedings'), 23, 24, 42 ('any available arbitral process'), 44, 56, 59, 68, 70, 73, 74, 79 and 82. One of the Act's themes is that if the rules of any relevant institution provide for some process which the courts might otherwise exercise, that process must be exhausted before the courts can be approached.

Agreements in writing

Whilst the 1950 Act only applied to arbitration agreements in writing, and this requirement is reproduced in s.5(1), now a wide definition of 'in writing' or 'evidenced in writing' is given in s.5(2)–(6).

Separability

The separability of an arbitration agreement from a main contract is confirmed by s.7. Although the concept of separability was one to which English common law had finally come, this had only happened shortly before the Act was passed.

Staying court proceedings

The position in relation to the court having discretion whether or not to stay proceedings when there was a domestic arbitration agreement has been substantially changed. S.86 of the Act would have allowed the court not to grant a stay in a domestic arbitration if it was satisfied that the arbitration agreement was null or void, inoperative or incapable of being performed (i.e. the same grounds as are provided for in s.9(4)) *or* if there were 'other sufficient grounds for not requiring the parties to abide by the arbitration agreement'. However, ss.85 to 87 were not brought into effect. Thus there is no distinction between domestic and non-domestic arbitrations, and in particular there is no discretion given to the court to hold that there are sufficient grounds for not holding parties to an arbitration agreement.

The abolition of the former distinction came about because of the responses to a consultation document which the DAC published on the issue, and because of the Court of Appeal's decision in *Philip Alexander Securities and Futures Limited* v. *Bamberger* [1996] CLC 1757 which held that any distinction between domestic and international arbitrations was incompatible with EU law. This limitation on the court's discretion has caused concern in some areas, particularly the construction industry, where it is no longer possible for the court to order that all issues arising between employers and contractors and contractors and sub-contractors be tried together if, in any of the relevant contracts, there is a binding arbitration clause. It has also caused concern that the courts may no longer refuse a stay when the only apparent dispute between the parties is the refusal by the defaulting party to pay the sum claimed and the claimant would, formerly, have wished to seek a summary judgment.

S.9 has been considered by the courts in a number of cases and these are dealt with in the commentary on that section. We would observe here, though, that because under Art.3 of the High Court and County Courts (Allocation of Proceedings) Order 1996, proceedings under s.9 are to be commenced in the court in which the legal proceedings are pending, there is a risk of inconsistency of approach when such applications are dealt with other than in the Commercial Court, which must be involved in (if not actually hearing) all other arbitration applications. This we feel to be undesirable.

Time limits
The concept of 'undue hardship' in relation to the extension of time by the court for the commencement of arbitration proceedings has been abolished and replaced, in s.12(3), by a requirement that the court shall only extend time if the circumstances were such as were outside the reasonable contemplation of the parties when they agreed the time limit, and that it would be just to extend the time; or where one party's conduct makes it unjust to hold the other to the strict terms of the time limit provisions. Cases where the courts have had to consider these provisions are mentioned in the commentary on this section.

New provisions on time limits arise in a different context in ss.16 and 17, namely in relation to the period of time to be allowed a party to make a responding appointment of an arbitrator (now 14 days under s.16(4)–(5)), and if a default appointment of a sole arbitrator is required to be made, a *further* seven clear days' notice must be given under s.17(2).

No majority decision
In a three-man tribunal, the possibility of there being no majority does exist in relation to some issues. Should that happen, the view of the chairman will now prevail under s.20(4).

Removal of arbitrators
Whilst the courts retain the power to remove an arbitrator, the rather restricted grounds on which they may do so are now spelt out in s.24. Then s.25 deals with the position of an arbitrator who resigns his appointment; in particular it creates a possibility of such an arbitrator applying to the court for relief from any liability thereby incurred and for an appropriate order in respect of the arbitrator's fees and expenses.

Repayment of excessive fees
A new provision in relation to arbitrators' fees and expenses occurs in s.28(3), where the court may now order the repayment of any such amount as can be shown to be excessive. Previously, at least where an award had been made and paid for without protest or the costs being paid into court, no such possibility existed.

Jurisdiction
Jurisdiction has given rise to a number of changes. There is now an express power given to tribunals to rule on their own jurisdiction, i.e. as to the existence of a valid arbitration agreement, the proper constitution of the tribunal and what matters have been submitted (s.30). S.31 then requires any party objecting to the substantive jurisdiction of a tribunal to do so at the outset of the proceedings, and any objection as to excessive jurisdiction to be made as soon as possible after the matter giving rise to the objection occurs. Further, ss.31 and 32 together provide for how arbitrators and the courts may deal with questions as to substantive jurisdiction, and for the possible loss of the right to object if objections are not raised timeously. Cases where the courts have had to consider these provisions are mentioned in the commentary on s.30.

Arbitrators' powers
S.34(2)(h) has already been mentioned, but the earlier sub-paragraphs of that sub-section list some of the procedural and evidential matters which arbitrators may decide (subject to any contrary agreement between the parties). These include where to hold proceedings, language matters, the form of statements of case (if any), disclosure of documents, evidence, and even the possibility of the tribunal itself taking the initiative to ascertain the facts or law (s.34(2)(g)). Following on from that, logically we suggest, s.37 empowers arbitrators to appoint experts, legal advisers or technical assessors and deals with the parties' rights in relation to any material provided by such individuals, as well as covering their fees and expenses.

Security for costs

The ordering of security for costs has long been a thorny problem, particularly where it was done by the courts although arbitrators or an institution had failed or refused to intervene. Now, under s.38(3) arbitrators alone may order the provision of security for the costs of an arbitration (subject to any agreement between the parties). It should be noted in particular that they may not exercise this power on the ground that the claimant party is either an individual or a corporation or similar body essentially outside the United Kingdom.

Provisional orders

Whilst arbitrators have always had the power to make interim awards dealing with parts of a reference, they will now have power under s.39 to make provisional orders – if the parties so agree – which may be subject to final adjudication at a later stage.

Peremptory orders

S.42 gives the courts the power to make orders requiring parties to comply with arbitrators' peremptory orders. However, given the extensive powers expressly set out for arbitrators in s.41 to cover the case where a party fails to comply with such an order or to attend a hearing, etc., it is perhaps not surprising that the courts have not apparently had to be troubled under s.42. Their powers under s.44, to make orders for the taking of evidence, the preservation of evidence and property and the like, and in particular under s.44(3) to act urgently for the purpose of preserving evidence or assets, are required to be used rather more often. Some cases on these powers are covered in the commentary on s.44.

Equity clauses

Turning to awards, s.46(1)(b) expressly provides for the possibility of the parties agreeing that arbitrators may decide cases other than in accordance with a specific law (e.g. on an equitable basis) and obliges arbitrators then so to act.

Reasons

Whatever the basis on which an award has been reached, it must contain reasons under s.52(4) unless the parties have agreed otherwise or unless it is an agreed award. This does not automatically mean that there will be a right of appeal, because parties are able – henceforth in all cases (the special categories having been abolished) – to agree to exclude such right (see the sub-section just mentioned and s.69(1)).

Supplementary awards

If it should happen that a tribunal omits to deal with a claim, it is able to make an additional award in respect of that claim under s.57(3)(b). This ought only to apply to any inadvertent omission since, otherwise, the tribunal will presumably have made an interim award quite deliberately.

Costs

Arbitrators have always been bound to follow the practice of the courts in awarding costs on the general principle that they should follow the event. This concept is now encapsulated in s.61(2) but with the saving that they may decline to follow that principle where it appears that in the circumstances it would not be appropriate to do so. This leaves open, of course, the question of what is 'the event', one to which arbitrators often have to address themselves. Plainly it is, in many cases, not simply the fact that one party recovers some money, whether on balance or otherwise.

The death of 'misconduct'

Happily, the concept of 'misconduct', whether technical or otherwise, has finally been laid to rest. That is not to say that awards may not be challenged in future on some of the bases previously adopted, but the term now used is 'serious irregularity', and that is defined in a limited way in s.68 which deals with challenges (other than on the grounds of substantive jurisdiction). The closed nature of the list given means that the courts are not able to invent other grounds for setting aside or remitting awards. The list itself sets out instances of failure by arbitrators to comply with their duties to which, we suggest, no one could reasonably object since they all plainly do amount to 'serious irregularity' in circumstances where, as the section requires, substantial injustice is caused to the party seeking the court's help. Cases where the courts have had to consider 'serious irregularity' are dealt with in the commentary on s.68.

Appeals

The courts' intervention may also be sought in relation to appeals on points of law (s.69) as before. That section now encapsulates the guidelines set out in *Pioneer Shipping Ltd* v. *BTP Tioxide Ltd* (*The 'Nema'*) [1982] AC 724. A particular innovation is found in s.69(3)(b) which provides that the court shall not give permission to appeal if the question of law is one which the tribunal was not asked to determine. Permission has, of course, been given in such circumstances in the past.

Waiver

An important new provision is contained in s.73 which provides for parties to lose the right to object to matters relating to substantive jurisdiction, the conduct of an arbitration, compliance with the arbitration agreement or Act or, indeed, any other irregularity affecting the tribunal or the proceedings, unless that objection is taken promptly.

Technical matters

Certain new matters of a 'technical' nature are dealt with in ss.76 and 77 which provide for the service of documents, and s.78 which explains how periods of time are to be counted under the Act. Further, s.82 contains a list of definitions. Of particular importance are those of 'dispute' which now includes *any difference* (a change from the previous law) and of 'substantive jurisdiction' which refers to the matters dealt with in s.30(1)(a)–(c).

The revolution

Why do we talk of a revolution at the beginning of this introduction? It seems to us that the changes wrought to arbitration law by this Act will, in the fullness of time, change the face of arbitration in England, Wales and Northern Ireland. The increased powers given to arbitrators, coupled with their new obligations to adopt suitable procedures for each case, avoiding unnecessary delay and expense (s.33), plus the new immunity and the promise of reduced judicial intervention, mean that far more responsibility now rests on arbitrators' shoulders than hitherto. That means that the importance of proper training is considerably enhanced. Arbitrators need to be far more pro-active and bold than hitherto: no longer can it suffice for them to sit back and wait for a party to ask them to do something (unless, of course, both parties agree that is how a matter should proceed). They must be more imaginative and professional. The idea that arbitration should be modelled on court procedures must be abandoned finally.

Practitioners too need to re-mould their thinking, not least because of this last point. There can no longer be any assumption that court-style pleadings are to be used, or that there is to be discovery (or even 'disclosure') as if in court, or even that there will be an oral hearing or that proceedings will be conducted adversarially. Far more thought must now be given to how most appropriately to conduct arbitrations; and to the possibility of costs being capped under s.65 and the effects of that. Accordingly, any temptation to think, following the Woolf reforms, that the CPR can simply be imported into arbitration should be resisted. Whilst the spirit informing those reforms is the same as that behind the Act, that does not mean that the procedures to be adopted are to be the same. And if some agreement as to procedure is thought to be appropriate between parties to a dispute, it is no longer possible to rely upon any tacit understanding: it must be in writing (s.5). Arbitrators too should bear this in mind, and not allow themselves to assume that because a matter is being conducted in a particular way by the parties or their representatives, it is bound to continue in that way: absent a written agreement to vary any of the non-mandatory provisions of the Act, arbitrators have all the 'default' powers it sets out and are *obliged* to use them in pursuance of their s.33 duties.

Parties now need to give more thought to their arbitration clauses and agreements, and to what (if any) agreements as to procedure should be entered into (which must be in writing, as observed above). Institutional and trade rules have mostly been looked at and revised as required.

Because the Act is so radical, and because human beings tend towards habit, the changes are likely to come about relatively slowly. But everyone involved in arbitration needs to be on their mettle henceforth, for otherwise a few brutal surprises are almost certainly in store for some, even though the Act has been in force for a few years. In cases to which the Act applies, thought must be given to the principles set out in ss.1 and 33. More detailed points that have to be focused upon include whether interest should be simple or compound, and the awarding of interest after the date of an award (s.49); the procedural options open to arbitrators (s.34 etc.); dealing with questions of jurisdiction (ss.31 and 32 etc.); the possible loss of rights to object (s.73); whether to exclude the right of appeal or the right to have reasons in an award; and the form, dating and notification of awards (ss.52–56).

These points, though, whilst important, are no more than a selection of those to which consideration must now be given by all involved in arbitration.

This book

The core of this book is a commentary on the Act. We have endeavoured to make it as 'user-friendly' as possible. Thus we have begun the consideration of each section – starting on a new page – with some basic information as to the 'status' of the section (whether *mandatory* or *non-mandatory*) and its derivation (whether new, Model Law, previous legislation or some combination of these). We have then set out a brief 'summary' of the effect of the section before moving on to consider in more detail the specific 'points' that seem to us to arise. We have endeavoured to explain concisely and clearly the various regimes established by the Act and have made, wherever relevant, cross-references between different sections. In this edition we have commented, where appropriate, on cases that have come before the courts under the new Act. Lastly, in most instances, we have included the text of the Model Law where there is a comparable provision, with notes as to differences between it and the Act. We have deliberately refrained from citing more than a few authorities decided before the Act came into force, principally because we believe that the courts will rarely now refer to those cases. There appears, however, to be at least one exception: see the *Trygg Hansa* case referred to in paragraph 6E.

In addition to the main commentary we have prepared certain materials designed to be useful to both arbitrators and practitioners. In a separate section following this introduction we have brought together drafting suggestions and checklists covering such matters as arbitration clauses, the agreements that may be made pursuant to the many sections of the Act which allow the parties a choice, the preliminary meeting, the position on resignation of an arbitrator and the operative parts of the award. The main text includes further drafting suggestions – coupled with commentary – as to peremptory orders (paragraph 41E); awards of interest (paragraph 49F); 'capping' orders (paragraph 65D); and costs awards (paragraph 61E). Comment is also made at various points as to how arbitration rules have been amended to operate under the Act.

Principal decisions

We note here some of the principal points decided by the courts since the Act came into effect.

S.3
Where parties agree the place of an arbitration pursuant to institutional rules, such place is to be the seat of the arbitration.

Ss.6 and 7
An arbitration in an illegal (gaming) agreement cannot be regarded as an arbitration agreement since any arbitration under it would be devoid of any legal consequence.

In relation to the incorporation of arbitration clauses by reference, it is appropriate to refer to cases decided before the 1996 Act; and in one case it has been held that an arbitration clause contained in standard conditions of contract incorporated by reference into a contract was binding, although no specific reference was made to it in the incorporating provisions.

S.9
This, together with a few other sections, has attracted most attention in the courts. Cases have dealt with the meaning of 'dispute', the effect of a time-bar on arbitrators' jurisdiction, the loss of a right to the stay of court proceedings, and the available options where an arbitration agreement is challenged.

S.12
The question of extensions of time for the commencement of arbitration proceedings has also come before the court on a number of occasions. The cases have dealt with the circumstances that must be considered when deciding whether to grant an extension, examples of what is not outside the mutual contemplation of parties for the purposes of this section, and what conduct may be taken as making it unjust to hold to strict time provisions.

S.14

There have also been a few cases on the commencement of arbitration proceedings, in particular what is required in terms of notice.

S.15

Of general significance, though in the particular circumstance it arose under this section, is a case in which it was held that an arbitration provision referring to the 1950 Act and any subsequent alterations effectively referred to the 1996 Act.

S.24

In the original text we gave the example (as did the DAC in its report) of the possibility of a challenge on the grounds of lack of independence where barristers appeared in front of a member of their own chambers; there has in fact been one (unsuccessful) challenge on precisely this basis.

Ss.30 and 67

If any provisions of the Act have given rise to problems the full extent of which were probably not anticipated, it is these. In the first place, the power of tribunals to rule on their own jurisdiction can impinge seriously on a s.9 application for a stay where there is a challenge to the arbitration agreement, and there may be many instances where it is more appropriate for a court to resolve the problems arising. In addition, because on hearing an appeal under s.67 against a ruling under s.30 the court may decide to hear the whole matter – evidence included – again, it may often be preferable for such an issue to go straight to court rather than having all the points tried twice.

S.44

There has been some assistance given as to the support that may be sought from the courts for arbitrators' orders, and as to the court's jurisdiction in respect of ancillary relief.

S.52 (and S.31)

Notwithstanding the formal requirements for an award set out in this section, it has been held that an arbitrator's letter dealing with his jurisdiction amounted to an award as to that subject even though it did not state the seat of the arbitration or his reasons.

Ss.67–70

A case under the 1950 Act relating to the prosecution of challenges to awards appears to lay down certain principles which would probably be applied to the 1996 Act.

In addition, there have been – sadly but perhaps not suprisingly – a number of challenges under s.68 on the grounds of 'alleged serious irregularity' and the courts have given some assistance in relation to these. Under s.69 the court has had to consider what amounts to an agreement consenting to a right of appeal.

Ss.85–88

These sections, which would have sustained distinctions between domestic arbitrations and international arbitrations, were not brought into effect. Accordingly there is now a single regime for all types of arbitration.

S.103

The courts have had to consider the exceptions which may allow the avoidance of enforcement of a New York Convention Award.

We very much hope that this second edition will prove a useful guide and a workable tool for arbitrators, practitioners and others interested in arbitration as they work with the new Act.

Part 2

Tables

Table A: Basic Information about the Act

This is a quick guide to some essential information about the Act and its application.

(1) *Scope*
Part I , that is ss.1 to 84, applies:

- where the seat of the arbitration is in England, Wales or Northern Ireland (and some sections apply even if the seat is elsewhere, or not yet designated) (s.2);
- to arbitral proceedings commenced on or after the date on which Part I comes into force, under an arbitration agreement *whenever* made (s.84).

(2) *When in force*
In relation to arbitrations commenced on and after 31 January 1997.

(3) *Existing arbitration agreements*
Part I applies to existing arbitration agreements.

(4) *Existing arbitrations*
Part I does *not* apply to on-going arbitrations commenced before it came into force.

(5) *Domestic arbitrations*
Part I applies both to domestic and international arbitrations.

(6) *International arbitrations*

- Part I applies to international as well as to domestic arbitrations.
- Part III covers the recognition and enforcement of foreign awards under the Geneva and New York Conventions.

(7) **Consumer arbitration agreements**
Ss.89–91 of Part II apply to consumer arbitration agreements.

(8) **Small claims arbitrations in the county court**
Part I does *not* apply to arbitration under s.64 of the County Courts Act 1984 (s.92) (which has now, in any event, been superseded by the small claims track).

(9) **Small-scale arbitrations**
There are no separate provisions, but there is intended to be sufficient flexibility in the procedures that may be followed (see s.34) to tailor proceedings to suit.

(10) **Statutory arbitrations**
Subject to ss.94 to 98 of Part II, Part I applies to all statutory arbitrations, whether the relevant enactment was passed before or after the commencement of the Act.

(11) **Enforcement of a domestic arbitration award**
See s.66

(12) **Enforcement of an international arbitration award**
See s.66 and Part III.

(13) **Unusual drafting**
The Act contains:

- the principles on which it is founded, which are of value to parties, tribunals and arbitral institutions construing its provisions (s.1);
- cross-references to other sections;
- user-friendly language.

Table B: Major Changes to the Law

This is a quick guide to some of the more significant changes to the law made by the Act.

- Stay of legal proceedings in favour of domestic arbitrations becomes mandatory (s.9).

- Excessive fees paid to arbitrators may be ordered to be repaid (s.28).

- Immunity for acts done in good faith is given to arbitrators (s.29) and arbitral institutions (s.74).

- The tribunal may rule on its own jurisdiction (s.30).

- The tribunal is placed under a general duty as to how to act (s.33).

- The tribunal is given increased powers to determine procedure (s.34).

- The tribunal is given power to order security for costs (s.38), and the court loses this power.

- The tribunal may make provisional orders (s.39).

- The tribunal is given express power to make peremptory orders (s.41), and the court is given power to order compliance (s.42).

- The concept of the court 'handing back' particular matters to the tribunal when the latter can act effectively is introduced (s.44).

- The tribunal may decide cases other than in accordance with the law (s.46).

- Compound interest may be awarded (s.49).

- Awards must contain reasons (s.54).

- The tribunal may make additional – or supplementary – awards (s.57).

- The tribunal is given the power to 'cap' recoverable costs (s.65).

- 'Misconduct' is abolished and 'serious irregularity' introduced as the basis for procedural challenge (s.68).

- Further limitations are placed on appeals on points of law (s.69).

Table C: Statutory and Similar Sources

This is an indicative table of derivations for the 1996 Act. It is no more than indicative because in many instances the 'sources' given do no more than deal with the subject covered by the relevant provision in the Act. On occasions the sources and provisions are comparable, sometimes they are similar, and yet others are almost totally different.

To produce anything more detailed would result in a cumbersome schedule, and would probably be of no help.

1996 Act S.	1950 Act S.	1975 Act S.	1979 Act S.	Model Law Article	Comments
1				5, 19(1)	
2				1(2)	
3					
4					
5	32	7(1)		7(2)	
6	32	7(1)		7(1),(2)	
7				16(1)	
8	2(1), (3)				
9	4(1)	1		8	
10	5				
11					Re-enacts Civil Jurisdiction and Judgments Act 1982, s.26
12	27				
13					Re-enacts (with modifications) s.34(1), (5) and (7)(b) of Limitation Act 1980, and Foreign Limitation Periods Act 1984, s.5.
14				21	Limitation Act 1980, s.34(3)
15	6,8,9			10(1)	
16				11(2),(3)	
17	7(b) and 10(3B)				
18	10(1), (2) and (3c)			11(4)	
19				11(5)	
20	9			29	
21	8(2), (3)				
22				29	

1996 Act S.	1950 Act S.	1975 Act S.	1979 Act S.	Model Law Article	Comments
23	1			14	
24	13(3), 23 and 24(1)			12–14	
25				14	
26(1)					
26(2)	2(2)				see also s.8
27	25			15	
28					
29					
30				16(1)	
31				16(2), (3)	
32					
33				18	
34	12(1)–(3)			19, 20, 22–24	
35					
36					
37				26(1)	
38	12(1)–(3), 6(a)			17	
39					
40					
41	13A			25	
42			5		
43	12(4), (5)			27	
44	12(6)			9	
45			2		
46				28	
47	14				
48	15				
49	19A and 20				
50	13(2)				
51				30	
52				31	
53				31(3)	
54					
55					
56	19				
57	17 and 18(4)			33	
58	16				
59					
60	18(3)				
61	18(1)				
62					
63	18(1), (2)				
64	19				
65					
66	26(1)			35	
67–68	22, 23			16, 34	
69			1, 3		
70	23(3)		1(5), (6)		
71	22(2)		1(8)		

1996 Act S.	1950 Act S.	1975 Act S.	1979 Act S.	Model Law Article	Comments
72					
73				4	
74					
75	18(5)				
76				3	
77					
78					
79					
80					
81					
82					
83					
84	33				
85		1(4)			
86	4(1)	1(1)			
87			3(6)		
88					
89–91					Consumer Arbitration Agreements Act 1988/Unfair Terms in Consumer Contract Regulations 1999
92			7(3)		
93	11				Administration of Justice Act 1970, s.4
94–98	31				
99	35–42	2			
100		7			
101		3			
102		4			
103		5			
104		6			
105					
106	30				
107					
108	34	8(4)	8(4)		

Table D: 'Destinations'

As with the table of sources (Table C), this table does no more than indicate where provisions related to those set out may be found in the 1996 Act. It is intended to provide a guide for those who wish to see how the Act deals with previous provisions and those of the Model Law.

1950 Act S.	1996 Act S.
1	23
2	8, 26(2)
3	Insolvency Act 1986, s.349A, inserted by Sch. 3, para. 46
4	9, 86
5	10
6	15
7	17
8	15, 21
9	15, 20
10	17, 18
11	93
12	34, 38, 43, 44
13	24, 41, 50
14	47
15	48(5)(b)
16	58
17	57
18	57, 60, 61, 63, 75
19	56, 64
19A	49
20	49
21	
22	67, 68, 71
23	24, 67, 68, 70
24(1)	24
24(2)	
24(3)	
25	27 (together with 16 and 18)
26	66
27	12
28	
29	109(1)
30	106
31	94–98
32	5(1) and 6
33	84
34	108
35–42	99

1975 Act S.	1996 Act S.
1	9, 85, 86
2	99
3	101
4	102
5	103
6	104
7	5, 6, 100

1979 Act S.	1996 Act S.
1	69–71
2	45
3	69, 87
4	
5	42
6	
7	92
8	108

Model Law Article	1996 Act S.
1	2
2	
3	76
4	73
5	1
6	
7	5, 6
8	9
9	44
10	15
11	16, 18, 19
12	24
13	24
14	23, 24, 25
15	27
16	7, 30–31, 67–68
17	38
18	33
19	1, 34
20	34
21	14
22	34
23	34
24	34
25	41

Model Law Article	1996 Act S.
26	37
27	43
28	46
29	20, 22
30	51
31	52, 53
32	
33	57
34	67, 68
35	66
36	

Part 3

Materials

A. Arbitration Clauses

Examples

We have prepared two sample arbitration clauses specifically designed for use with the Act. In each case they are intended simply to establish the arbitration tribunal and do not embark upon the possible powers that the tribunal might have.

It should be noted that both these clauses anticipate another clause in the relevant contract which provides for a notice to be given that will determine when the arbitration starts. In the absence of such a provision, the default provisions in s.14(3) to (5) will apply.

Clause (1) – Sole arbitrator
'Any dispute or difference arising out of or in connection with this contract shall be referred to the arbitration of a sole arbitrator to be appointed in accordance with s.16(3) of the Arbitration Act 1996 ('the Act'), the seat of such arbitration being hereby designated as London, England. In the event of failure of the parties to make the appointment pursuant to s.16(3) of the Act, the appointment shall be made by the President for the time being of the Chartered Institute of Arbitrators. The arbitration will be regarded as commenced for the purposes set out in s.14(1) of the Act when one party sends to the other the notice described in clause [   ] of this contract. The arbitrator shall decide the dispute according to the substantive laws of England and Wales.'

Clause (2) – Two arbitrators and a chairman
'Any dispute or difference arising out of or in connection with this contract shall be referred to the arbitration of two arbitrators and a chairman to be appointed in accordance with s.16(5) of the Arbitration Act 1996 ('the Act'), the seat of such arbitration being hereby designated as London, England. S.17 of the Act shall not apply. In the event

of failure of either of the parties to make the appointment pursuant to s.16(5) of the Act, or in the event of failure by the arbitrators to appoint a chairman, such appointment shall be made by the President for the time being of the Chartered Institute of Arbitrators who shall have the powers otherwise given to the court under s.18(3) of the Act. The arbitration will be regarded as commenced for the purposes set out in s.14(1) of the Act when one party sends to the other the notice described in clause [   ] of this contract. Save that in respect of matters of procedure (other than where the parties are agreed) decisions or orders may be made by the chairman acting alone, s.20(3) and (4) of the Act shall apply. The arbitrators shall determine the dispute in accordance with the substantive laws of England and Wales.'

B. Agreements Prior to or on Constitution of Arbitration

Introduction

This is one of a number of checklists of points arising for consideration or agreement designed to assist with implementing the Act.

Whilst the lists have been drawn up broadly by reference to the different stages or aspects of an arbitration, such an exercise is arbitrary since many such points may be addressed at more than one stage or in relation to more than one aspect. We have therefore tried to identify those matters which we think *most likely* to be addressed at certain stages.

Dealing with preliminary meetings and awards, the checklists are, of course, primarily addressed to the arbitrator.

Checklist for agreements prior to or on constitution of the arbitration

Seat of the arbitration – (s.3)

- The parties may designate the seat;
- The parties may empower an arbitral institution or other person to designate the seat;
- The parties may authorise the tribunal to designate the seat;
- Otherwise, the court will determine the seat.

General means of providing for non-mandatory provisions – (s.4)

- The parties may make their own agreement or agreements dealing with specific non-mandatory provisions;
- The parties may adopt institutional rules;
- The parties may adopt the laws of another state;
- The parties may use a combination of the above;
- Otherwise, default provisions apply.

Separability of arbitration agreement – (s.7)

- The parties may agree that an arbitration clause is not separable from the main agreement;
- Otherwise, it is so separable.

Death of a party – (s.8)

- The parties may agree that an arbitration agreement is discharged by the death of a party and so may not be enforced by or against the personal representatives of that party;

- Otherwise, it is not so discharged.

Commencement of arbitral proceedings – (s.14)

- The parties may agree when arbitral proceedings are to be regarded as commenced for the purposes of Part I of the Act and of the Limitation Acts;
- Otherwise, default provisions apply.

Arbitral tribunal – (s.15)

- The parties may agree on the number of arbitrators to form the tribunal and whether there is to be a chairman or umpire;
- Where the parties have agreed on an even number of arbitrators, they may agree that there should be *no* additional arbitrator as chairman (otherwise an agreement for two arbitrators is understood as requiring the appointment of a third, as chairman);
- In default of agreement as to number, there will be a sole arbitrator.
- (Note, therefore, that if the parties require more than one arbitrator, agreement is essential.)

Procedure for appointment of arbitrators – (s.16)

- The parties may agree on the procedure for appointing the arbitrator or arbitrators, including the procedure for appointing any chairman or umpire;
- Default provisions apply if there is no agreement, or only an agreement covering some of these matters.

Appointment of sole arbitrator in case of default – (s.17)

- The parties may agree that s.17 (appointment of first party's arbitrator as sole arbitrator where second party fails to make an appointment) does *not* apply.

Failure of appointment procedure – (s.18)

- The parties may agree what is to happen in the event of a failure of the procedure for the appointment of the arbitral tribunal;
- Default provisions apply if or to the extent that there is no agreement.

Chairman – (s.20)

- Where there is to be a chairman, the parties may agree his functions as to the making of decisions, orders and awards;
- Default provisions apply if or to the extent that there is no agreement.

Umpire – (s.21)

- Where there is to be an umpire, the parties may agree his functions;
- Default provisions apply if or to the extent that there is no agreement.

Decision making where there is no chairman or umpire – (s.22)

- The parties may agree how the tribunal is to make decisions, orders and awards where they have previously agreed that there shall be two or more arbitrators with no chairman or umpire;
- In the absence of such an agreement, decisions are made by all arbitrators or a majority.

Revocation of arbitrator's authority – (s.23)

- The parties may agree in what circumstances the authority of an arbitrator may be revoked;
- Default provisions apply to the extent that there is no agreement.

Jurisdiction of the arbitral tribunal – (s.30)

- The parties may agree that the tribunal *cannot* rule on its own substantive jurisdiction;
- Otherwise, it can.

Consolidation – (s.35)

- The parties may agree on consolidation of arbitrations or concurrent hearings, or that the tribunal shall have power so to order;
- Otherwise, the tribunal has no such power.

Representation – (s.36)

- The parties may agree that rights to representation be limited;
- Otherwise, they may be represented by a lawyer or other person.

Appointing experts, legal advisers or assessors – (s.37)

- The parties may agree that the tribunal may *not* appoint experts, etc;
- The parties may agree they should *not* be given a reasonable opportunity to comment on the output of such an expert;
- Otherwise, the tribunal has such power of appointment.
- (Note that s.37(2), in relation to the fees and expenses of such experts, is mandatory.)

General powers of tribunal – (s.38)

- The parties may agree on the powers exercisable by the tribunal;
- *But* the powers in subss.(3) to (6) *will* apply unless a contrary agreement can be spelt out.

Power to make provisional orders – (s.39)

- The parties may agree that the tribunal shall have power to order on a provisional basis any relief which it would have power to grant in a final award;
- If there is no such agreement, the tribunal has no such power.

Powers on default of party – (s.41)

- The parties may agree on the powers of the tribunal in case of a party's failure to do something necessary for the proper and expeditious conduct of the arbitration;
- *But* the powers in subss.(3) to (7) *will* apply unless a contrary agreement can be spelt out.

Court powers in support of arbitral proceedings – (s.44)

- The parties may agree to exclude the court's powers in relation to any or all of the matters listed in s.44(2)(a) to (e);
- *But* its powers in respect of those matters *will* remain unless a contrary agreement can be spelt out.
- (Note, we do not think that the parties can make an agreement that excludes the effect of subss.(3) to (7) in respect of any matter within subs.(2) as to which the court has powers.)

Determination of preliminary point of law – (s.45)

- The parties may agree, in advance of any arbitration or dispute as well as after an arbitration has commenced, to exclude the court's power to determine any question of law arising in the proceedings;
- An agreement to dispense with reasons for the tribunal's award is equivalent to an agreement to exclude this jurisdiction;
- *But* the provisions of the section *will* apply unless a contrary agreement can be spelt out.

Rules applicable to substance of dispute – (s.46)

- The parties may choose the substantive law that is to apply;
- Or, the parties may agree as to what other considerations should be applied;
- Or, the parties may agree that the tribunal should determine the considerations to be applied;
- In default, the tribunal will apply the substantive law determined by the applicable conflict rules.

Awards on different issues – (s.47)

- The parties may agree to *exclude* the power to make more than one award at different times on different aspects of the matters to be determined;
- Otherwise the tribunal has this power.

Remedies – (s.48)

- The parties may agree on the powers the tribunal may exercise as regards remedies;
- *But* the tribunal *will* have the powers set out in subss.(3) to (5) unless a contrary agreement can be spelt out.

Interest – (s.49)

- The parties may agree on the powers the tribunal may exercise as regards interest;
- *But* the tribunal *will* have the powers set out in subss.(3) to (5) unless a contrary agreement can be spelt out.

Form of award – (s.52)

- The parties may agree on the form of the award;
- Note that the parties may agree to dispense with reasons;
- Default provisions apply if or to the extent that there is no agreement.

Place where award made – (s.53)

- The parties may agree that the award shall *not* be treated as made at the seat of the arbitration, where that seat is in England, Wales or Northern Ireland;
- Otherwise, it will be treated as so made.

Date of award – (s.54)

- The parties may agree that the tribunal shall *not* decide on the date of the award;
- If they so agree, or the tribunal fails to decide the date, subs.(2) applies.

Notification of award – (s.55)

- The parties may agree on the requirements as to notification of the award to the parties;
- In the absence of agreement, subs.(2) applies.

Correction of award or additional award – (s.57)

- The parties may agree on the powers of the tribunal to correct an award or make an additional award;
- Default provisions apply if or to the extent that there is no agreement.

Effect of award – (s.58)

- The parties may agree on the effect of the award;

- In the absence of contrary agreement, the award is final and binding.

Appeal on point of law – (s.69)

- The parties may agree, in advance of any arbitration or dispute as well as after an arbitration has commenced, to exclude the court's power to determine any question of law arising out of an award;
- An agreement to dispense with reasons for the tribunal's award is equivalent to an agreement to exclude this jurisdiction;
- *But* the provisions of the section *will* apply unless a contrary agreement can be spelt out.

Service of notices – (ss.76 and 77)

- The parties may agree on the manner of service of any notice or other document required or authorised to be given or served in pursuance of an arbitration agreement or for the purposes of arbitral proceedings;
- Default provisions apply if or to the extent there is no agreement under s.76;
- The parties may agree that the court should *not* have the powers set out in s.77 relating to service.

Time periods – (s.78)

- The parties may agree on the method of reckoning periods of time for the purposes of any of their own agreed provisions or any non-mandatory provisions of the Act having effect in default of other agreement;
- Default provisions apply if or to the extent that there is no agreement.

Extension of time limits by court – (s.79)

- The parties may agree to *exclude* the power of the court to extend time limits.

C. Agreements as to Tribunal and Court Powers

This is a checklist for the more significant powers of the tribunal and the court which the parties may affect by their agreement. The list has been drawn up in the format of draft agreements that could be entered into, so as to indicate some of the options that are available.

Of course, where the parties' wishes concur with the default provisions of the Act, no agreement will be necessary since, in the absence of *contrary* agreement, the default provisions will apply.

'The parties have agreed as follows (all references to 'the Act' being references to the Arbitration Act 1996):'

Jurisdiction (s.30)	'The tribunal may not rule on its own substantive jurisdiction.'
Consolidation (s.35)	'The tribunal shall have the following power of consolidation/power to order concurrent hearings, namely...'
Appointment of experts etc. (s.37)	'The tribunal shall not have the power to appoint its own experts, legal advisers or assessors.' (s.37 (1)(a)) 'In relation to any expert appointed by the tribunal, the parties shall not be given any opportunity to comment on the opinion expressed by him.' (s.37(1)(b))
General powers (s.38)	'The tribunal has the following powers (and only those powers) for the purposes of and in relation to these proceedings, namely...' 'S.38(3)/s.38(4)/s.38(5)/s.38(6) of the Act does/ do not apply to these proceedings.'
Provisional order (s.39)	'The tribunal shall have power to order on a provisional basis any relief which might be granted in the final award [limited to...]...'
Power on party's default (s.41)	'In the event that either party fails to do something necessary for the proper and expeditious conduct of the arbitration, the tribunal shall have the following powers (and only those powers) namely...' 'S.41(3)/s.41(4)/s.41(5)/s.41(6)/s.41(7) of the Act does/ do not apply to these proceedings.'
Court's power to enforce peremptory orders (s.42)	'The court shall not have the power to require any party to comply with any peremptory order made by the tribunal in these proceedings pursuant to s.41(5) of the Act, and s.42 of the Act shall not apply to these proceedings.'

Court's power in support of arbitral proceedings (s.44)	'For the purposes of and in relation to these proceedings, the court shall not have the power to make orders about the matter[s] listed in s.44(2)(a)/(b)/(c)/(d)/(e) of the Act.'
Determination of preliminary point of law (s.45)	'Neither party shall be entitled to apply to the court for the determination of any question of law arising in the course of these proceedings, and the court's jurisdiction under s.45 of the Act is hereby excluded.'
Awards on different issues (s.47)	'The tribunal shall not have the power to make awards on different issues, and it will make only one substantive award in respect of the matters referred to it in these proceedings.'
Remedies (s.48)	'The tribunal may make an award including some or all of the following remedies (and only those remedies), namely ...' '**S.48(3)/s.48(4)/s.48(5)** of the Act does/do not apply to these proceedings.'
Interest (s.49)	'The tribunal shall have the following power (and no further power) to award interest, namely ...' 'S.49(3)/s.49(4) of the Act does/do not apply to these proceedings.'
Extension of time for making award (s.50)	'The power of the court to extend the period of time provided in clause [ ] of this agreement within which the tribunal is to make its award shall not apply, and the court's jurisdiction under s.50 of the Act is hereby excluded.'
Power to correct awards or make additional awards (s.57)	'The tribunal shall have the following powers (and only those powers) to correct awards or make additional awards, namely ...' 'S.57(3)(a)/s.57(3)(b) of the Act does/do not apply to these proceedings.' 'S.57(4) to (6) of the Act applies to these proceedings with the following amendments, namely ...'
Power to limit recoverable costs (s.65)	'The tribunal shall not have the power to direct that the recoverable costs of these proceedings or any part of them be limited to a specified amount, and its jurisdiction pursuant to s.65 of the Act is hereby excluded.'
Determination of question of law arising out of an award (s.69)	'Neither party shall be entitled to appeal to the court on any question of law arising out of an award made in these proceedings, and the court's jurisdiction under s.69 of the Act is hereby excluded.'

D. Checklist for Preliminary Meetings

This is a checklist of some of the more significant points that might come up for consideration at a preliminary meeting. It has been drawn up bearing in mind some of the requirements – and also possibilities – created by the Act.

It should be remembered that in dealing with *all* the points below, the tribunal should have its s.33 duty – 'to adopt procedures suitable to the circumstances of the particular case, avoiding unnecessary delay or expense, so as to provide a fair means for the resolution of the matters falling to be determined' – very firmly in mind.

Appointment – (ss.16–18)

- Are all matters in relation to the appointment of the arbitrators resolved?
- Is a chairman or umpire to be appointed?
- Are there any outstanding matters relating to fees?

Seat – (s.3)

- Has the seat of the arbitration been designated?
- If not, how is it to be designated?

Provision for non-mandatory sections – (s.4)

- Are institutional rules to be adopted?
- Have the parties made any other agreement or arrangement in respect of such sections?

Jurisdiction – (ss.30–32)

- Does the tribunal obviously have jurisdiction?
- Are there likely to be issues as to substantive jurisdiction requiring resolution?
- If so, is the tribunal to rule on the matter in an award as to jurisdiction, or deal with the objection in its award on the merits?
- Or is there to be an application to the court for a determination? If so, should the proceedings continue in the meantime?

Procedure and evidence – (s.34)

- What form should the proceedings take so as to be best suited to the case?
- Are written statements of claim and defence to be used? If so, what form should they take – statements of case, annexing documents, witness statements and the like, or some other form?

- Should there be disclosure of documents, and if so, to what extent?
- Are strict rules of evidence to apply?
- When and how should proofs of evidence be exchanged?
- Should the tribunal proceed inquisitorially, itself taking the initiative in ascertaining facts and law?
- Should there be an oral hearing or only (and if so, what) written evidence and/or submissions?

Consolidation – (s.35)

- Have the parties made some agreement in this respect?
- Or does the tribunal have power to consolidate or hold concurrent hearings, and is it being asked to use it?

Representation – (s.36)

- Are the parties to be represented, and if so, by whom?

Experts – (s.37)

- Should the tribunal consider appointing its own experts, legal advisers or assessors?

General powers – (s.38)

- Does the tribunal have all the powers set out in s.38?
- If not, what are its general powers?
- Should the tribunal make any order in respect of security for costs? (Directions for dealing with an appropriate application, possibly at a separate hearing, will probably be necessary.)
- Should the tribunal give directions in relation to any property? (Directions for dealing with this may again be necessary.)
- Are any directions necessary in respect of witnesses and their examination?

Provisional orders – (s.39)

- Is there, or is there likely to be, an application for any relief on a provisional basis? (Directions for dealing with this would almost certainly be necessary.)

Substantive law – (s.46)

- Is there any dispute or doubt about what substantive law (or which other considerations) the tribunal is to apply?
- If yes, how is this to be resolved?

Awards on different issues – (s.47)

- Is this a case in which it would be appropriate to make awards on different issues at different stages?

Remedies/Interest – (ss.48 and 49)

- Are there any agreements of which the tribunal should be aware as to its powers to include different remedies and provisions as to interest in its award?

Reasons – (s.52)

- Do the parties choose *not* to have a reasoned award?
- Do they otherwise wish to exclude the powers of the court under s.45 to determine a preliminary point of law, and under s.69 to entertain an appeal on a point of law?

Costs capping – (s.65)

- Are there any aspects of the arbitration that suggest themselves as appropriate for a ceiling on recoverable costs?

Writing – (s.5)

- Have any agreements made by the parties as to procedure or otherwise been put into writing?
- Does the tribunal have authority to record any agreements reached at the preliminary meeting on behalf of the parties?

E. Agreement with Arbitrator on Resignation

Resignation

Under s.25, the parties may agree with the arbitrator the consequences of his resignation as to his fees and any liability he may have incurred for breaching his contract with them. In the absence of agreement, the arbitrator may apply to the court, which may grant him relief from liability if his resignation was reasonable and may rule on his entitlement to fees.

These provisions could well become of considerable importance if arbitrators, regarding a particular course or procedure as appropriate pursuant to their s.33 duty, find themselves in conflict with the parties who have agreed on a course which the arbitrator would (reasonably) regard as unacceptable. Resignation may then be the preferred option.

Whilst an agreement as to the consequences of resignation may be made at any time, our model is a simple form of agreement made after resignation.

The parties may agree whether the vacancy left by the resigning arbitrator is to be filled, and if so, how. They may also agree whether and to what extent the proceedings prior to the resignation should stand, and what effect there is on any appointment in which the resigning arbitrator has had a part. Default provisions apply to the extent there is no such agreement, (s.27).

Model agreement with arbitrator on his resignation

'Mr Arbitrator having resigned his appointment as arbitrator in the aforesaid arbitration, the parties and Mr Arbitrator hereby agree as follows:

(a) Mr Arbitrator may retain such fees and expenses as have been paid to him at the date of this agreement, but his invoice of 31st July 1997 is cancelled and he shall not be entitled to the fees sought therein;

(b) as between the parties payment of the fees and expenses of Mr Arbitrator will be dealt with as part of the costs of the arbitration, and the tribunal as reconstituted shall make an award allocating such costs as part of the costs of the arbitration under s.61 of the Act;

(c) Mr Arbitrator will incur no further liability to the parties, save that he will bear the expenses of the Chartered Institute of Arbitrators [or some other arbitral institution] in the appointment of a replacement arbitrator.'

F. Checklist for Awards

Checklist for arbitrators on awards

We have tried here to provide some guidance in short form for arbitrators making awards under the regime contained in the Act.

When are formal awards required?

Apart, of course, from an award on the merits, awards are also required in the following circumstances:

S.31(4)(a)	Specific award as to the tribunal's substantive jurisdiction
S.41(3)	Dismissal of claim for inordinate and inexcusable delay
S.41(6)	Dismissal of claim for failure to comply with peremptory order to provide security for costs
S.47	Awards on different issues
S.49	Awards as to interest
S.51(2)	Agreed award on settlement
S.57(3)(b)	Additional award in respect of any claim presented to the tribunal but not dealt with in the original award
S.61	Allocation of costs as between the parties
S.63(3)	Determination of the recoverable costs of the arbitration

Should the award contain reasons?

Yes, in our view, in all cases unless there is agreement to dispense with reasons under s.52(4) or it is an agreed award under s.51. The court may refer an award back to the tribunal for sufficient reasons to be given to enable a challenge or appeal to be properly considered (s.70(4)).

Other determinations, decisions and orders, and in particular provisional orders under s.39, need not be given or made with reasons.

Form of the award

(1) The parties may have agreed on formal requirements. If so, they must be followed (s.52(1)).

(2) In the absence of agreement, every award:

  (a) must be in *writing*;
  (b) must be *signed*, at least by all those assenting to its contents;
  (c) must contain *reasons* (unless the award is agreed, or the parties have agreed to dispense with reasons);

    (d)   must state the *seat* of the arbitration;

    (e)   must state the *date* when the award is made (see s.54: the power of the tribunal to decide this date may be taken away from them by the parties).

(3)   If the award is made under s.47, it must additionally specify the issue, or the claim or part of a claim, which is its *subject-matter*.

(4)   Where applicable, the award must deal with *interest* either in accordance with the power so to award conferred by the parties, or in accordance with the default provisions of s.49. There may be a contractual provision for interest which should be dealt with.

       Interest is normally awarded by reference to the periods leading up to the date of the award and after the award. See our drafting suggestion at paragraph 49F of the main text.

       Note that the tribunal must now positively provide for interest after the date of the award. It will not automatically accrue (as it did previously).

(5)   Where applicable, the award must allocate the *costs* of the arbitration as between the parties (subject to any agreement by them), following the general principle that costs should follow the event (s.61).

       The costs of the arbitration are defined in s.59 and are:

    (a)   the arbitrators' fees and expenses;

    (b)   the fees and expenses of any arbitral institution concerned;

    (c)   the legal and other costs of the parties.

The award should deal with each of these in turn. See our drafting suggestion at paragraph 61E of the main text.

Form of agreed award

(6)   An agreed award will follow the form set out above with certain differences:

    (a)   it will not contain reasons;

    (b)   it must state that it is an award of the tribunal (s.51(3));

    (c)   it will probably deal with interest and costs in accordance with the parties' request, because these matters will have been agreed. If they have not been agreed, the award must deal with the recoverable costs of the arbitration (s.51(5)). These include the arbitrators' own fees.

G. Agreements as to Costs

We have prepared here a list of possible agreements that may be made concerning costs at various stages of the arbitration. For our drafting suggestion as to an award of costs, see paragraph 61E of the main text.

Agreements before dispute

(1) The parties may agree there should be no allocation of costs, or may make some other agreement as to how costs should be allocated (e.g. that each party will bear its own) that does not infringe s.60.

(2) Note the effect of s.60: an agreement which has the effect that a party is to pay the whole or part of the costs of the arbitration in any event is only valid if made *after* the dispute in question has arisen.

(3) The parties may agree the principles applicable to the award of costs (if different from 'costs following the event' under s.61(2)).

(4) The parties may agree that the allocation of costs extends to something other than the 'recoverable costs' referred to in s.63 – see s.62.

(5) The parties may agree the nature or classes of the costs of the arbitration which are, in principle, recoverable – see ss.63(1) and 64(1), but not (at this stage) their quantum.

(6) The parties may agree to exclude the power of the tribunal to limit the recoverable costs of the arbitration to a specified amount – see s.65.

Agreements after dispute and before award on merits

(7) The parties may agree the matters set out in (1) and (3) to (6) above. Additionally they may make an agreement which has the effect that a party is to pay the whole or part of the costs of the arbitration in any event.

Agreements after award on merits

(8) The parties may agree the allocation of costs as between themselves.

(9) The parties may agree that the tribunal allocate costs, but on a basis other than 'costs following the event' under s.61(2).

(10) The parties may agree that the allocation of costs under an agreement between themselves or under an award extends to

something other than 'recoverable costs' referred to in s.63 – see s.62.

(11) The parties may agree the nature or classes of the costs of the arbitration which are, in principle, recoverable.

(12) The parties may agree the quantum of recoverable costs.

H. Arbitration Applications
General Note on Arbitration Applications

The procedure for applications to the court in relation to arbitration matters is now governed by Part 49 of the Civil Procedure Rules (CPR) and, specifically, the Practice Direction on Arbitrations which is supplemental to Part 49. The Practice Direction is subdivided. Part I concerns applications to which the Act applies (but not enforcement proceedings or a claim on the award): its provisions are summarised in this note. Part II concerns applications to which the law prior to the Act applies and is not dealt with here. Part III concerns enforcement proceedings (other than by a claim on the award) and is dealt with primarily at paragraph 66G.

General principles

It is noteworthy that the general principles set out in s.1 of the Act, emphasising party autonomy and restraint in respect of court intervention, are applied to arbitration applications. The overriding objective otherwise applicable to the CPR under Part 1, with its more specific emphasis on active case management by the court, was not, presumably, thought appropriate.

Claim form

Arbitration applications are begun by issuing an arbitration claim form. This must be on Form 8A which is to be found in volume 2 of the Supreme Court Practice. It is simple to complete, but to comply with the Practice Direction the completed form must:

(1) include a concise statement of
    (a) the remedy claimed, and
    (b) (where appropriate) the questions on which the applicant seeks the determination or direction of the court;
(2) give details of any arbitration award that is challenged by the applicant, showing the grounds for any such challenge;
(3) where the applicant claims an order for costs, identify the respondent against whom the claim is made;
(4) (where appropriate) specify the section of the Act under which the application is brought;
(5) show that any statutory requirements have been satisfied (we have dealt with these under the heading 'Application' in our commentary on the sections affected);

(6) state whether the application is made on notice or without notice and, if made on notice, give the names and addresses of the persons on whom it is to be served, stating their role in the arbitration and whether they are made respondents to the application;

(7) state whether the application will be heard by a judge sitting in public or in private;

(8) state the date and time when the application will be heard or that such date has not yet been fixed; and

(9) be indorsed with the applicant's address for service.

Issue of claim form

The allocation of arbitration proceedings between different courts is dealt with by the High Court and County Courts (Allocation of Arbitration Proceedings) Order 1996, SI 1996 No.3215 as amended by SI 1999 No.1010, together with paragraph 5 of the Practice Direction.

Applications for a stay under s.9 of the Act must be made by way of application in the court in which the legal proceedings which it is sought to stay are pending. (The application is made in the legal proceedings.) As suggested in the Introduction to this edition, this could lead an inconsistency of approach to the interpretation of the relevant provisions of the Act, given the situation outlined in the next paragraph.

In respect of other types of arbitration application, there is an overriding provision that applications must be commenced in the Commercial Court, but they may also be commenced in a Mercantile Court or in the Central London County Court (entered into the Business List). In relation to the latter courts, the judge in charge of the list must as soon as practicable consult with the judge in charge of the commercial list and consider whether the application should be transferred to the Commercial Court or any other list. The criteria to which regard must be had in considering transfer are the financial substance of the dispute including the value of any claim or counterclaim (but disregarding considerations of interest, costs or contributory negligence), the nature of the dispute, the importance of the proceedings (and, in particular, their importance to non-parties) and whether the balance of convenience points to having the proceedings taken other than in the Commercial Court. Where the financial substance of the dispute exceeds £200,000 the proceedings must be commenced and will be retained in the Commercial Court unless they do not raise questions of general importance to non-parties.

Conversely, the Commercial Court may transfer arbitration applications commenced there to another list, court or division of the High Court, adopting the same criteria. In practice, the Commercial Court seeks to exercise a supervisory jurisdiction over the Act and its development, reserving to itself applications in commercial arbitrations and those concerning significant points of arbitration law and practice. Thus, if no significant point of arbitration law or practice is raised, rent-review arbitrations will normally be referred to the Chancery Division; building and civil engineering arbitrations will normally be referred to the Technology and Construction Court; and ship salvage arbitrations will normally be transferred to the Admiralty Court.

Service within the jurisdiction

An arbitration claim form is valid for service within the jurisdiction for one month. Generally, service is effected as for other types of court proceeding in accordance with CPR Part 6. The court may also, following an application without notice, authorise service on the solicitor or agent within the jurisdiction of a respondent who is outside the jurisdiction. Moreover, service in a second or subsequent application arising out of the same arbitration may be effected at the addresses for service given in relation to the first application.

Note that there are particular provisions in relation to service of an application notice (which will be in the same form as an arbitration claim form) for a stay of legal proceedings under s.9 of the Act: see paragraph 9G.

Service outside the jurisdiction

For the purpose of service outside the jurisdiction, a claim form is valid for such period as the court may fix. The permission of the court is required. Permission may be granted if the application seeks to challenge or appeal against an arbitration award made within the jurisdiction; if the application is for an order under s.44 of the Act (court powers exercisable in support of arbitral proceedings) whether the arbitral proceedings are taking place within or outside the jurisdiction; and, in relation to other applications, where the seat of the arbitration is or will be within the jurisdiction or, where no seat has been designated, if there is an appropriate connection with the jurisdiction. The application for permission must be supported by evidence. If granted, service may be effected overseas in the same way as for other types of claim form.

Evidence

The applicant must file evidence in support of the application and serve it with the claim form. This may be an affidavit or a witness statement. CPR Rule 32.6 provides that generally, in applications of this kind, evidence is to be by witness statement. Presumably, a witness statement should comply with the requirements of CPR Part 32 as to format and content, so that, for example, it must include a statement by the intended witness that he believes the facts in it are true. Unless ordered otherwise, affidavits and witness statements may contain hearsay. Where an application is made with the agreement of the other parties or the permission of the tribunal, the evidence must include details and exhibit copies of any document evidencing those matters.

Notice

In most instances, the Act requires applications to be made on notice. We have noted these in our commentary on individual sections. This is generally done by making the other parties and, in the case of applications under ss.24, 28 or 56 of the Act, the arbitrator or arbitrators concerned respondents to the application and serving on them the claim form and evidence in support.

Otherwise, notice to the arbitrator may be effected simply by sending a copy of the claim form and supporting evidence to the arbitrator at his last known address for his information. In such a case, the arbitrator may request, without notice, to be made a respondent to the application. Alternatively (and, we suggest, normally) the arbitrator may make representations to the court by filing an affidavit or witness statement or making representations in writing. Such representations must be sent to all the parties to the application and the court may admit them and give weight to them as it sees fit.

Note that different provisions concerning notice apply to applications for a stay under s.9: see paragraph 9G.

Acknowledgment of service

A respondent to an arbitration application may acknowledge service by filing and serving on the other parties the appropriate form (No. N15A) within 14 days of service of the claim form. Failure to do so will debar the respondent from contesting the application without the permission of the court and the court will not notify such a respondent of the hearing date. In relation to applications under s.9 of the Act (as to which see paragraph 9G) and subsequent applications in the same arbitration this rule as to acknowledgment of service does not apply.

Directions

Unless the court otherwise directs, certain directions take effect automatically:

(1) a respondent may serve evidence in response to the application within 21 days after the time for acknowledging service or, if

acknowledgment of service is not required, within 21 days after service of the claim form;

(2) any further evidence in response by the applicant must be served within seven days after service of the respondent's evidence;

(3) where a date has not been fixed for the hearing, the applicant must, and the respondent may, apply to the court for a hearing date within 14 days after the time limit within which a respondent may serve evidence;

(4) the applicant (with the co-operation of the respondent) is responsible for the preparation of agreed, indexed and paginated bundles of all the evidence and other documents for use at the hearing;

(5) not less than five clear days before the hearing date time estimates and a complete set of the bundles must be lodged with the court;

(6) not less than two days before the hearing date the applicant must lodge with the court and send to the respondent –
   (a) a chronology cross-referenced to the bundles;
   (b) (where necessary) a list of the persons involved;
   (c) a skeleton argument listing succinctly the issues; the grounds relied upon for seeking or opposing relief; submissions of fact cross-referred to the evidence; and submissions of law cross-referred to relevant authorities;

(7) not less than the day before the hearing the respondent must lodge with the court and send to the applicant a similarly configured skeleton argument.

If the court gives specific directions then these will be such as it thinks best adapted to secure the just, expeditious and economical disposal of the application. The CPR rules which allocate cases in court to certain procedural tracks do not apply to arbitration applications. Nevertheless, the court may give directions for the attendance of witnesses for cross-examination; for hearing the application on oral evidence or partly on oral and partly on written evidence if there is or may be a dispute as to fact; and indeed any other directions which the court could give in other types of proceedings.

Accordingly, although the majority of applications are dealt with simply on written evidence and documentation, in some instances the hearing of the application may take on the characteristics of a small trial.

The sanctions for the applicant who defaults or delays are that the court may dismiss the application or make such other order as may be just. In relation to the defaulting respondent the court may determine the application without having regard to his evidence or submissions.

Hearing

Generally, arbitration applications are heard in private with the exception of the substantive determination of a preliminary point of law under s.45 and the substantive appeal on a question of law under s.69, which are heard in public. However the court may order differently.

Security for costs

The court may order any applicant (including one granted permission to appeal) to provide security for costs of any arbitration application. See also s.70(6) of the Act in this regard.

Specific provisions

There are specific provisions in relation to a number of different kinds of application under the Act which are dealt with in our commentary on those sections.

Part 4

The Arbitration Act 1996

Text of the Act and Commentary

An Act to restate and improve the law relating to arbitration pursuant to an arbitration agreement; to make other provision relating to arbitration and arbitration awards; and for connected purposes. [17 June 1996]

The Arbitration Act 1996: Arrangement of Sections

PART II
OTHER PROVISIONS RELATING TO ARBITRATION
Domestic arbitration agreements

Consumer arbitration agreements

Part I

Arbitration Pursuant to an Arbitration Agreement

Introductory

Section 1 – General Principles

1. The provisions of this Part are founded on the following principles, and shall be construed accordingly–

(a) the object of arbitration is to obtain the fair resolution of disputes by an impartial tribunal without unnecessary delay or expense;
(b) the parties should be free to agree how their disputes are resolved, subject only to such safeguards as are necessary in the public interest;
(c) in matters governed by this Part the court should not intervene except as provided by this Part.

Definitions

'agree': s.5(1).
'the court': s.105.
'dispute': s.82(1).
'party': ss.82(2), 106(4).

[1A] ## Status

So far as English law is concerned, this is a new provision, but s.1(b) and (c) reflect Arts.19(1) and 5 respectively of the Model Law.

Whilst ss.1 to 6 are not expressly made mandatory by s.4(1) and Sched.1 and are therefore, in theory at least, capable of being supplanted by agreement of the parties in accordance with s.4(2), we think that these sections are so fundamental to the operation of the Act that their application could not be affected by contrary agreement.

[1B] ## Summary

This section sets out the general principles on which this Part of the Act is founded and by reference to which the courts must construe the Act.

Points

[1C]   *Interpretation*

It is unusual for a statute expressly to formulate its founding principles in this way. Having done so, these principles will be essential to its construction. Where, for example, the meaning of any section in Part I has to be considered because it leaves room for doubt or is open to more than one interpretation, these principles will provide the basis for interpretation.

In another respect, too, the Act breaks new ground. If the meaning of any section has to be considered in connection with the 'mischief' which it was designed to remedy, the Report of the Departmental Advisory Committee on Arbitration Law of February 1996, which gives a succinct account of how and why each section was formulated, provides a readily accessible external source of interpretative material. Hereafter we refer to this Report as the DAC Report.

Although not a statute, the Civil Procedure Rules (CPR) have similarly now set out an 'Overriding Objective' in Rule 1.1. Rule 1.2 of the CPR requires the court to seek to give effect to that objective when exercising any powers under the CPR and in the interpretation thereof.

[1D]   *Object of arbitration – S.1(a)*

Whilst there is no attempt to define arbitration, its object is set out, in the first principle, as being the fair resolution of disputes by an impartial tribunal without unnecessary delay or expense. This aim is taken up at s.33 where positive obligations are placed on the tribunal to act fairly and impartially and to adopt procedures which provide a fair means for the resolution of the matters to be determined.

There is, deliberately, no express requirement of independence on the part of the tribunal. (As to the independence of the tribunal in the context of statutory arbitrations, see paragraph 96D.) However an arbitrator may be removed on several grounds, including circumstances giving rise to justifiable doubts as to his *impartiality* and a failure or refusal on his part to conduct the proceedings properly (see s.24).

At s.40 there is a corresponding duty placed on the parties to do all things necessary for the proper and expeditious conduct of the proceedings. This latter section is backed up by new powers to make peremptory orders (ss.41 and 42), and by sections such as s.73 which provide for the loss of certain rights in the event of delay.

In similar vein, Rule 1.1 of the CPR requires the court to deal with cases 'justly'. 'Justly' includes

(1)   ensuring that the parties are on an equal footing;
(2)   saving expense;
(3)   dealing with the case in ways that are proportionate to the amount in dispute, the importance of the case, the complexity of the issues and the financial position of each party;
(4)   ensuring that the case is dealt with expeditiously and fairly.

In our view the intention of both the Act and the CPR is similar, and the arbitrator of the new millennium should have at least half an eye on the provisions of the CPR.

[1E]   *Party autonomy – S.1(b)*
The second principle is that of party autonomy, or leaving the initiative, as far as possible, with the parties themselves. The principle that parties are free to choose their procedure is also a fundamental basis of the Model Law, and is equally to be found in international arbitration, (see International Chamber of Commerce (ICC) Rules Art.15 and London Court of International Arbitrators (LCIA) Rules Art.14). Art.19(1) of the Model Law provides:

> 'Subject to the provisions of this Law, the parties are free to agree on the procedure to be followed by the arbitral tribunal in conducting the proceedings.'

This principle is reflected in the Act in a number of ways. Most significantly, apart from certain mandatory provisions (which apply in all cases, and to that extent limit party autonomy) the provisions of Part I of the Act are non-mandatory, giving the parties the opportunity to make their own choices wherever possible, and only coming into play if they do not do so, (see s.4). To a large degree, the parties can control the format of the arbitration, and consequently the cost and speed with which it is to be conducted.

We should draw particular attention to s.34 which provides that all procedural and evidential matters are for the tribunal to decide 'subject to the right of the parties to agree any matter'. This section alone gives the parties considerable scope for imposing procedures on the tribunal. We comment in paragraph 33H on the potential conflict between party autonomy, as expressed through the exercise of s.34 rights by the parties, and the tribunal's duty to adopt procedures suitable to the circumstances of the particular case, as expressed in s.33.

In order further to encourage party autonomy, the Act has been drafted in a 'user-friendly' way, using simple language, avoiding obscure terminology and adopting provisions that are complete in themselves and do not require reference to other sources to be fully understood.

[1F]   *Court intervention – S.1(c)*
The third principle deals with the role of the court. Art.5 of the Model Law provides that, 'In matters governed by this Law, no court shall intervene except where so provided in this Law'. S.1(c) therefore follows the Model Law in restricting the circumstances in which the court may intervene in arbitrations. These are broadly confined to where the court's powers support the arbitral process, for instance under s.44, or correct manifest injustice, for instance in the retention of a limited appeal process.

As indicated in the Introduction, we believe that 'should' is used here, in contrast to 'shall' in the Model Law, because the section is setting out principles, and the subjunctive is therefore required. In our view, the court in fact has no discretion: it *must* not intervene except as provided elsewhere in Part I of the Act.

It is also to be noted that the principle of non-intervention is restricted to 'this Part' of the Act. It would not appear, therefore, to apply to consumer arbitration agreements, as set out in Part II. This is plainly consistent with the wide-ranging protection given to consumers under the Unfair Terms in Consumer Contracts Regulations 1999 (as to which see the general note at paragraph 89GN).

(1G)   *Rules*

Consistent with the style of drafting that contains an objective, CIMAR Rule 1.2 provides that the objective of the rules is to provide for the fair, impartial, speedy, cost-effective and binding resolution of construction disputes. By Rule 1.7 the Rules do not exclude the powers of the court, nor any agreement between the parties concerning those powers. By Rule 1.3, after an arbitrator has been appointed under the Rules, the parties may not, without the agreement of the arbitrator, amend the Rules or impose procedures in conflict with them. For our part, in the light of section 1(b) of the Act, we think the enforceability of Rule 1.3 is open to question. Art 1.2 of the Chartered Institute of Arbitrators (CIArb) Rules contains a similar provision and must therefore be subject to the same reservation.

Rule 1.1 of the Institution of Civil Engineers (ICE) Arbitration Procedure recites the object of arbitration in the terms of s.1(a). There is no attempt *per se* to limit s.1(b), but Rule 24.4 provides that

> 'If after the appointment of the Arbitrator any agreement is reached between the parties which is inconsistent with this Procedure the Arbitrator shall be entitled upon giving reasonable notice to terminate his appointment, and shall be entitled to payment of his reasonable fees and expenses incurred up to the date of termination.'

As we have intimated above, the principle of party autonomy prevails in international arbitration. Thus Art.14.1 of the LCIA Rules provides that the parties are encouraged to agree on the conduct of their arbitral proceedings (within the confines of their s.33 duties) and the ICC Rules (Art.15.1) permit the parties to settle on other rules (the IBA (International Bar Association) Rules of Evidence, for example) where the ICC Rules are silent.

Section 2 – Scope of Application of Provisions

2.–(1) The provisions of this Part apply where the seat of the arbitration is in England and Wales or Northern Ireland.

(2) The following sections apply even if the seat of the arbitration is outside England and Wales or Northern Ireland or no seat has been designated or determined–

(a)   sections 9 to 11 (stay of legal proceedings, &c.), and
(b)   section 66 (enforcement of arbitral awards).

(3) The powers conferred by the following sections apply even if the seat of the arbitration is outside England and Wales or Northern Ireland or no seat has been designated or determined–

(a)   section 43 (securing the attendance of witnesses), and
(b)   section 44 (court powers exercisable in support of arbitral proceedings);

but the court may refuse to exercise any such power if, in the opinion of the court, the fact that the seat of the arbitration is outside England and Wales or Northern Ireland, or that when designated or determined the seat is likely to be outside England and Wales or Northern Ireland, makes it inappropriate to do so.

(4) The court may exercise a power conferred by any provision of this Part not mentioned in subsection (2) or (3) for the purpose of supporting the arbitral process where–

(a)   no seat of the arbitration has been designated or determined, and
(b)   by reason of a connection with England and Wales or Northern Ireland the court is satisfied that it is appropriate to do so.

(5) Section 7 (separability of arbitration agreement) and section 8 (death of a party) apply where the law applicable to the arbitration agreement is the law of England and Wales or Northern Ireland even if the seat of the arbitration is outside England and Wales or Northern Ireland or has not been designated or determined.

Definitions

'arbitration agreement': ss.6, 5(1).   'party': ss.82(2), 106(4).
'the court': s.105.   'seat of the arbitration': s.3.
'legal proceedings': s.82(1).

[2A]   ## Status

This is a new provision in the sense that the 1950 Act contains no equivalent. It reflects Art.1(2) of the Model Law.

See our comment as to the mandatory/non-mandatory status of ss.1 to 6 at paragraph 1A.

[2B] Summary

This section sets out the scope of application of Part I of the Act. The section was evidently difficult to draft, and there were several attempts to reach a satisfactory solution. In the end, the scope of this Part of the Act is based on the concept of the 'seat' of the arbitration, so that in general Part I applies to arbitrations where the 'seat' of the arbitration is in England, Wales or Northern Ireland. The 'seat' is defined in s.3. It is designated either by the parties to the arbitration agreement, or by an institution or third party with such power, or by the arbitral tribunal if authorised by the parties, or lastly (in the absence of a designation) by the court.

Certain sections of Part I apply even if the seat is outside England, Wales or Northern Ireland, but the Act refrains from trespassing upon other jurisdictions by exercising powers in relation to arbitrations abroad other than in a very limited and necessary way, even if the parties have chosen English law as the substantive law.

Points

[2C] *Importance of identifying the seat – Subs.(1)*
Part I of the Act applies without exception where the seat of the arbitration is in England, Wales or Northern Ireland.

Since 'seat' is carefully defined, and to avoid complexities arising under s.3, it seems to us important that parties to an arbitration agreement to which the Act is intended to apply should specifically provide that the seat of the arbitration is agreed to be in one of those countries.

[2D] *Stay and enforcement – Subs.(2)*
Certain sections of Part I concerning, for example, a stay of legal proceedings and enforcement of the arbitral award, apply irrespective of the location of the seat. Thus any legal proceedings brought in England, Wales or Northern Ireland may be stayed, pursuant to s.9, in favour of an arbitration agreement that provided for arbitration in, say Switzerland; and any resulting Swiss award may, pursuant to s.66, be enforced in England, Wales and Northern Ireland; notwithstanding that the seat of the arbitration was designated as Switzerland.

This subsection ensures that the United Kingdom's obligations under the New York Convention on the Recognition and Enforcement of Foreign Awards are fulfilled. As to these generally, see the commentary on Part III of the Act.

[2E] *Powers in support of foreign arbitrations – Subs.(3)*
The powers referred to in this subsection (which supersedes parts of s.25 of the Civil Jurisdiction and Judgments Act 1982) apply notwith-

standing that the seat is not in England, Wales or Northern Ireland. This means that if the seat were in, say, Switzerland, and subject to the discretion set out in the latter part of subs.(3), the court might order United Kingdom-based witnesses to attend a hearing being conducted in England, Wales and Northern Ireland, (s.43). It might also exercise powers under s.44, for example as to the preservation of evidence and assets, the inspection (etc.) of property, and the issuing of interim injunctions.

The exercise of these powers is, by the proviso, a matter of 'appropriateness'. Thus we would, for example, think it quite appropriate that the court accede to a request by a party to an arbitration, the seat of which is in Switzerland, that there be an inspection of property in the hands of one of the parties in England in the event that the Swiss courts could not act effectively, (s.44(2)(c) and (5)). We would also think it possible that the court might issue an injunction restraining one of the parties in this country from dealing with property that might be dissipated in this country, when the seat was elsewhere.

[2F]   *No seat – Subss.(2), (3)*
The provisions in subss.(2) and (3) apply notwithstanding that no seat has been designated or determined. Unless the seat is designated in the arbitration agreement, there may be a period, which could be considerable, between the commencement of the arbitration, (s.14), and identification of the seat. Alternatively, the arbitration may not have started at all but the court's assistance may be required, for instance under s.44. Hence this provision.

It leads, of course, to a complexity in the proviso to subs.(3) in that where no seat has been designated or determined, the court, in making its assessment of appropriateness, must prejudge as best it can the likely outcome of the designation or determination process.

[2G]   *Powers where no seat elsewhere – Subs.(4)*
The court may also exercise any other power it has within Part I of the Act in support of an arbitration *before* the seat has been designated or determined, where it is appropriate to do so by virtue of a connection with England and Wales or Northern Ireland. If there is no seat elsewhere, then there is no risk of a conflict with other jurisdictions. The required domestic connection may well be the likelihood that the seat of the arbitration will be in England, Wales or Northern Ireland.

One of the most likely uses of this power is where an appointment procedure has begun, but failed. Thus the powers likely to be relevant prior to designation or determination of the seat would appear to be those in s.12 (power of court to extend time for beginning arbitral proceedings); s.17(3) (power of court to set aside appointment of sole arbitrator in default); s.18 (failure of appointment procedure); s.24 (power of court to remove arbitrator); and s.32 (power to determine preliminary point of jurisdiction).

[2H]   ***Provisions where English law applies – Subs.(5)***
S.7 (separability of the arbitration agreement) and s.8 (death of a party)
apply where the law applicable to the *arbitration agreement* is the law of
England and Wales or Northern Ireland, regardless of where the seat of
the arbitration may be, and regardless of whether or not a seat has yet
been identified. It is important to note that what is relevant here is the
law applicable to the arbitration agreement. This may differ from the
law applicable to the substance of the dispute, (as to which, see s.46).

The subsection has the effect that wherever an arbitration agreement
may be litigated (or arbitrated), if, according to that other state's conflict of laws rules, it is subject to the law of England and Wales or
Northern Ireland, the arbitration agreement shall be treated as a distinct agreement that is separable from the main agreement, (s.7), and
that will not be discharged by the death of a party, (s.8). The application of these sections regardless of the location of the seat is designed to
preserve the original choice of the parties to submit their disputes to
arbitration. The importance of the former is obvious; the latter prevents
the revival of the common law rule that an arbitration agreement is
discharged by the death of a party. Since the Act repeals earlier legislation abrogating this rule, it would revive unless specifically dealt
with.

[2I]   *Model Law*
The Model Law has a similar, but less detailed provision. Art.1(2)
provides:

> 'The provisions of this Law, except articles 8, 9, 35 and 36, apply only if the
> place of arbitration is in the territory of this State.'

Art.8 provides for a stay of legal proceedings to arbitration (as in s.9);
Art.9 provides for interim measures of protection (as in s.44); and
Arts.35 and 36 provide for recognition and enforcement.

Section 3 – The Seat of the Arbitration

3. In this Part 'the seat of the arbitration' means the juridical seat of the arbitration designated–

(a) by the parties to the arbitration agreement, or
(b) by any arbitral or other institution or person vested by the parties with powers in that regard, or
(c) by the arbitral tribunal if so authorised by the parties,

or determined, in the absence of any such designation, having regard to the parties' agreement and all the relevant circumstances.

Definitions

'agreement': s.5(1).
'arbitration agreement': ss.6, 5(1).
'party': ss.82(2), 106(4).
'seat of the arbitration': s.3.

[3A] Status

This is a new provision.

See our comment as to the mandatory/non-mandatory status of ss.1 to 6 at paragraph 1A.

[3B] Summary

This section defines 'the seat of the arbitration' which is used as the springboard for other important sections of the Act, including the scope and applicability of Part I. It is defined as the juridical seat as designated or, in the absence of designation, as determined. The 'juridical seat' is a concept independent of the place where the hearings or other parts of the arbitral process occur, which may not be the same (see paragraph 3C below).

Points

[3C] *Location of the seat*
The seat may well be synonymous with the place where the arbitration is conducted. However, not all international arbitrations are conveniently conducted in one place, and not all awards are conveniently signed in one location. The Act thus provides for a constant juridical seat regardless of where the arbitration may be conducted (s.34(2)(a) in effect allows the tribunal freedom of choice as to the location of hear-

ings) or where the award is signed, despatched or delivered, as to which see ss.53 and 100(2)(b).

[3D]   *Statutory arbitrations*
Note that statutory arbitrations are deemed to have their seat in England and Wales or Northern Ireland, as appropriate (s.95(2)). Thus for their purposes, the possibility of Part I not applying because of a designation or determination of the seat outside England and Wales or Northern Ireland is eliminated.

[3E]   *Designation*
The seat of the arbitration may be designated by the parties themselves, or by an authorised third party (such as an institution), or by the arbitral tribunal, if so authorised. It will clearly be most convenient if the parties designate the seat in their arbitration agreement, or agree its location once an arbitration has been commenced, (see s.14). It has been held that where parties agree the place of an arbitration pursuant to institutional rules (e.g. Art.7.1 of the 1985 LCIA Rules) it follows that such place is to be the seat of the arbitration: *ABB Lummus Global Ltd* v. *Keppel Fels Ltd* [1999] 2 Lloyd's Rep. 24.

Note that the position under the 1998 LCIA Rules is quite clear: Art.16 specifically refers to the seat of an arbitration. See also our comments in paragraph 3I below.

In the absence of agreement by the parties it will next lie with an arbitral or other institution vested by the parties with appropriate powers to make the designation. For instance, Art.14 of the ICC Rules provides:

> 'The place of arbitration shall be fixed by the [International] Court [of Arbitration] unless agreed upon by the parties.'

In the absence of either of these two methods of designation, the task will fall to the tribunal, if so authorised. For instance, Art.16(1) of the UNCITRAL Arbitration Rules provides: 'Unless the parties have agreed upon the place where the arbitration is to be held, such place shall be determined by the arbitral tribunal, having regard to the circumstances of the arbitration.'

[3F]   *Determination*
If there is a failure of the process of designation, determination by the court may become necessary. For instance, the parties may need to know if Part I applies at all, or they may need to make an application to the court in respect of a power given to it in Part I. If so, and if no seat has been designated, it is likely to be necessary for a determination to be made by the court.

Certainly, by the time an award is made, the seat must have been identified, since the award is required to state the seat, (s.52). Internationally, for enforcement purposes, the seat is essential.

[3G] *Delay*

Whilst we endorse the principle of identifying a 'seat' for the arbitration which operates as a springboard for other provisions and as a cornerstone of enforcement, the provision is complex and may be a source of delay, although we are not yet aware of any instances where this has proved to be the case. As we have suggested above, much the best course is for the parties to make an agreement. Otherwise, an alternative mechanism avoiding reference to the court is to be preferred.

[3H] *Model Law*

Art.20(1) of the Model Law provides:

> 'The parties are free to agree on the place of arbitration. Failing such agreement, the place of arbitration shall be determined by the arbitral tribunal having regard to the circumstances of the case, including the convenience of the parties.'

The nature and purpose of this Article is similar to s.3, but the options are clearly more limited.

[3I] *Rules*

CIMAR Rule 1.6 states that the Rules apply where the seat of the arbitration is in England and Wales or Northern Ireland. It does not follow that this is a designation of the seat. In contrast, Rule 24.3 of the ICE Arbitration Procedure states that where the seat is in a country other than England and Wales or Northern Ireland the Procedure applies to the extent that the applicable law permits. We have covered the international rules in paragraph 3E above.

We have cited Art.16(1) of the UNCITRAL Arbitration Rules in that paragraph. Use of this Article will clearly satisfy the needs of s.3.

Section 4 – Mandatory and Non-mandatory Provisions

4.–(1) The mandatory provisions of this Part are listed in Schedule 1 and have effect notwithstanding any agreement to the contrary.

(2) The other provisions of this Part (the 'non-mandatory provisions') allow the parties to make their own arrangements by agreement but provide rules which apply in the absence of such agreement.

(3) The parties may make such arrangements by agreeing to the application of institutional rules or providing any other means by which a matter may be decided.

(4) It is immaterial whether or not the law applicable to the parties' agreement is the law of England and Wales or, as the case may be, Northern Ireland.

(5) The choice of a law other than the law of England and Wales or Northern Ireland as the applicable law in respect of a matter provided for by a non-mandatory provision of this Part is equivalent to an agreement making provision about that matter.

For this purpose an applicable law determined in accordance with the parties' agreement, or which is objectively determined in the absence of any express or implied choice, shall be treated as chosen by the parties.

Definitions

'agreement', 'agree': s.5(1).
'party': ss.82(2), 106(4).

[4A] Status

Although the 1950 Act contained some provisions which might fall away where there was contrary agreement, in this form this is a new provision.

See our comment as to the mandatory/non-mandatory status of ss.1 to 6 at paragraph 1A.

[4B] Summary

This section has the effect of rendering certain sections of Part I of the Act 'mandatory', in the sense that they cannot be overriden by agreement of the parties, whilst leaving others 'non-mandatory', so that agreement to the contrary may be made. It is fundamental to the scheme of Part I of the Act.

Points

[4C]   *Effect*
The section reflects the principle of party autonomy declared in s.1(b), so far as practicable. Thus the majority of sections in Part I are in the 'non-mandatory' category. The parties have been given the freedom to make their own arrangements by agreement. Where they do not do so, rules set out in each of these sections apply by default.

The balance of the sections (listed in Sched.1) are, however, 'mandatory'. Mandatory status means that these provisions cannot be overriden by contrary agreement. The scope of party autonomy is thereby restricted. The mandatory sections are generally those which are fundamental to the Act's regime or its founding principles, as set out in s.1. Agreement contrary to such principles would undermine the contemplated arbitral process unacceptably.

Mandatory sections include such important supportive and regulatory provisions as the court's powers to stay legal proceedings (ss.9 to 11); to remove an arbitrator (s.24); to secure the attendance of witnesses (s.43); to enforce the award (s.66); and to deal with challenges to the award (ss.67 and 68). They also include provisions reflecting the basic tenets of the Act such as the duty of the tribunal to act fairly and expeditiously, (s.33), and the corresponding duty of the parties, (s.40).

[4D]   *Application*
It should be borne in mind that the mandatory provisions of Part I only apply if Part I itself applies. It does so if the seat of the arbitration is in England and Wales or Northern Ireland, (s.2(1)). Thus, for instance, Part I would not be applicable to an arbitration concerning a contract the proper law of which was English, but where the seat of the arbitration was Scotland.

[4E]   *Mandatory provisions – Subs.(1)*
It is vital to an understanding of the scheme of the Act to appreciate that mandatory provisions have effect 'notwithstanding any agreement to the contrary'.

Subs.(3) points to the adoption of rules as a possible means of making arrangements in respect of non-mandatory provisions. We anticipate that, at least initially, some parties may have adopted existing institutional rules not specifically adapted to the Act's provisions. Any provision of such rules that is contrary to a mandatory provision of the Act will automatically be ineffective. We also assume that any ancillary provision of the ineffective rule would also be rendered ineffective, although the extent of that ineffectiveness may be debatable in each instance. We consider, however, that the ineffectiveness of some rules would not render the balance ineffective, and they would then form the arrangement contemplated by subs.(2).

[4F] *Non-mandatory provisions – Subs.(2)*

In respect of non-mandatory provisions, the parties may make their own arrangements by agreement, but the Act provides rules that apply in the absence of such agreement.

We originally thought that this system was, at least initially, likely to present considerable difficulty for lay parties and considerable scope for lawyers. However, we have seen no evidence that this has in fact been the case. Nonetheless, until custom-made agreements are universally available and are being fully used, the only truly satisfactory and thorough course is for the parties to consider each non-mandatory section of Part I and decide whether or not the default rules should apply. If not, then the parties will have to make an agreement to the contrary. We have compiled a number of checklists which appear in the Materials section earlier in this book to assist in identifying the provisions that require this attention.

A comparison of the operative wording of some of the non-mandatory sections reveals that different formulae are used, with different consequences. For example, s.14 provides: 'The parties are free to agree when arbitral proceedings are to be regarded as commenced ... If there is no such agreement, the following provisions apply ...' Here, *any* agreement will oust *all* the default rules. By contrast, s.16 provides: 'The parties are free to agree on the procedure for appointing the arbitrator ... If or to the extent that there is no such agreement, the following provisions apply ...' Here the default rules continue to apply save to the extent that they are expressly contradicted or, presumably, obviously rendered inapplicable by contrary agreement.

[4G] *Institutional rules – Subs.(3)*

The first part of this subsection expressly contemplates the use of institutional rules as a means by which the parties may make their own arrangements. We would caution against the use of rules which have not been specifically drafted or adapted to suit the provisions of the Act, unless the parties or their representatives have carried out a close analysis of them against both mandatory and non-mandatory sections. There are now a wide range of rules drafted since 1996 which take the Act into account.

An important question arises as to the effect of incorporating into an arbitration agreement rules which are silent on one or more of the topics covered by some provisions of the Act, e.g. ss.38 and 41. The answer depends on the proper reading to be given to s.4(2) and (3). In our view unless the incorporated rules or the parties' particular agreement expressly address the matter in question, the Act's default provisions apply. Otherwise, in the unlikely but possible situation where parties incorporated a set of rules that did not deal at all with any of the default powers given under the Act a tribunal would find itself without any of these powers. In our view that was neither the intention of Parliament, nor the intention that should be inferred from a proper reading of the words used.

[4H]   *Agreement in writing*
The effect of s.5(1) is to require any agreement between the parties to be in writing if it is to be 'effective for the purposes of this Part'. It follows that agreements between the parties to change or make substitutions for the non-mandatory provisions of this Part must be in writing, as defined by s.5.

[4I]   *Law applicable to agreement – Subs.(4)*
The law applicaole to the agreement making the parties' arrangement in respect of a non-mandatory provision is immaterial.

[4J]   *Choice of foreign law – Subs.(5)*
The choice of a foreign law (which may be made expressly or impliedly, or which may be objectively determined) in respect of a matter provided for by a non-mandatory provision of Part I is the equivalent of an agreement about that matter.

   Again, we consider that where parties wish a foreign law to apply, they should specify by agreement exactly which non-mandatory provisions it is intended to replace, and to what extent. We think that the use of the expression 'in respect of *a matter* . . .' suggests such a specific approach. We foresee considerable difficulties arising where the parties simply provide that their arbitration procedure shall be subject to the arbitration law of a particular state. We consider that with non custom-made institutional rules there are difficulties in ascertaining the extent to which they would replace non-mandatory default rules, even though they provide for those matters. The same difficulties would be inherent in the global selection of a foreign law. Again, it is important to remember that matters in the foreign law which conflict with the *mandatory* provisions of the Act would not be effective at all.

[4K]   *Conclusion*
We consider that the implications of this section are not at all easy. We caution against the use of non custom-made institutional rules, as also against the global adoption of a foreign law. The most comprehensive course is for the parties to make an agreement that considers each non-mandatory section. A more pragmatic approach would be for the parties to accept the regime put in place by the default rules of the non-mandatory provisions, making such amendments as may occur to them as necessary. We suspect that the latter may be the course most commonly adopted. Indeed in some fields where little thought is given to arbitration clauses and provisions – for example, maritime – this course is likely to be followed by default.

Section 5 – Agreements to be in Writing

5.–(1) The provisions of this Part apply only where the arbitration agreement is in writing, and any other agreement between the parties as to any matter is effective for the purposes of this Part only if in writing.

The expressions 'agreement', 'agree' and 'agreed' shall be construed accordingly.

(2) There is an agreement in writing–

(a) if the agreement is made in writing (whether or not it is signed by the parties),
(b) if the agreement is made by exchange of communications in writing, or
(c) if the agreement is evidenced in writing.

(3) Where parties agree otherwise than in writing by reference to terms which are in writing, they make an agreement in writing.

(4) An agreement is evidenced in writing if an agreement made otherwise than in writing is recorded by one of the parties, or by a third party, with the authority of the parties to the agreement.

(5) An exchange of written submissions in arbitral or legal proceedings in which the existence of an agreement otherwise than in writing is alleged by one party against another party and not denied by the other party in his response constitutes as between those parties an agreement in writing to the effect alleged.

(6) References in this Part to anything being written or in writing include its being recorded by any means.

Definitions

'agreement', 'agree', 'agreed': s.5(1).   'party': ss.82(2), 106(4).
'agreement in writing': s.5(2) to (5).   'written', 'in writing': s.5(6).
'arbitration agreement': ss.6, 5(1).

[5A]   ## Status

This section derives from the Model Law, Art.7(2), and preserves the requirement for writing found in the Arbitration Act 1950, s.32 and the Arbitration Act 1975, s.7(1).

See our comment as to the mandatory/non-mandatory status of ss.1 to 6 at paragraph 1A.

[5B]   ## Summary

This section sets out the formal requirements of an arbitration agreement and indeed of all agreements between the parties concerning an arbitration if they are to be effective for the purposes of this Part of the Act.

For the substantive definition of an arbitration agreement, see s.6.

Points

[5C]   *Effect – Subs.(1)*
The effect of the section is that an arbitration agreement must be in writing in order for this Part of the Act to apply.

Similarly, other agreements between the parties concerning the arbitration such as variations to the arbitration agreement, agreements as to procedural matters, agreements to opt out of non-mandatory provisions (in accordance with s.4(2)), and so on must also be in writing to be effective for the purposes of this Part of the Act. 'Agreement in writing' is generously interpreted (subss.(2) to (5)), and 'writing' is very broadly defined (subs.(6)).

Note that a purely oral arbitration agreement is not totally ineffective, since the common law applying to such agreements is expressly preserved by section 81(1)(b).

[5D]   *Agreement in writing – Subs.(2)*
An agreement may be made – that is to say itself embodied – in writing, in which case its form may be a document that is signed or unsigned, or it may be an exchange of communications, such as letters, faxes or e-mails (but see paragraph 5H in this context).

Alternatively, the writing may not itself embody, but may only evidence, or record, the agreement. In that case the writing may take the form of a memorandum written by one of the parties, or even a duly authorised third party.

In any of these cases separate written terms or a distinct arbitration clause may be incorporated into the agreement by a sufficient reference (as to which, see s.6(2)).

[5E]   *Agreement not in writing referring to writing – Subs.(3)*
A non-written agreement that incorporates by reference the terms of a written form of agreement containing an arbitration clause constitutes an arbitration agreement in writing.

This subsection is primarily designed to cover situations such as ship salvage operations, where parties make a purely oral agreement, but by reference to the terms of a written form containing an arbitration clause. However it would also cover an agreement by conduct, where one of the parties proposes to contract on written terms containing an arbitration clause, and the other party, without acknowledging those terms in writing, nevertheless accepts them by performing the contract in accordance with them.

The 'terms which are in writing' could include, for example, a standard form of agreement containing an arbitration clause, or a specific written agreement containing such a clause, or a set of written arbitration rules. To incorporate the written terms the reference must be sufficient, (as to which see s.6(2)).

[5F]   *Agreement not in writing evidenced by recording – Subs.(4)*
This sub-section allows a single party to record a binding arbitration agreement by some form of writing or in some other way, such as, we think, a tape recording, (see subs.(6)). The arbitration agreement is then effective, without more. It is also possible for a third party to make such a record. In either case the recording must have the authority of both parties. The authority can presumably be given orally.

The London Maritime Arbitrators Association (LMAA) has made use of this provision in its new Terms, on the basis of which all its members accept appointment. Relatively few LMAA arbitrations arise from arbitration clauses which expressly refer to the Terms. Although the Terms themselves say that they apply whenever arbitrators are appointed on the basis of the Terms, to avoid any possibility (however remote) of this being held not to bind parties to arbitrate under the Terms, they expressly authorise the parties' respective arbitrators to confirm in writing on behalf of the parties that the arbitration agreement governing the dispute has been made or varied so as to incorporate the Terms.

The sub-section will incidentally have the effect of facilitating flexibility during hearings. Variations to the arbitration agreement or agreements as to procedural matters may be made orally, but will still be effective for the purposes of this Part of the Act as long as they have been duly recorded, with authority. The recording may well be carried out by the tribunal as the authorised third party.

[5G]   *Agreement in submissions – Subs.(5)*
The mere allegation of an oral agreement made by one party in an exchange of written submissions in an arbitration or an action suffices to make an agreement in writing if the other party responds, but does not controvert the allegation. This only applies as between the parties to the exchange, and to the effect alleged.

We believe that 'written submissions' must be intended to have a wider application than merely to formal statements of case and pleadings. It is interesting to note that s.34(2)(c), dealing with procedural matters, uses the different terminology 'statements of claim and defence'. The precise scope of the phrase is unclear. We doubt if an uncontradicted statement concerning the alleged existence of an arbitration agreement in a letter simply relating to an appointment would amount to a 'written submission'. On the other hand, such an uncontradicted statement in a letter seeking a direction, to which the other party responds, could (we think) be covered.

[5H]   *Definition of writing – Subs.(6)*
The definition of 'writing' includes any means by which an agreement may be recorded. Such a wide definition would seem to include all forms of electronic transmission and communication, and will almost certainly embrace a sound recording.

In the computer age, there is, of course, clear scope for arguing over the exact meaning of 'recorded'. Thus for instance an electronic mail transmission may not, perhaps, provide a 'record' unless the recipient chooses to save the message to a file.

[5I]   *Exception*

The only exception to the requirement for writing concerns an agreement to terminate an arbitration, (as to which see s.23(4)). The exception is permitted because of the impracticality of imposing a requirement for writing in certain of the circumstances in which an arbitration may be mutually allowed to determine, for example where both parties simply abandon proceedings, or allow them to lapse.

[5J]   *Model Law and New York Convention*

Art.7(2) of the Model Law provides:

> 'The arbitration agreement shall be in writing. An agreement is in writing if it is contained in a document signed by the parties or in an exchange of letters, telex, telegrams or other means of telecommunications which provide a record of the agreement, or in an exchange of statements of claim and defence in which the existence of an agreement is alleged by one party and not denied by another...'

Art.II of the New York Convention similarly provides:

> '1. Each Contracting State shall recognise an agreement in writing under which the parties undertake to submit to arbitration...
> 2. The term 'agreement in writing' shall include an arbitral clause in a contract or an arbitration agreement, signed by the parties or contained in an exchange of letters or telegrams.'

Over the years, and as systems of communication have changed, there has been considerable difficulty over the requirements of a signature in these, and other, international arbitration laws. It is reasonably clear that there is no requirement either under the Model Law or under the New York Convention that letters, telexes or telegrams must be signed. The position is less clear where there is a formal arbitration agreement or contract containing an arbitration clause.

In this respect the new Act significantly differs, providing very clearly in s.5(2) that no signature is necessary.

The Arbitration Agreement

Section 6 – Definition of Arbitration Agreement

6.–(1) In this Part an 'arbitration agreement' means an agreement to submit to arbitration present or future disputes (whether they are contractual or not).

(2) The reference in an agreement to a written form of arbitration clause or to a document containing an arbitration clause constitutes an arbitration agreement if the reference is such as to make that clause part of the agreement.

Definitions

'agreement': s.5(1).
'arbitration agreement': ss.6, 5(1).
'dispute': s.82(1).
'written': s.5(6).

[6A] Status

This section corresponds to the Model Law, Art.7(1) and – in part – Art.7(2), and it substantially derives from the Arbitration Act 1950, s.32 and the Arbitration Act 1975, s.7(1).

See our comment as to the mandatory/non-mandatory status of ss.1 to 6 at paragraph 1A.

[6B] Summary

This section contains the substantive definition of 'arbitration agreement'. No change in the law is apparently intended, but the definition makes it clear that an arbitration agreement may relate to disputes that are not contractual. For the formal requirements of an arbitration agreement, see s.5.

The Court of Appeal has held that the hallmark of arbitration is that it is a procedure to determine legal rights and obligations, with binding effect, which determination is enforceable in law. Consequently an arbitration clause in an illegal gaming agreement could not be regarded as an arbitration agreement since any arbitration in the circumstances would be devoid of any legal consequences whatsoever and could not therefore be characterised as an arbitration: *O'Callaghan v. Coral Racing Limited* (The Times, 26 November 1998). This case was decided under the 1950 Act. It must be open to question whether the same decision would be reached under the 1996 Act, although at the end of the day the outcome would be the same since no arbitrator could make a valid monetary award.

Note that in the context of statutory arbitrations it is the enactment under which the arbitration is conducted that comprises the 'agreement', (s.95(1)).

Points

[6C] *Dispute – Subs.(1)*
'Dispute' includes 'any difference' between the parties by virtue of the definition in s.82(1) and thus has a broader meaning than hitherto.

In *Sykes* v. *Fine Fare* [1967] 1 Lloyd's Rep. 53, Lord Justice Danckwerts, at p.60, considered that 'difference' would be apt to cover a wider situation than a dispute. In that case the difference comprised a failure by the parties to agree detailed terms for the future performance of a substantive agreement that had come into effect, and which itself clearly indicated that such future performance and the agreement of appropriate terms for that performance was anticipated.

[6D] *Model Law and Subs.(1)*
Art.7(1) of the Model Law, which is reflected in subs.(1), provides:

> '"Arbitration agreement" is an agreement by the parties to submit to arbitration all or certain disputes which have arisen or which may arise between them in respect of a defined legal relationship, whether contractual or not. An arbitration agreement may be in the form of an arbitration clause in a contract or in the form of a separate agreement.'

It will be noted that both the Act and the Model Law define an arbitration agreement by reference to existing and future disputes, and specifically go beyond contractual disputes into (at least) tortious ones. The New York Convention equally recognises arbitration clauses which refer to future disputes.

The classification in the section is necessary because some civil law countries do not recognise agreements to refer future disputes to arbitration. Such an agreement takes away the right of the aggrieved party to have recourse to his own local courts and, so it is said, he should only agree to relinquish such a right in the face of an identified dispute.

[6E] *Incorporation by reference – Subs.(2)*
An arbitration agreement may refer to a written arbitration clause to be found elsewhere, provided the latter is effectively incorporated by reference into the agreement. The question of whether there is an effective incorporation by reference in any particular case will depend upon the facts and the construction of the agreement in that case.

What the subsection should resolve is that there need not be specific reference to the arbitration clause as a pre-requisite to incorporation, as Lord Justice Megaw considered in *Aughton* v. *M.F. Kent Services* (1991) 57 BLR 1, and as subsequent cases such as *Ben Barrett* v. *Henry Boot Management Ltd* [1995] CILL 1026 have emphasised. Rather, as Lord Justice Ralph Gibson considered in *Aughton*, a more general reference to a document containing such a clause may be sufficient to effect incorporation. However, whether Lord Justice Ralph Gibson's approach is to apply without limitation (as the DAC clearly thought it should) has been left to the courts to decide.

The authorities were comprehensively reviewed by Judge Bowsher QC in *Secretary of State for Foreign and Commonwealth Affairs* v. *The Percy Thomas Partnership* [1998] CILL 1342. The judge held himself (gladly) bound by the Court of Appeal decision in *Modern Building Wales* v. *Limmer* [1975] 1 WLR 1281 and that of Brandon J in *The 'Annefield'* [1971] P 168 and concluded that an arbitration clause contained in standard conditions of contract that were incorporated by reference into a contract was binding, though no specific reference was made to the clause in the incorporating provision.

It has been held that it is appropriate to refer to cases decided before the 1996 Act when deciding whether a reference in an agreement to a written form of arbitration clause or to a document containing an arbitration clause is such as to make that clause part of the agreement, and that in the absence of special circumstances general words of incorporation are not effective: *Trygg Hansa Insurance Co. Ltd* v. *Equitas* [1998] 2 Lloyd's Rep. 439.

[6F]   *Model Law and Subs.(2)*

Subs.(2) is in almost identical terms to the latter part of Art.7(2) of the Model Law:

> 'The reference in a contract to a document containing an arbitration clause constitutes an arbitration agreement provided that the contract is in writing and the reference is such as to make that clause part of the contract.'

Section 7 – Separability of Arbitration Agreement

7. Unless otherwise agreed by the parties, an arbitration agreement which forms or was intended to form part of another agreement (whether or not in writing) shall not be regarded as invalid, non-existent or ineffective because that other agreement is invalid, or did not come into existence or has become ineffective, and it shall for that purpose be treated as a distinct agreement.

Definitions

'agreement', 'agreed': s.5(1).
'arbitration agreement': ss.6, 5(1).
'party': ss.82(2), 106(4).

[7A] **Status**

This section corresponds to the Model Law, Art.16(1), and codifies existing case law, (see *Harbour Assurance* v. *Kansa General International Insurance* [1993] QB 701). It is *non-mandatory*, (s.4(2)).

[7B] **Summary**

This section confirms the separability of the arbitration clause from the main contract (or 'agreement', since the Act uses this expression rather than the Model Law's 'contract'). Operation of the section may only be excluded by contrary agreement. Such agreement must be in writing, in accordance with s.5.

Points

[7C] *Effect*

Thus an arbitration clause will remain valid despite an allegation of illegality affecting the substantive agreement (which allegation, if proved, would render the substantive agreement void). Similarly, a decision by an arbitral tribunal that a main agreement is null and void, or the termination of a main agreement by performance, will not of itself entail a similar consequence for the arbitration clause. The validity of the latter, regarded as a separate, collateral agreement, must be examined as a separate issue.

However, these comments should be read in the light of the decision in *O'Callaghan* v. *Coral Racing Ltd* (see paragraph 6B). There, an arbitration clause in an illegal gaming agreement was held to be of no effect. It could not be severed from the agreement. Any purported arbitrator would have had to hold that the gaming transaction was void and that he was debarred from awarding any sum of money alleged to have been won; consequently the clause had to be treated as an integral part of the terms on which the parties agreed to do business, and it could not be separated from them and survive independently. In any event, as mentioned in paragraph 6B above, the clause could not be considered as an arbitration agreement. This case was decided under the 1950 Act and it is unclear what the court's approach would have been had the 1996 Act applied.

This section is particularly significant when read in conjunction with the express provision in s.30 conferring competence on the arbitral tribunal to rule on its own jurisdiction, (see paragraph 7E below).

[7D]   *Specific phrases*

The words 'whether or not in writing' make it plain that the section applies where the arbitration clause forms part of a substantive agreement made orally or other than in writing (as envisaged, for example, by s.5(3)).

The words 'for that purpose' confine the separability principle to situations concerning the impact of invalidity, non-existence or ineffectiveness of the substantive agreement on the arbitration agreement. They do not therefore affect the question of whether an assignment of rights under the substantive agreement carries with it the right or obligation to submit to arbitration in accordance with the arbitration agreement. (The latter is an area in respect of which the Act has stopped short of making any provision because the DAC thought it impractical – given the complexity to which the application of different laws can give rise – and probably of no real use to try to do so).

[7E]   *Model Law*

In the Model Law, the principle of separability of the arbitration agreement is combined with the question of the tribunal's power to rule on its own jurisdiction. The Act deliberately separates these two issues, regarding them as distinct questions.

Art.16(1) of the Model Law provides:

> 'The arbitral tribunal may rule on its own jurisdiction, including any objections with respect to the existence or validity of the arbitration agreement. For that purpose, an arbitration clause which forms part of a contract shall be treated as an agreement independent of the other terms of the contract. A decision by the arbitral tribunal that the contract is null and void shall not entail *ipso jure* the invalidity of the arbitration clause.'

The principle of separability of the arbitration clause is well enshrined internationally.

Section 8 – Whether Agreement Discharged by Death of a Party

8.–(1) Unless otherwise agreed by the parties, an arbitration agreement is not discharged by the death of a party and may be enforced by or against the personal representatives of that party.

(2) Subsection (1) does not affect the operation of any enactment or rule of law by virtue of which a substantive right or obligation is extinguished by death.

Definitions

'agreed': s.5(1).
'arbitration agreement': ss.6, 5(1).
'party': ss.82(2), 106(4).

[8A] Status

This section substantially reproduces the Arbitration Act 1950, s.2(1), (3). It is *non-mandatory*, (s.4(2)), so that the parties may agree that death terminates an arbitration agreement.

[8B] Summary

This section deals with the consequences of the death of a party on the arbitration agreement. Note that this section does not apply in relation to a statutory arbitration, (s.97(a)).

For the effect of the death of the arbitrator, or the person appointing him, see s.26, which complements this section.

Operation of the section may only be excluded by contrary agreement. Such agreement must be in writing, in accordance with s.5.

Points

[8C] *Effect*

The general rule is that on the death of any person all causes of action, except those for defamation, subsisting against or vested in him survive against, or for the benefit of, his estate, see the Law Reform (Miscellaneous Provisions) Act 1934, s.1(1). An arbitration agreement relating to the resolution of a dispute in respect of such a cause of action would, unless the parties agree otherwise, similarly survive, (subs.(1)).

However, where the substantive cause of action is itself extinguished by death (such as a claim for damages for defamation), the arbitration agreement will be discharged with it, (subs.(2)).

[8D]   **Bankruptcy**

Note that the substance of the Arbitration Act 1950, s.3, concerning the effect of bankruptcy on an arbitration agreement, is re-enacted by a new s.349A of the Insolvency Act 1986, inserted by paragraph 46 of Sched. 3 to this Act (where the new section is set out).

Essentially, if a trustee in bankruptcy adopts a contract containing an arbitration agreement, then the agreement is enforceable by or against the trustee. If the trustee does not adopt the contract then either the trustee, with the consent of the creditors' committee, or the other party to the arbitration agreement can apply to the court with jurisdiction in the bankruptcy proceedings for an order referring a matter to arbitration in accordance with the arbitration agreement.

[8E]   **Dissolution**

Where, however, a corporate body ceases to exist, any arbitration in which it was involved lapses absolutely and cannot be revived: *Morris v. Harris* [1927] AC 252 and *Baytur v. Finagro* [1992] QB 610.

There are situations where a corporate body may be struck off a register (e.g. for failing to file a return) and then be re-instated expressly with effect as if it had continued on the register all along. What the position is in such circumstances is unclear: we suspect a court would strain to hold that the arbitration continued in existence, as in effect would the corporate body in such a situation. In a case where a corporation was dissolved following a merger with another, it was held that arbitration proceedings initiated by the dissolved corporation did not lapse or come to an end: *Eurosteel Ltd* v. *Stinnes AG* (unreported, Longmore J, 8 December 1999).

[8F]   **Model Law**

The Model Law has no corresponding provision.

[8G]   **Rules**

Of those rules we have considered, we are not aware of provisions contrary to s.8.

Stay of Legal Proceedings

Section 9 – Stay of Legal Proceedings

9.–(1) A party to an arbitration agreement against whom legal proceedings are brought (whether by way of claim or counterclaim) in respect of a matter which under the agreement is to be referred to arbitration may (upon notice to the other parties to the proceedings) apply to the court in which the proceedings have been brought to stay the proceedings so far as they concern that matter.

(2) An application may be made notwithstanding that the matter is to be referred to arbitration only after the exhaustion of other dispute resolution procedures.

(3) An application may not be made by a person before taking the appropriate procedural step (if any) to acknowledge the legal proceedings against him or after he has taken any step in those proceedings to answer the substantive claim.

(4) On an application under this section the court shall grant a stay unless satisfied that the arbitration agreement is null and void, inoperative, or incapable of being performed.

(5) If the court refuses to stay the legal proceedings, any provision that an award is a condition precedent to the bringing of legal proceedings in respect of any matter is of no effect in relation to those proceedings.

Definitions

'agreement': s.5(1).
'arbitration agreement': ss.6, 5(1).
'the court': s.105.
'legal proceedings': s.82(1).

'notice (or other document)': s.76(6).
'party': ss.82(2), 106(4).
'upon notice' (to the parties or the tribunal): s.80.

[9A] ## Status

This section corresponds to the Model Law, Art.8, and derives from the Arbitration Act 1950, s.4(1), and the Arbitration Act 1975, s.1. It is a *mandatory* provision, (s.4(1) and Sched.1).

[9B] ## Summary

The section provides for the stay of legal proceedings in respect of a matter that is the subject of an arbitration agreement. Because ss.85 to 87 were never brought into effect, subs.(4) applies in respect of both international and domestic arbitrations.

Perhaps not surprisingly this section has been the subject of more court decisions than any other.

Points

[9C]   *Party who may apply – Subs.(1)*
The section makes it clear that it is only the party against whom the legal proceedings are brought (whether as respondent to the claim or counterclaim) who may apply for a stay. This is a change in wording from s.4(1) of the 1950 Act and s.1 of the 1975 Act, under which 'any party to the proceedings' (including, arguably, a plaintiff) could apply. Thus a plaintiff who takes proceedings to arrest a ship and then wishes to apply for a stay so as to send the matter to arbitration will now, beyond doubt, be unable to do so. Whether he could have done so before is not clear, as it could be said that he had taken a step in the proceedings, and thus lost any right to a stay.

The stay may extend to so much of the proceedings as form common matter with the subject of the arbitration agreement. It follows that a stay may be sought in respect of part only of the legal proceedings.

The section may, in appropriate circumstances, affect a third party under the Contracts (Rights of Third Parties) Act 1999.

[9D]   *Matter to be referred to arbitration – Subs.(1)*
Where an agreement refers disputes to arbitration, the court has only to consider whether there is a dispute within the meaning of the arbitration agreement, not whether in fact there is a dispute between the parties. Hence there is a dispute once money is claimed unless and until it is admitted that the sum is due and payable, and there is a dispute if a party refuses to pay a sum which is claimed, or denies that it is owing: *Halki Shipping Corporation v. Sopex Oils Ltd* [1998] 1 Lloyd's Rep. 465 (Court of Appeal). In *Secretary of State for Foreign and Commonwealth Affairs v. The Percy Thomas Partnership* [1998] CILL 1342 it was held that the law as to what constitutes a dispute is well summarised in *Mustill and Boyd* (2nd Edition) at pages 127/8, every case being heavily dependent on its facts. Further it was held that once a dispute has arisen it remains a dispute unless and until something happens to indicate that the dispute is either resolved or abandoned.

To obtain a stay it is not necessary to show that there is an arbitration agreement relevant to the cause of action pleaded: it is sufficient to show that there is an arbitration agreement and that there is a relevant dispute concerning that agreement. That dispute may be one as to whether the arbitration agreement is relevant to the cause of action: *Ahmad Al-Naimi v. Islamic Press Agency Inc.* (1998) 16-CLD-08-08.

It has also been held that where a contractual time-bar operates to bar a claim, it does not deprive an arbitration tribunal of jurisdiction. In the absence of the agreement of the parties to the court determining the question whether the claim was time-barred, it would be for the arbitrators to deal with that issue, and proceedings designed to have the court decide it would be stayed: *Grimaldi Compagnia di Navigazione SpA v. Sekihyo Lines Ltd (The 'Siki Rolette')* [1999] 1 WLR 708.

Where a contract contains a provision making it subject to the exclusive jurisdiction of the English Courts, but also contains an arbitration clause, it has been held (following *Paul Smith Ltd* v. *H&S International Holdings Inc.* [1991] 2 Lloyd's Rep. 127) that the arbitration clause is effective and thus a stay must be granted. In the same case the judge held that a badly drafted shortened version of clause 18 of the 1984 edition of the Federation of Civil Engineering Contractors (FCEC) form of subcontract should not be declared null and void, inoperative or incapable of being performed, at least in the abstract; whilst expressing the view *obiter* that a provision abrogating an arbitration agreement and leaving a party to be bound by the decision of a tribunal in proceedings in which it had no right to participate must be void as contrary to public policy: *McNicholas plc* v. *AEI Cables Ltd* (unreported, Judge Humphrey Lloyd QC, 25 May 1999).

[9E] *Summary judgment*

Before the Act, it was common for a plaintiff who believed there was no defence to his claim to issue a writ and at the same time apply for summary judgment under the former Rules of the Supreme Court (RSC) Order 14. If convinced there was no defence and therefore, in effect, no dispute, the court did not stay the action but would give judgment for the plaintiff. The effect of *Halki Shipping Corporation* v. *Sopex Oils Ltd* (see paragraph 9D above) has been to render such applications pointless since there is nonetheless a dispute between the parties within the meaning of the arbitration clause. The arbitrator, it is thought, has a sufficient armoury to deal with such obvious cases on a speedy basis. The situation remains the same under Part 24 of the CPR.

[9F] *Other dispute resolution procedures – Subs.(2)*

An application under this section may be made even if other dispute resolution procedures (such as reference to a panel of experts, an adjudicator or mediator) would have to be exhausted before the reference to arbitration, and those other procedures have not been exhausted at the date of the application.

[9G] *The application*

Under the CPR the application for a stay is made under the procedure set out in the Practice Direction on Arbitrations supplementary to CPR Part 49. Prior to the application, the claimant will have issued and served a claim form under CPR Part 7. The claim form may contain particulars of the claim, or they may be served separately, within 14 days. Under Part 9 the defendant need not respond until the particulars of claim have been served on him. When the particulars of claim are served on him (as part of the claim form or separately) the defendant (who does not admit the claim) must, if he wishes to seek a stay, file an acknowledgment of service under Part 10 within 14 days. Thereafter, within the period for filing a defence, which will normally be 28 days *after service of the particulars of claim* (see Rule 15.4(1)(b)), the defendant must apply to the court by application notice for a stay of the proceedings under s.9. Form 8A is used (as it would be for an arbitration application). He must not take any other step in the proceedings, and the claim form must state that he has not done so or otherwise demonstrate the same. The application must be supported by evidence. By Rule 32.6 evidence would normally be by witness statement: for general details on the format of a witness statement, see the Practice Direction to Part 32. The application should be served in accordance with CPR Rule 6.5 on the party bringing the relevant legal proceedings and on any other party to those proceedings who has given an address for service, and should be sent for information to any party who has not given such an address at his last known address or at a place where it is likely to come to his attention.

[9H]   *Time for application – Subs.(3)*
The application may not be made by a defendant before taking the appropriate step to acknowledge the legal proceedings against him. Under the CPR, this step is the filing of the acknowledgment of service under Part 10. Where the applicant is the claimant, who seeks to stay a counterclaim, there will be no such appropriate step, and the Act accommodates this by including the words '(if any)'.

Under the 1950 Act there was a considerable body of law as to what constituted a 'step in the proceedings'. For a summary see pp. 472/3 of *Mustill and Boyd* (2nd Edition). Since the new Act came into force a party has been held to have taken a 'step in … proceedings to answer the substantive claim', and accordingly not to be entitled to a stay where, unrepresented, he had written a letter accompanying an acknowledgment of service form in respect of a generally indorsed writ, purporting to set out a 'defence' but also making it clear that full details of the claim were awaited and mentioning that the preparation of the 'defence' was restricted: *London Central and Suburban Developments Ltd* v. *Banger* (1999) ADRLJ 119. On the other hand the issuing of a summons seeking leave to defend has been held by the Court of Appeal not to amount to a 'step … in proceedings to answer the substantive claim': *Patel* v. *Patel* [1999] 1 All ER (Comm.) 923.

[9I]   *Grant of a stay – Subs. (4)*
It is important to note that the regime for a stay provided for by s.9 is applicable to both domestic and international arbitration. The stay is mandatory unless the arbitration agreement is null and void, inoperative or incapable of being performed.

The exceptions which applied under the 1950 Act, and might have continued to apply had s.86 been enacted, do not apply. In particular this has serious implications for a claimant who wishes to bring proceedings against more than one party who may alternatively or jointly have been responsible for his loss. A typical example is a claimant who seeks damages for repair work to a building. The defect may have been caused by bad workmanship, bad design or bad supervision, or a combination of any of them. Prior to the Act the claimant could have safely issued a writ against the builder and any of his design team. Since the Act he is unable to do so. If he wishes to take proceedings against both, he must proceed in different fora, an expensive and time-consuming exercise. It is not surprising that there are calls to amend the Act to re-instate the old position – which almost certainly would have continued to prevail had s.86 been enacted. On the other hand, this problem was anticipated and considered by the DAC which was unable to find any solution that it considered satisfactory.

Arbitral joinder provisions to be found in various arbitral rules may, to a very limited extent, mitigate this difficulty. But such provisions are likely to depend on the same arbitrator being appointed in more than one arbitration under the same rules (or rules that allow some form of joinder) and do not assist the situation where one set of proceedings will be dealt with in arbitration, and the second in court.

The absence of any provision for multi-party arbitration causes further injustice to the defendant who considers another person liable with him for the loss caused to the claimant. Under the previous law the defendants could easily claim contribution between them in the one set of proceedings, so that the court would apportion liability appraised of all the relevant factors. Now the defendant must consider whether he is able to take separate proceedings for a contribution or indemnity towards his liability to the claimant. This works particularly harshly on a designer who has supervised a contractor's work, and the workmanship is found to be defective. In multi-party proceedings, the court would typically have apportioned liability 85% (builder), 15% (supervisor). Under the new regime, the supervisor will have to pay 100% unless he has such a route to contribution.

In the case of both domestic and international arbitration the onus is placed on the party bringing the legal proceedings (and therefore resisting the stay) to challenge the arbitration agreement (or make out other grounds for refusing a stay) and persuade the court to allow the legal proceedings to continue. To this extent the section ties in with the later sections (ss.30–32) concerning the jurisdiction of the arbitral tribunal. In the ordinary case, and subject to agreement of the parties to the contrary, the *tribunal* will rule on the validity of the arbitration clause, subject to final determination by the court as a preliminary point (s.32), or on a challenge to the award (s.67). On an application for a stay, which is limited to a challenge on the grounds set out in subs.(4), the *court* has the first opportunity. Further, where on an application for a stay a question arises as to whether an arbitration agreement has been concluded or as to whether the dispute falls within the terms of such an agreement, by para. 6.2 of the Practice Direction 49G on Arbitrations supplementary to CPR Part 49, the court may determine that question or give directions for its determination.

For a case in which there was a challenge to the existence of an arbitration agreement and in which the court considered the various procedural options available, see *Birse Construction Ltd* v. *St David Ltd* [1999] BLR 194 expanded on below at paragraph 30F.

[9J]   **Scott *v.* Avery *clause – Subs.(5)***
This subsection deals with the effect of a clause in the arbitration agreement which makes an award a condition precedent to legal proceedings to enforce the contract (known as a *Scott* v. *Avery* clause, after the case of that name, reported at [1856] 5 HL Cas. 811). The court's refusal of a stay will deprive such a clause of any effect in relation to those proceedings. A party that is refused a stay may not, therefore, subsequently seek to defend the proceedings (relying on a *Scott* v. *Avery* clause) on the grounds that no award has been made.

Note that this subsection does not apply in relation to a statutory arbitration, (s.97(1)(c)).

[9K]   *Appeals to Court of Appeal*
The Court of Appeal has jurisdiction to hear appeals from a judge or a court under this section, and a party may apply to the judge or to the Court of Appeal for permission to appeal and such judge and the court have power to grant that permission: *Inco Europe Ltd* v. *First Choice Distribution* [1999] 1 WLR 270.

[9L]   *Model Law and New York Convention*
Art.8 of the Model Law provides:

> '(1) A court before which an action is brought in a matter which is the subject of an arbitration agreement shall, if a party so requests not later than when submitting his first statement on the substance of the dispute, refer the parties to arbitration unless it finds that the agreement is null and void, inoperative, or incapable of being performed.
> (2) Where an action referred to in paragraph (1) of this article has been brought, arbitral proceedings may nevertheless be commenced or continued, and an award may be made, while the issue is pending before the court.'

Art.II.3 of the New York Convention provides:

> 'The court of a Contracting State, when seized of an action in a matter in respect of which the parties have made an agreement within the meaning of this article [shall,] at the request of one of the parties, refer the parties to arbitration unless it finds that the said agreement is null and void, inoperative or incapable of being performed.'

The regime contemplated by the Model Law in respect of international arbitrations is thus implemented in s.9 for both international and domestic arbitrations.

Section 10 – Reference of Interpleader Issue to Arbitration

10.–(1) Where in legal proceedings relief by way of interpleader is granted and any issue between the claimants is one in respect of which there is an arbitration agreement between them, the court granting the relief shall direct that the issue be determined in accordance with the agreement unless the circumstances are such that proceedings brought by a claimant in respect of the matter would not be stayed.

(2) Where subsection (1) applies but the court does not direct that the issue be determined in accordance with the arbitration agreement, any provision that an award is a condition precedent to the bringing of legal proceedings in respect of any matter shall not affect the determination of that issue by the court.

Definitions

'agreement': s.5(1).
'arbitration agreement': ss.6, 5(1).
'claimant': s.82(1).

'the court': s.105.
'legal proceedings': s.82(1).

[10A] Status

This section derives from the Arbitration Act 1950, s.5. It is a *mandatory* provision, (s.4(1) and Sched.1).

[10B] Summary

The section provides for the reference of interpleader issues to arbitration.

Points

[10C] *Effect*
An interpleader arises, for example, where two persons claim the same asset which is held by a third person. The person in possession of the asset acknowledges that one or other of the claimants is entitled to the asset, but is unable to determine which. That person may turn to the court and ask it to determine the issue.

In such a case, if the rival claimants are parties to an arbitration agreement covering the issue between them, the court is bound to direct that the issue be referred to arbitration, unless the circumstances are such that an application to stay the proceedings would be refused (as to which, see s.9.

This is a change from the law prior to this Act, which provided (in s.5

of the 1950 Act) that the court *might* direct the issue as between the claimants to be determined in accordance with the agreement, giving the court a wide discretion.

It should be noted that this section is a necessary addition to s.9, since the party beginning the proceedings in which interpleader relief is granted is not necessarily a party to an arbitration agreement.

[10D]   **Scott *v.* Avery *clause – Subs.(2)***
This subsection deals with the effect of a *Scott* v. *Avery* clause in the arbitration agreement, as to which see paragraph 9J above. The court's refusal to direct an issue to be referred to arbitration deprives such a clause of any effect in relation to its subsequent determination of that issue.

Note that this subsection does not apply in relation to statutory arbitrations, see s.97(c).

[10E]   *Appeals to Court of Appeal*
The Court of Appeal has jurisdiction to hear appeal from a judge or a court under this section, and a party may apply to the judge or to the Court of Appeal for permission to appeal and such judge and the court have power to grant that permission: *Inco Europe Ltd* v. *First Choice Distribution* [1999] 1 WLR 270.

[10F]   *Model Law*
There is no corresponding provision in the Model Law.

It should be noted, however, that the change to a mandatory stay of court proceedings brings English law, so far as international arbitration is concerned, into line with Art. II of the New York Convention, which requires a reference to arbitration unless the agreement to arbitrate is null and void, inoperative or incapable of being performed.

Section 11 – Retention of Security
Where Admiralty Proceedings Stayed

11.–(1) Where Admiralty proceedings are stayed on the ground that the dispute in question should be submitted to arbitration, the court granting the stay may, if in those proceedings property has been arrested or bail or other security has been given to prevent or obtain release from arrest–

(a) order that the property arrested be retained as security for the satisfaction of any award given in the arbitration in respect of that dispute, or
(b) order that the stay of those proceedings be conditional on the provision of equivalent security for the satisfaction of any such award.

(2) Subject to any provision made by rules of court and to any necessary modifications, the same law and practice shall apply in relation to property retained in pursuance of an order as would apply if it were held for the purposes of proceedings in the court making the order.

Definitions

'the court': s.105.

[11A] Status

This section reproduces, and simply re-enacts, so much of the Civil Jurisdiction and Judgments Act 1982, s.26, as relates to arbitration. It is a *mandatory* provision, (s.4(1) and Sched.1).

[11B] Summary

The section is concerned with the situation where Admiralty proceedings are stayed in order for the dispute in question to be submitted to arbitration, and the vessel has been arrested, or security has been given to prevent, or obtain its release from, arrest. On granting the stay, the court may order the retention of the arrested vessel or the provision of equivalent security as a means of ensuring the satisfaction of any award in respect of that dispute.

Points

[11C] *Practice*
Since there is no change in the law, there is no reason to suppose the practice will change. As in most cases in the past, the Admiralty proceedings will remain in existence in case they are later needed for the enforcement of any award made. And security – usually in the form of

91

a letter of undertaking or guarantee – will be given in respect of any award(s) that may be rendered in the arbitration.

[11D]   *Appeals to Court of Appeal*
The Court of Appeal has jurisdiction to hear appeals from a judge or a court under this section, and a party may apply to the judge or to the Court of Appeal for permission to appeal and such judge and the court have power to grant that permission: *Inco Europe Ltd* v. *First Choice Distribution* (1999) 1 WLR 270.

[11E]   *Model Law*
There is no corresponding provision in the Model Law.

Commencement of Arbitral Proceedings

Section 12 – Power of Court to Extend Time for Beginning Arbitral Proceedings, etc.

12.-(1) Where an arbitration agreement to refer future disputes to arbitration provides that a claim shall be barred, or the claimant's right extinguished, unless the claimant takes within a time fixed by the agreement some step–

(a) to begin arbitral proceedings, or
(b) to begin other dispute resolution procedures which must be exhausted before arbitral proceedings can be begun,

the court may by order extend the time for taking that step.

(2) Any party to the arbitration agreement may apply for such an order (upon notice to the other parties), but only after a claim has arisen and after exhausting any available arbitral process for obtaining an extension of time.

(3) The court shall make an order only if satisfied–

(a) that the circumstances are such as were outside the reasonable contemplation of the parties when they agreed the provision in question, and that it would be just to extend the time, or
(b) that the conduct of one party makes it unjust to hold the other party to the strict terms of the provision in question.

(4) The court may extend the time for such period and on such terms as it thinks fit, and may do so whether or not the time previously fixed (by agreement or by a previous order) has expired.

(5) An order under this section does not affect the operation of the Limitation Acts (see section 13).

(6) The leave of the court is required for any appeal from a decision of the court under this section.

Definitions

'agreement': s.5(1).
'arbitration agreement': ss.6, 5(1).
'available arbitral process': s.82(1).
'claimant': s.82(1).
'the court': s.105.
'dispute': s.82(1).

'Limitation Acts': s.13(4).
'notice (or other document)': s.76(6).
'party': ss.82(2), 106(4).
'upon notice' (to the parties or the tribunal): s.80.

[12A]   **Status**

This section derives from the Arbitration Act 1950, s.27, but has its origins in s.16 of the 1934 Act. It is a *mandatory* provision, (s.4(1) and Sched.1).

[12B]   **Summary**

The section gives the court power to extend time for beginning arbitral proceedings. S.12 operates independently of the general power given to the court by s.79 (1) of the Act to extend other time limits relating to arbitral proceedings. That power does not apply to a time limit to which this section applies.

Note that this section does not apply in relation to statutory arbitrations, (s.97(b)).

Points

[12C]   *Effect*

Whilst retaining the existing philosophy that a time limit may be extended, this section makes important changes to the tests to be applied, and certain other minor alterations.

It is worth noting that these provisions apply *only* to requirements that arbitral proceedings or other dispute resolution proceedings must be started within a specific time. Contractual provisions barring claims unless other steps are taken (e.g. the making of a claim, or the provision of documents) cannot be affected by this section, see e.g. *The Medusa* (1986) 2 Lloyd's Rep. 328.

In appropriate cases third parties will have to bear in mind this section having regard to the Contracts (Rights of Third Parties) Act 1999.

[12D]   *The test – Subs.(3)*

The most significant change introduced by the section is that the test to be applied by the court is no longer whether 'undue hardship would otherwise be caused' if an extension were not granted. There are now two alternatives. The first test is whether the court is satisfied that circumstances unforeseen at the time of agreeing the time bar have arisen, and it would be just to grant an extension, (subs.(3)(a)). In the absence of such unforeseen circumstances, the parties are held to their bargain, unless the second test can be satisfied, namely whether the court is satisfied that one party has so conducted itself as to make it unjust to apply the time bar to the other, (subs.(3)(b)). Inequitable conduct by one party may thereby be remedied.

This section follows on the heels of s.9 in the popularity stakes for court challenges, again not surprisingly. It has been held that the circumstances to be considered are all those placed before the court, which then has to focus on those which appear particularly relevant (*Cathiship SA* v. *Allansons Ltd (The 'Catherine Helen')* [1998] 2 Lloyd's Rep. 511), and that the relevant time to consider is that when the parties agree the arbitration clause: *Cathiship* v. *Allansons*; *Fox & Widley* v. *Guram* [1998] 3 EG 142, and *Grimaldi Compagnia di Navigazione SpA* v. *Sekihyo Lines Ltd (The 'Siki Rolette')* [1999] 1 WLR 908.

The court has to consider what the parties would reasonably have contemplated and thus must consider the relevant transaction, ordinary practices within that type of transaction and the reasonable expectation of parties involved in such a transaction (*Cathiship* v. *Allansons* (above)). Thus, an extension of time was refused to tenants who, if they had wished to object to a proposed rent increase had to serve a counter-notice within three months of the landlords' notice, notwithstanding that the landlords proposed a substantially exaggerated rent. This could not be said to have been outside the reasonable contemplation of the parties when they agreed the rent review clause (*Fox & Widley* v. *Guram* (above)). Similarly, the fact that a contractual time-bar did not occur to either party until a late stage was quite insufficient to show that its existence or application on the facts which occurred was outside the reasonable contemplation of the parties when they agreed a clause in their contract incorporating the time-bar. Further, the circumstances to be considered were those in which the consensual time-bar applied, but did not include the circumstances in which s.12 came to replace s.27 of the 1950 Act; if that was wrong, it still could not be suggested that the introduction of s.12 came without ample warning and consultation (*Grimaldi* v. *Sekihyo* (above)).

In another case, it was held impossible to characterise a negligent omission to comply with a time-bar, however little delay is involved, as – without more – being outside the mutual contemplation of parties. Narrowly overlooking a time-bar due to an administrative oversight is far from being so uncommon as to be treated as being beyond the parties' reasonable contemplation and it will not normally be a relevant exercise to weigh the prejudice to the time-barred party against the degree of fault on its part in order to determine whether there are circumstances beyond the reasonable contemplation of the parties (*Harbour and General Works Ltd* v. *The Environment Agency*) [1999] 1 All ER (Comm.) 953. The Court of Appeal in the same case ([1999] 2 All ER (Comm.) 686) held that the aim of subs.(3) is 'to allow the court to consider an extension in relation to circumstances where the parties would not reasonably have contemplated them as being ones where the time bar would apply'.

Where reliance is placed on s.12(3)(b) there must be shown some conduct which is proved somehow to have led the claimant to omit to give notice in time. Mere silence on the part of the respondent, or failure to alert the claimant to the need to comply with the time-bar cannot make the barring of the claim unjust (*Harbour and General Works*

v. *The Environment Agency* (above)) and the proposal of a substantially exaggerated rent by landlords could not make it unjust to hold the tenant to the strict terms of the contract under which the tenant was barred from challenging a rental increase (*Fox & Widley* v. *Guram* (above)).

In shipping arbitrations, 12 month time limits are not uncommon in charterparty arbitration clauses, whilst cargo claims under bills of lading must almost invariably be brought within 12 months by statute. Accordingly, we opined in the first edition, an owner seeking an indemnity from a charterer in respect of a cargo claim the subject of proceedings brought just at the end of the 12 month period might have problems in obtaining an extension since such a situation ought to be quite foreseeable, and so it has been held.

In *Cathiship* v. *Allansons* (above), shipowners ought reasonably to have contemplated the possibility of cargo claims in respect of which they would wish to be indemnified by charterers, but they made a mistake as to the operation of an arbitration clause both with regard to making a claim and with regard to appointing an arbitrator to cover such an indemnity. Such mistakes were held not to be outside the reasonable contemplation of the parties when they concluded their contract, and an extension was refused.

[12E] *Other dispute resolution procedures – Subs.(1)(b)*
The power to extend time may also be exercised in relation to the time for beginning other dispute resolution procedures (such as reference to a panel of experts) which are a pre-condition of a reference to arbitration, mediation or adjudication. This is, of course, a change from the law prior to the Act, but makes allowance for the growing practice of allowing for other pre-arbitration steps. It should be noted that it will not normally be the case under adjudication clauses inserted in contracts pursuant to the Housing Grants, Construction and Regeneration Act 1996 that the adjudication process must be exhausted before the arbitration can be commenced.

[12F] *Other recourse – Subss.(2), (4)*
An applicant for an extension of time must first have exhausted any possible application for an extension under institutional rules, or other available arbitral process, before applying to the court. Again, this is a change from the law prior to this Act which will inevitably limit the number of applications, as well as make some fall later in time.

Subs.(4), which provides that an extension may be granted even where the time limit which it is sought to extend has itself expired, acquires considerable importance in this regard. Where a time-consuming arbitral process for obtaining an extension of time must be exhausted before application to the court can be made, the relevant time limit is all the more likely to have expired by the time of the application.

Where there is an arbitral process to be exhausted and where that process has produced a negative response, it seems unlikely that an application to the court will be successful. Assuming that the arbitral tribunal or institution has acted properly, the court would seldom reach a contrary conclusion.

[12G]   *Limitation – Subs.(5)*
Extensions of time granted under this section do not affect the working of enactments relating to limitation (as to which, see s.13). Thus a limitation period affecting a substantive claim will not be enlarged by an extension of time granted under this section.

[12H]   *Appeal – Subs.(6)*
Any appeal from a decision of the court under this section may only be made with leave. Presumably, leave would only be granted if some question of difficulty or principle appeared to justify it. In most cases, the decision of the court may be expected to be final.

[12I]   *The application*
The application made under s.12(2) is an 'arbitration application' falling within the terms of the Practice Direction issued under CPR Part 49. We deal generally with arbitration applications in Part 3, Materials, Section H. Specifically in respect of an application under s.12:

(i)   the application must be on notice to the other party or parties to the arbitration agreement;
(ii)   the requirement to give notice to other parties is to be met by making those parties respondents to the application and serving on them the arbitration claim form and any evidence in support;
(iii)   the claim form must show that a claim has arisen and that any available arbitral process for obtaining an extension of time has been exhausted;
(iv)   the application may include as an alternative an application for a declaration that an order under s.12 is not needed; and
(v)   unless otherwise ordered, the hearing will be in private.

[12J]   *Model Law*
There is no corresponding provision in the Model Law.

Section 13 – Application of Limitation Acts

13.–(1) The Limitation Acts apply to arbitral proceedings as they apply to legal proceedings.

(2) The court may order that in computing the time prescribed by the Limitation Acts for the commencement of proceedings (including arbitral proceedings) in respect of a dispute which was the subject matter–

(a)   of an award which the court orders to be set aside or declares to be of no effect, or

(b)   of the affected part of an award which the court orders to be set aside in part, or declares to be in part of no effect,

the period between the commencement of the arbitration and the date of the order referred to in paragraph (a) or (b) shall be excluded.

(3) In determining for the purposes of the Limitation Acts when a cause of action accrued, any provision that an award is a condition precedent to the bringing of legal proceedings in respect of a matter to which an arbitration agreement applies shall be disregarded.

(4) In this Part 'the Limitation Acts' means–

(a)   in England and Wales, the Limitation Act 1980, the Foreign Limitation Periods Act 1984 and any other enactment (whenever passed) relating to the limitation of actions;

(b)   in Northern Ireland, the Limitation (Northern Ireland) Order 1989, the Foreign Limitation Periods (Northern Ireland) Order 1985 and any other enactment (whenever passed) relating to the limitation of actions.

Definitions

'commencement' (in relation to arbitral proceedings): s.14.
'the court': s.105.
'dispute': s.82(1).

'enactment': s.82(1).
'legal proceedings': s.82(1).
'Limitation Acts': s.13(4).

[13A]   ## Status

This section derives from the Limitation Act 1980, s.34, and the Foreign Limitation Periods Act 1984, s.5. It is a *mandatory* provision, (s.4(1) and Sched.1).

[13B]   ## Summary

The section provides for the application of the Limitation Acts, as defined by subs.(4), to arbitral proceedings. It reflects the law prior to this Act.

Points

[13C] *Effect*

Limitation periods apply to substantive claims in relation to arbitrations in the same way as they apply in relation to legal proceedings. In appropriate cases, under the provisions of the Foreign Limitation Periods Act 1984, where foreign law is taken into account in the determination of any matter, then foreign limitation law may apply in respect of that matter.

So far as most arbitrations are concerned, the relevant limitation periods are those applicable to contract and tort. In relation to contract claims, the limitation period is six years from the accrual of the cause of action (which will normally be the date of the alleged breach of contract), see the Limitation Act 1980, s.5. In relation to claims in tort, the limitation period is similarly six years from the accrual of the cause of action (which will normally be the date of the damage), see the Limitation Act 1980, s.2.

[13D] *Disregarding abortive time – Subs.(2)*

The court has power to order that where an award, or part of an award, has been set aside or declared to be of no effect and fresh proceedings in respect of the dispute are commenced, the period of time between the commencement of the original arbitration and the date of the order setting aside or declaring of no effect the original award should be disregarded for limitation purposes.

In this way, time expended on abortive arbitral proceedings that have not produced a valid award may be disregarded so that it does not, itself, prejudice further proceedings.

[13E] **Scott *v.* Avery** *clause – Subs.(3)*

This subsection negates the effect of a clause in the arbitration agreement which makes an award a condition precedent to legal proceedings to enforce the contract (a *Scott* v. *Avery* clause, as to which see paragraph 9J).

Such a clause (which might be argued to have the effect of deferring the accrual of a cause of action until the making of the award) is to be disregarded in determining when a cause of action accrued for limitation purposes. It will not, therefore, influence the considerations that would otherwise determine that date.

[13F] *Appeals to Court of Appeal*

The Court of Appeal has jurisdiction to hear appeals from a judge or a court under this section, and a party may apply to the judge or to the Court of Appeal for permission to appeal and such judge and the court have power to grant that permission: *Inco Europe Ltd* v. *First Choice Distribution* (1999) 1 WLR 270.

Section 14 – Commencement of Arbitral Proceedings

14.-(1) The parties are free to agree when arbitral proceedings are to be regarded as commenced for the purposes of this Part and for the purposes of the Limitation Acts.

(2) If there is no such agreement the following provisions apply.

(3) Where the arbitrator is named or designated in the arbitration agreement, arbitral proceedings are commenced in respect of a matter when one party serves on the other party or parties a notice in writing requiring him or them to submit that matter to the person so named or designated.

(4) Where the arbitrator or arbitrators are to be appointed by the parties, arbitral proceedings are commenced in respect of a matter when one party serves on the other party or parties notice in writing requiring him or them to appoint an arbitrator or to agree to the appointment of an arbitrator in respect of that matter.

(5) Where the arbitrator or arbitrators are to be appointed by a person other than a party to the proceedings, arbitral proceedings are commenced in respect of a matter when one party gives notice in writing to that person requesting him to make the appointment in respect of that matter.

Definitions

'agreement', 'agree': s.5(1).
'arbitration agreement': ss.6, 5(1).
'arbitrator': s.82(1).
'commencement' (in relation to arbitral proceedings): s.14.
'Limitation Acts': s.13(4).
'notice (or other document)': s.76(6).

'party': ss.82(2), 106(4).
'serve' (notice or other document): s.76(6).
'upon notice' (to the parties or the tribunal): s.80.
'in writing': s.5(6).

[14A] Status

This section derives in part from the Model Law, Art.21, and from the Limitation Act 1980, s.34(3). It is *non-mandatory*, (s.4(2)).

[14B] Summary

The section explains when arbitral proceedings are to be regarded as commenced. It essentially provides that the parties are free to agree as to the time of deemed commencement of the arbitration for the purposes of this Part of the Act and limitation purposes, (subs.(1)). Certain rules apply in the absence of such agreement, as follows.

Points

[14C]   *Commencement – Subss.(3), (4), (5)*

If the arbitrator is already identified in the arbitration agreement, arbitral proceedings commence when one party serves on the other a written notice requiring the matter to be referred to that arbitrator, (subs.(3)). The word 'matter', used throughout this section rather than 'dispute', is a neutral term that encompasses both disputes and claims, see *Commission for the New Towns* v. *Crudens* (1995) CILL 1035.

Where the arbitrator has yet to be appointed by the parties, arbitral proceedings commence when one party serves on the other a written notice requiring the appointment, or agreement to the appointment, of an arbitrator, (subs.(4)).

Where the arbitrator has yet to be appointed by a third party (such as an institution), arbitral proceedings commence when one party gives written notice to that third party requesting the appointment, (subs. (5)). This is an important provision, and one of which practitioners making claims in the last days before expiry of a limitation period need to be careful. It is a change to the law prior to the Act, which required notice to the other party in all cases. It does, however, reflect the requirements of institutions such as the ICC and LCIA. Thus, Art.4.2 of the ICC Rules provides: 'The date on which the Request is received by the Secretariat shall, for all purposes, be deemed to be the date of commencement of the arbitral proceedings'; and Art.1.2 of the LCIA Rules is to similar effect.

[14D]   *Notice*

Where an arbitration clause provided, by virtue of s.15(3), for a sole arbitrator it has been held sufficient to commence arbitration proceedings for a claimant to write:

> 'In view of your refusal to pay hire under the above charterparty, we hereby give you formal notice of our commencement of arbitration against you under the terms of the above charterparty which requires the appointment of a sole arbitrator.'
> (*Villa Denizcilik Sanayi ve Ticaret AS* v. *Longen S.A. (The 'Villa')*
> [1998] 1 Lloyd's Rep. 195)

It has been held by a court construing s.14(4) that reference should only be made to pre-1996 Act cases when the Act does not cover a point or such reference is otherwise necessary: *Seabridge Shipping AB* v. *AC Orssleff EFT's A/S* [1999] 2 Lloyd's Rep. 685. Nevertheless it seems likely that in deciding what amounts to a notice for the purposes of this section, the courts will follow the approach in *Nea Agrex SA* v. *Baltic Shipping Co. Ltd* [1976] 2 Lloyd's Rep. 47. In a recent case decided under s.27(3) of the Limitation Act 1939 (now replaced by s.14(4) of the present Act, which is in materially the same terms) it was held that it is necessary to look at the real meaning of the notice purporting to commence arbitration rather than merely at the form of words used. A notice in writing which, read in its context, makes it clear by whatever language that the sender is invoking the arbitration agreement and is requiring the recipient to take steps in response to enable the tribunal to be constituted is sufficient to satisfy the requirements: *Allianz Versicherungs-AG* v. *Fortuna Co Inc (The 'Baltic Universal')* [1999] 1 Lloyd's Rep. 497. Two days after judgment was given in that case another judge held, under the 1950 Act, that the question whether a notice is good is whether it sufficiently, or in substance, makes clear that the respondent is expected to act on the claimant's submission of a dispute to arbitration on his invocation of the arbitration clause so as to participate in the submission of the relevant dispute to arbitration: *Charles M. Willie and Co. (Shipping) Ltd* v. *Ocean Laser Shipping Ltd (The 'Smaro')* [1999] 1 Lloyd's Rep. 225. And in *Seabridge Shipping* v. *Orssleff* (above) a message to an arbitrator asking him to accept appointment, which was copied to the opposite party and which contained a specific request to that party to indicate if they would accept the arbitrator as sole arbitrator, or alternatively appoint their own arbitrator, was held to be a good notice.

[14E] *'Gives notice' – Subs.(5)*
It should be noted that there may be a distinction between the word 'gives' in this sub-section and the word 'serves' in others. 'Serve' is defined in s.76(4) by reference to delivery, and, in our view, service (for limitation purposes) is only effective upon delivery. It is likely that 'gives' in this sub-section also requires delivery. The word is probably used because any third party that has to make an appointment is likely to be an institution which has rules that specify how a valid notice is to be given to it, and such provision may sit ill with the definition of 'serve' in s.76(4). In this context it is useful to note that the Model Law refers to 'receipt' but it does not specifically address the question of a notice to a third party.

[14F] *Agreement before default appointment – Subss.(4), (5)*
There are many arbitration clauses which provide for the parties to attempt agreement on an arbitrator before a default appointment by an institution. The Act does not specifically provide for the inter-relationship between subss.(4) and (5), but it is, we think, clear that in the case of such a clause, subs.(4) would apply even if the parties are subsequently unable to agree an arbitrator.

[14G] *Relevance to exclusion agreements*
The date of commencement of the proceedings is relevant to the effectiveness of agreements to exclude the court's jurisdiction under s.45 (determination of preliminary point of law) and s.69 (appeal on a point of law).

[14H] *Model Law*

The section reflects Art.21 of the Model Law, in part. This provides:

> 'Unless otherwise agreed by the parties, the arbitral proceedings in respect of a particular dispute commence on the date on which a request for that dispute to be referred to arbitration is received by the respondent.'

The clause is not as complex as the new Act. It is similar in that the parties may agree something different; it uses the word 'dispute', rather than 'matter', and is thus more confined; and where the arbitrator is appointed by a third party, notice is still to the other party, rather than the appointing body.

[14I] *Rules*

Some arbitration rules have taken up the invitation under s.14(1) to define when arbitral proceedings are commenced. For example, CIMAR Rule 2.1 provides:

> 'Arbitral proceedings are begun in respect of a dispute when one party serves on the other a written notice of arbitration identifying the dispute and requiring him to agree to the appointment of an arbitrator.'

And the ICE Arbitration Procedure states that the relevant date is the date upon which the Notice to Refer under that Procedure is served. The CIArb Rules provide in Art.2.1 for the arbitration to 'be regarded as commenced in accordance with the provisions of section 14'. Of the international rules, we have noted the ICC and LCIA rules above. In contrast, the UNCITRAL Arbitration Rules, providing for ad hoc rather than institutional arbitration, set the date for commencement as the date on which the notice of arbitration is received by the respondent.

The Arbitral Tribunal

Section 15 – The Arbitral Tribunal

15.–(1) The parties are free to agree on the number of arbitrators to form the tribunal and whether there is to be a chairman or umpire.

(2) Unless otherwise agreed by the parties, an agreement that the number of arbitrators shall be two or any other even number shall be understood as requiring the appointment of an additional arbitrator as chairman of the tribunal.

(3) If there is no agreement as to the number of arbitrators, the tribunal shall consist of a sole arbitrator.

Definitions

> 'agreement', 'agree', 'agreed': s.5(1).
> 'arbitrator': s.82(1).
> 'party': ss.82(2), 106(4).

[15A] ## Status

This section corresponds in part to the Model Law, Art.10(1); and derives in part from the Arbitration Act 1950, ss.6, 8, 9. It is *non-mandatory*, (s.4(2)).

[15B] ## Summary

This section deals with the composition of the arbitral tribunal. As well as the role of umpire (that is to say, an arbitrator who, as long as the other arbitrators are in agreement, has no real function or status, but who may come to replace the other arbitrators as the deciding tribunal), this section refers to the role of chairman (that is to say, an arbitrator who has the same functions and obligations, and whose views rank equally with those of the other members of the tribunal, save where there is no majority). For the functions of a chairman, see s.20; and for those of an umpire, see s.21.

Points

[15C] *Agreement – Subs.(1)*
It is for the parties to agree (if they so wish or are able) on the composition of the tribunal with regard both to number and

structure, (subs.(1)). The agreement must be in writing in accordance with s.5.

In most cases the agreement of the parties as to the structure of the tribunal is likely to be found within the body of the arbitration agreement. It may well be by reference to institutional rules. The parties may need to take care in their use of such rules.

We note, for instance, that Art.7 of the UNCITRAL Arbitration Rules refers to a 'presiding arbitrator'. The parties are, of course, free to have a 'presiding arbitrator' rather than a 'chairman'. It may be that a court would decide they were the same thing. Subject to that, however, without a 'chairman' the parties would lose the protection of the chairman's view prevailing, under s.20(4). The tribunal's decision-making would be determined by s.22 – unanimous or majority decisions in the absence of other agreement – since the UNCITRAL Arbitration Rules do not themselves provide for how decisions are to be reached.

[15D]   *Additional arbitrator as chairman – Subs.(2)*

If the agreement of the parties is for a tribunal consisting of an even number of arbitrators, then (in the absence of contrary agreement) it will be implied that the appointment of a further arbitrator as chairman is also required. It follows that if the parties wish to have a tribunal consisting of an even number of arbitrators and an umpire, or simply an even number of arbitrators alone, they must specifically so opt.

In practice, where the parties have not opted for a single arbitrator, they are likely to have agreed that they shall each appoint an arbitrator. If there is no provision for a chairman or umpire, this subsection will operate, so that a chairman is appointed.

It is useful to note here that in the absence of agreement, or an agreed procedure, and where there are more than two parties, the appointment of the chairman will be made by the court under s.18.

[15E]   *Sole arbitrator – Subs.(3)*

In the absence of any agreement, the tribunal will consist of a sole arbitrator. This preserves the position prior to this Act in English law, which is equally common in many other countries, in preference to the Model Law's default provision for a tribunal of three arbitrators, (see Art.10(2), below). The DAC decided not to opt for the Model Law solution because the cost of three arbitrators was likely to be 'three times the cost of employing one' and it seemed right that 'this extra burden should be available if the parties so choose, but not imposed on them'.

In one case, an arbitration clause provided for any matters in dispute to be referred to arbitration in London in accordance with the 1950 Act and any subsequent alterations. It was not suggested that this referred to anything other than the 1996 Act. It was held to be plain that the agreement required a sole arbitrator, and the court had no jurisdiction to appoint a tribunal differently constituted: *Villa Denizcilik Sanayi ve Ticaret AS* v. *Longen SA (The 'Villa')* [1998] 1 Lloyd's Rep. 195.

[15F] **Model Law**

As we have indicated, in contrast to the Act, Art.10 of the Model Law provides:

> '(1) The parties are free to determine the number of arbitrators.
> (2) Failing such determination, the number of arbitrators shall be three.'

[15G] **Rules**

Domestic rules generally provide for a sole arbitrator. Of the main international rules, Art.5.4 of the LCIA Rules provides that the LCIA will appoint a sole arbitrator 'unless the parties have agreed in writing otherwise, or unless the LCIA Court determines that in view of all the circumstances of the case a three-member tribunal is appropriate'. And Art.8.2 of the ICC Rules has a similar effect:

> 'Where the parties have not agreed upon the number of arbitrators, the Court shall appoint a sole arbitrator, save where it appears to the Court that the dispute is such as to warrant the appointment of three arbitrators.'

In contrast, but in line with the Model Law itself, Art.5 of the UNCITRAL Arbitration Rules provides:

> 'If the parties have not previously agreed on the number of arbitrators (i.e. one or three), and if within fifteen days after the receipt by the respondent of the notice of arbitration the parties have not agreed that there shall be only one arbitrator, three arbitrators shall be appointed.'

We would venture to suggest that the more recent rules have 'defaulted' towards a sole arbitrator, and that this will continue to be the trend as other rules (such as UNCITRAL) come to be revised.

Section 16 – Procedure for Appointment of Arbitrators

16.–(1) The parties are free to agree on the procedure for appointing the arbitrator or arbitrators, including the procedure for appointing any chairman or umpire.

(2) If or to the extent that there is no such agreement, the following provisions apply.

(3) If the tribunal is to consist of a sole arbitrator, the parties shall jointly appoint the arbitrator not later than 28 days after service of a request in writing by either party to do so.

(4) If the tribunal is to consist of two arbitrators, each party shall appoint one arbitrator not later than 14 days after service of a request in writing by either party to do so.

(5) If the tribunal is to consist of three arbitrators–

(a) each party shall appoint one arbitrator not later than 14 days after service of a request in writing by either party to do so, and
(b) the two so appointed shall forthwith appoint a third arbitrator as the chairman of the tribunal.

(6) If the tribunal is to consist of two arbitrators and an umpire–

(a) each party shall appoint one arbitrator not later than 14 days after service of a request in writing by either party to do so, and
(b) the two so appointed may appoint an umpire at any time after they themselves are appointed and shall do so before any substantive hearing or forthwith if they cannot agree on a matter relating to the arbitration.

(7) In any other case (in particular, if there are more than two parties) section 18 applies as in the case of a failure of the agreed appointment procedure.

Defintions

'agreement', 'agree': s.5(1).
'arbitrator': s.82(1).
'party': ss.82(2), 106(4).

'service' (of notice or other document): s.76(6).
'in writing': s.5(6).

[16A] Status

This section derives from the Model Law, Art.11(2) and (3). It is a *non-mandatory* provision, (s.4(2)).

[16B] Summary

This section deals with the procedure for appointing arbitrators. Essentially, the parties are at liberty to agree on the procedure for appointing the tribunal. This will often be by reference to institutional rules, but in any event must be in writing, to accord with s.5.

The section sets out the rules that will apply to the extent that there is no such agreement. It should be read in conjunction with s.17 (which provides for the appointment of a sole arbitrator in default of the appointment of a second arbitrator); and s.18 (which makes provision for the failure of the appointment procedure).

Points

[16C] *Default provisions – Subss.(3) to (7)*

These subsections set out the rules for appointment that apply in the absence of agreement, depending upon whether the tribunal is to comprise a sole arbitrator, (subs.(3)); two arbitrators, (subs.(4)); three arbitrators, one of whom is a chairman, (subs.(5)); or three arbitrators, one of whom is an umpire, (subs.(6)). For the different functions of chairman and umpire, see ss.20 and 21. It is worth noting, particularly in relation to the default provisions of s.17, that (in the absence of contrary agreement) it is not necessary for the first party to have appointed an arbitrator in order to start running the other party's time for making an appointment: subs.(4) requires only a written request for an appointment.

Note, in relation to subs.(6), the new requirement that two arbitrators must appoint an umpire before any substantive hearing, even if at that stage there is no disagreement between them. This of course only reflects what was the practice before: for two arbitrators to sit at a substantive hearing without an umpire, and then to disagree, would mean a complete re-hearing before an umpire who would then have to be appointed.

Note also that because it is often more convenient (and cheaper) for two arbitrators to deal with interlocutory matters without appointing a third arbitrator (where one is required), parties often agree that no third arbitrator need be appointed until a substantive hearing. The LMAA Terms, for instance, expressly deal with this in para.8(b) which provides that the party-appointed arbitrators 'may at any time [after their appointments] appoint a third arbitrator so long as they do so before any substantive hearing or forthwith if they cannot agree on any matter relating to the arbitration'.

However, we now see some possible difficulty in the relationship between this sub-section and s.21(3) which provides that where parties have not agreed as to what an umpire's functions are to be, the umpire 'shall attend the proceedings'. It must be very strongly arguable that 'the proceedings' in that sub-section are not merely the 'substantive hearing' referred to here, and if that is right it means that arbitrators should always appoint an umpire to sit with them even at interlocutory hearings, unless (which in most instances will be highly desirable) they can get the parties' agreement not to do so.

The court has powers under s.18 to deal with any situations not covered by these provisions, (see subs.(7)).

For the inter-relationship between s.16(1) and s.27(3), see the discussion at paragraph 27D below.

[16D]   *Time*

Time limits imposed by the default provisions are expressed in multiples of seven days in order to limit the possibility of their expiring on a weekend. The court has power to extend the time limits pursuant to s.79, but the DAC was of the view that this power should not normally be exercised in ordinary cases.

Periods of time are dealt with generally in s.78. We give some examples at paragraphs 78C to E. In summary, the parties are free to agree on a method of reckoning periods of time and could thus (for example) agree not to count weekends, or agree that any period ending on a weekend was automatically extended to the next working day. Where they have not so provided, 14 or 28 day periods are counted from (but including) the next day, so that a period commencing on a Monday expires on a Monday, and so on. Where periods exceed seven days, weekends and public holidays are included, (s.78(5)), so particular care must be taken when time expires on a public holiday.

[16E]   *Model Law*

Art.11(2) of the Model Law provides, as does the Act, that the parties are 'free to agree on a procedure of appointing the arbitrator or arbitrators'. The balance of Art.11 covers matters dealt with by ss.16 to 18 of the Act, and we refer to them in paragraph 18K.

[16F]   *Rules*

Generally arbitration rules set out the procedure for appointment. Since the default provisions in s.16 apply to 'the extent' that there is no agreement, in certain cases some or all of s.16 may nonetheless apply. Of those rules we have considered, all provide detailed appointment mechanisms. See CIMAR Rule 2, ICE Arbitration Procedure Rules 3 and 4, CIArb Rules, Arts.2 and 4, LCIA Rules Art.5, ICC Arts.8 to 10, and UNCITRAL Arbitration Rules Arts.6 to 8. In the ordinary course of events, none of these leave much, if any, scope for the application of the default provisions.

Section 17 – Power in Case of Default to Appoint Sole Arbitrator

17.–(1) Unless the parties otherwise agree, where each of two parties to an arbitration agreement is to appoint an arbitrator and one party ('the party in default') refuses to do so, or fails to do so within the time specified, the other party, having duly appointed his arbitrator, may give notice in writing to the party in default that he proposes to appoint his arbitrator to act as sole arbitrator.

(2) If the party in default does not within 7 clear days of that notice being given–

(a)   make the required appointment, and
(b)   notify the other party that he has done so,

the other party may appoint his arbitrator as sole arbitrator whose award shall be binding on both parties as if he had been so appointed by agreement.

(3) Where a sole arbitrator has been appointed under subsection (2), the party in default may (upon notice to the appointing party) apply to the court which may set aside the appointment.

(4) The leave of the court is required for any appeal from a decision of the court under this section.

Definitions

'agreement', 'agree': s.5(1).
'arbitration agreement': ss.6, 5(1).
'arbitrator': s.82(1).
'the court': s.105.
'notice (or other document)': s.76(6).

'party': ss.82(2), 106(4).
'upon notice' (to the parties or the tribunal): s.80.
'in writing': s.5(6).

[17A]   ## Status

This section replaces the Arbitration Act 1950, ss.7(b) and 10(3B). It is a *non-mandatory* provision, (s.4(2)).

[17B]   ## Summary

This section allows for the possible appointment of a sole arbitrator in default of the appointment of a second arbitrator. It applies to a two-party case where each party is to appoint an arbitrator (under an agreed procedure or under s.16 (4) to (6)) and one of them fails to do so. Once again the parties are free to agree otherwise, so that they might themselves, for instance, provide for a third party to make the appointment in place of the defaulting party. Alternative remedies are set out in s.18.

To oust the effect of this section, there must be specific contrary agreement, or specific agreement that the section does not apply. Such agreement must be in writing, in accordance with s.5.

Points

[17C]   *Changes – Subss.(1), (2)*
The law prior to this Act is preserved, but with certain changes. Thus, by subs.(1), notice may be given to a defaulting party either if that party *fails* to appoint his arbitrator within the time specified in the arbitration agreement (that is to say, after the time has been allowed to elapse), or if the defaulting party *refuses* to make the appointment, in which case notice may be given immediately. This second alternative is new, and plainly desirable since there is no reason why a party expressly refusing to act should be allowed any further tolerance.

In addition, as noted in paragraph 16C, in the absence of any contrary agreement, 'the time specified' in s.17(1) can be 14 days after service of a written request to appoint an arbitrator, without the party giving that notice having had to appoint his own arbitrator at that stage. However, before giving seven clear days' notice under s.17(2) that party must have appointed his arbitrator and notice of that appointment must have been given.

A further possible change introduced by subs.(2) is that in order to avoid the consequences of the section, the defaulting party must not only make the appointment within the time specified, but must also notify the other party that they have done so. Under the law prior to this Act it was not, in our view, clear whether or not an appointment was complete prior to notification to the other side. It was the DAC's view that under the 1950 Act a defaulting party was under no obligation to say that an appointment had been made. Whatever the position, the clarification is to be welcomed.

It is also worth noting that the time periods have effectively been extended. Under s.7(b) of the 1950 Act it was possible to appoint one's own arbitrator and *simultaneously* call for a counter-appointment within seven clear days, any default appointment duly following after that period. Now, the combined effect of ss.16 and 17 is that there must first be a 14 days' notice (s.16(4)), and if there is then default, a further seven clear days' notice under s.17(2), before appointing a sole arbitrator. The seven day period must allow for the inclusion of a weekend at least (see paragraph 17D below), and the day of giving the default notice would not be counted in either period. Thus (unless there is a prompt refusal to appoint by the defaulting party) we calculate that appointment on the 25th actual day from the date of the first notice would be the shortest possible time scale.

[17D]   *Seven clear days – Subs.(2)*
Subs.(2) refers to a time period of seven clear days. Clear days are defined in s.78(4) such that at least seven days must intervene between

the starting date and the date the following event happens. Thus if the starting date were a Monday, this would suggest that the Tuesday to the following Monday would be clear days, and the next event might occur on the Tuesday. However, by s.78(5), where the Act prescribes a period of seven days or less, weekends and public holidays are not counted. Thus in an ordinary week, the weekend would affect our example. Accordingly, the seven clear days would be Tuesday to Friday of the first week and Monday to Wednesday of the next, making the following Thursday the earliest day on which the following event might happen.

[17E]   *The court – Subss.(3), (4)*

The court has power pursuant to s.79 to extend the time limit in subs.(2) within which the defaulting party must act. Alternatively it may set aside the sole appointment under subs.(3). In either case, notice must be given (s.80) and leave would be required from the court's decision for an appeal, (s.79(6); and subs.(4)).

No grounds are prescribed upon which the court should exercise its discretion under subs.(3) to set aside the appointment of a sole arbitrator pursuant to this section. It will therefore have an unfettered discretion to deal with each case on its own merits, in the light of the principles set out in the Act (as to which, see s.1 and CPR Part 49G (Practice Direction), para.1).

S.7(b) of the 1950 Act did, of course, provide that an appointment under that section might be set aside by the court. However, the reported cases are of little help as to how that power might have been used. At page 183 of *Mustill and Boyd* (2nd edition) it is suggested that 'the most likely occasions will be those where there is manifest bias on the part of the person appointed as sole arbitrator, or excessive haste by the person making the appointment'.

[17F]   *The application*

We deal with applications to extend time under s.79 in the commentary on that section. The application under s.17(3) is an arbitration application falling within the terms of the Practice Direction on Arbitrations issued under CPR Part 49. We deal generally with arbitration applications in Part 3, Materials, Section H. Specifically in respect of an application under s.17(3):

(i)   the application must be on notice to the appointing party;
(ii)   the requirement to give notice to the appointing party is to be met by making that party a respondent to the application and serving on it the arbitration claim form and any evidence in support;
(iii)   unless otherwise ordered, the hearing will be in private.

[17G]   *Model Law*

There is no similar provision in the Model Law. We set out in paragraph 18K the regime under the Model Law where the appointment procedure fails.

[17H]   ***Rules***

The section is only of limited application, applying to arbitration agreements (or to rules) where each of two parties are to appoint an arbitrator. It applies 'unless the parties otherwise agree'. The section will not generally apply to domestic arbitrations and applicable rules where a single arbitrator is contemplated. Of the international rules we have considered, under the LCIA Rules, by Art.7, and under the ICC Rules, by Art.8.4, the respective institutional courts may make the appointment. In our view these provisions oust s.17. Under the UNCITRAL Rules, by Art.7 the party who has appointed his arbitrator may request the appointing authority designated to make the appointment. If there is no such authority, the Secretary-General of the Permanent Court of Arbitration at The Hague may be asked to designate such an authority. In our view, again, s.17 would be ineffective.

Section 18 – Failure of Appointment Procedure

18.–(1) The parties are free to agree what is to happen in the event of a failure of the procedure for the appointment of the arbitral tribunal.

There is no failure if an appointment is duly made under section 17 (power in case of default to appoint sole arbitrator), unless that appointment is set aside.

(2) If or to the extent that there is no such agreement any party to the arbitration agreement may (upon notice to the other parties) apply to the court to exercise its powers under this section.

(3) Those powers are–

(a) to give directions as to the making of any necessary appointments;
(b) to direct that the tribunal shall be constituted by such appointments (or any one or more of them) as have been made;
(c) to revoke any appointments already made;
(d) to make any necessary appointments itself.

(4) An appointment made by the court under this section has effect as if made with the agreement of the parties.

(5) The leave of the court is required for any appeal from a decision of the court under this section.

Definitions

'agreement', 'agree': s.5(1).
'arbitration agreement': ss.6, 5(1).
'arbitrator': s.82(1).
'the court': s.105.

'notice (or other document)': s.76(6).
'party': ss.82(2), 106(4).
'upon notice' (to the parties or the tribunal): s.80.

[18A] ## Status

This section derives from the Model Law, Art.11(4), and the Arbitration Act 1950, s.10. It is a *non-mandatory* provision, (s.4(2)).

[18B] ## Summary

This section explains what is to happen in the event of the failure of the appointment procedure, whether the procedure is prescribed by the parties or imposed by the Act, and assuming s.17 has not been operated so as to produce a sole arbitrator. The parties are at liberty to agree what is to happen in the event that the appointment procedure fails to produce a tribunal. Any agreement must be in writing to accord with s.5.

To the extent that there is no such agreement, the court is given powers to make good the deficiency. These powers are available to deal with any situations not met by the procedures for appointment provided in default of agreement by s.16(3) to (6), and for any other cases, (see s.16(7)).

Points

[18C]   *Default of agreement – Subs.(2)*
The court only has power to the extent that any agreement between the parties does not cover the events which have, or have not, occurred.

[18D]   *Powers of the court – Subs.(3)*
Whereas the powers of the court were previously limited to making appointments itself, it now has greater flexibility. Thus it may confirm appointments already made as constituting the tribunal; or revoke appointments already made (in order to arrive at an appropriate tribunal). It may also give directions. We consider these powers individually in paragraphs 18E to H below.

In considering whether, and if so how, to exercise its powers under this section, either in relation to the appointment procedure or on the failure of such procedure, the court is required to have due regard to any agreement of the parties as to the qualifications required of the arbitrators, (s.19).

[18E]   *The power to give directions – Subs.(3)(a)*
The extent of this power is not obvious, but we can see that it might prove very useful. On a strict analysis, it would seem confined to giving directions to the parties. Plainly the court could direct one of the parties either to initiate some process for making an appointment, or to make an appointment which it had failed to make.

The court could not direct an appointing person or body to make an appointment which it had failed to make, because that person or body would not be before the court. However, if an appointing person or body no longer existed, the court might well direct that another such body (which had indicated that it would comply) should make the appointment in its place. Equally, we suppose, if an appointing authority's procedures required something of a putative respondent before it could act, e.g. confirmation of a request for an appointment, the court might direct that such steps should be deemed to have been taken.

We see this power as one which the court may use prior to making an appointment itself, which it would do as a last resort.

[18F]   *The power to direct the tribunal to be constituted by appointments already made – Subs.(3)(b)*
Ordinarily, where there is to be more than one arbitrator and only two parties, and where the second party fails to make an appointment, the procedure under s.17 is likely to be employed, so that a sole arbitrator

results. The scope for use of this subsection seems, therefore, limited, and it is only likely to be used where there are a number of arbitrators. It might well be used where, for instance, two arbitrators have failed to appoint a third arbitrator or umpire, or where there are more than two parties, and one party has failed to make an appointment.

[18G]   *The power to revoke appointments made – Subs.(3)(c)*
This power was inserted to ensure fair treatment of the parties following the *BKMI Industrieanlagen* v. *Dutco* case in France, [1994] ADRLJ 36. Where one party chooses its own arbitrator and the other (through no fault of its own) has an arbitrator imposed upon it by the court, it may be said that there is unfairness. This provision allows the court to redress the balance.

   In our view, the power could only be invoked in the context of a failure of the appointment procedure, and as part of a more general structuring of the tribunal by the court. It could not be invoked by a mischievous party seeking the removal of an arbitrator when the appointment procedure had otherwise been successfully implemented.

[18H]   *The power to make necessary appointments itself – Subs.(3)(d)*
We believe that the previous practice, by which the party seeking the assistance of the court suggested names to the court in its affidavit, continues. It is important to obtain an indication of willingness to act from the potential arbitrators whose names are put forward, together with an indication of their availability and suitability. The court (and therefore any proposing party) must take into account any agreement of the parties as to the qualifications required of the arbitrators, see s.19.

[18I]   *Appeals – Subs.(4)*
By contrast with the Model Law (which provides that the corresponding court decisions should be subject to no appeal, see Art.11(5)), an appeal will lie against the court's decision under this section, although only with leave. Thus the court's exercise of its powers, and in particular, its new powers, under this section are subject to appellate review in appropriate cases.

[18J]   *The application*
The application made under s.18(2) is an arbitration application falling within the terms of the Practice Direction on Arbitrations issued under CPR Part 49. We deal generally with arbitration applications in Part 3, Materials, Section H. Specifically in respect of an application under s.18:

(i)   the application must be on notice to the other party or parties to the potential arbitration;
(ii)   the requirement to give notice to other parties is to be met by making those parties respondents to the application and serving on them the arbitration claim form and any evidence in support;

(iii) the arbitration claim form must show that there has been a failure in the procedure for the appointment of the tribunal;

(iv) unless otherwise ordered, the hearing will be in private.

[18K] *Model Law*

Art.11 of the Model Law provides:

> '(2) The parties are free to agree on a procedure of appointing the arbitrator or arbitrators, subject to the provisions of paragraphs (4) and (5) of this article.'

This clause corresponds to s.16(1) of the Act, save that the Act provides for a means of appointing a chairman or an umpire. (UNCITRAL does not provide for an umpire; it provides for a 'presiding arbitrator' through Art.7 of its Arbitration Rules.)

> '(3) Failing such agreement,
> (a) in an arbitration with three arbitrators, each party shall appoint one arbitrator, and the two arbitrators thus appointed shall appoint the third arbitrator; if a party fails to appoint the arbitrator within thirty days of receipt of a request to do so from the other party, or if the two arbitrators fail to agree on the third arbitrator within thirty days of their appointment, the appointment shall be made, upon request of a party, by the court or other authority specified in article 6;
> (b) in an arbitration with a sole arbitrator, if the parties are unable to agree on the arbitrator, he shall be appointed, upon request of a party, by the court or other authority specified in article 6.
> (4) Where, under an appointment procedure agreed upon by the parties,
> (a) a party fails to act as required under such procedure, or
> (b) the parties, or two arbitrators, are unable to reach an agreement expected of them under such procedure, or
> (c) a third party, including an institution, fails to perform any function entrusted to it under such procedure,
> any party may request the court or other authority specified in article 6 to take the necessary measure, unless the agreement on the appointment procedure provides other means for securing the appointment.'

[18L] *Rules*

Of the domestic rules, CIMAR specifically provides that s.18 applies, subject to its own provisions applicable to related arbitrations. The CIArb Rules provide in Art.4.1 for ss.15–17 to apply 'to the procedure for the appointment of the arbitrator but if no appointment is made under these provisions the arbitrator shall be appointed by the Appointing Authority on the application of either party'. The appointing authority is defined in Art.3 as any authority agreed by the parties, and the President or a Vice President for the time being of the Institute in the absence of agreement.

The failure of the parties to make their own appointments or nominations (where, as in the case of the LCIA and ICC, the institutional court makes the appointment) will normally result in an institutional appointment. Thus the ICE Arbitration Procedure so provides in Rule 4, the LCIA Rules so provide in Arts.5 and 7, and the ICC Rules so provide in Art.8. Under UNCITRAL Rules the matter will usually be resolved by the appointing authority. Presumably if the institution or appointing authority failed to act, there would then be a failure of the appointment procedure not provided for by the parties sufficient to justify intervention of the court under s.18.

Section 19 – Court to have Regard to Agreed Qualifications

19. In deciding whether to exercise, and in considering how to exercise, any of its powers under section 16 (procedure for appointment of arbitrators) or section 18 (failure of appointment procedure), the court shall have due regard to any agreement of the parties as to the qualifications required of the arbitrators.

Definitions

'agreement': s.5(1).
'arbitrator': s.82(1).
'the court': s.105.
'party': ss.82(2), 106(4).

[19A]   Status

This section derives from the Model Law, Art.11(5). It is classed as *non-mandatory*, (s.4(2)).

[19B]   Summary

The section requires the court to have due regard to any agreement of the parties as to the qualifications required of the arbitrators in deciding whether, and if so how, to exercise its powers under s.18 in relation to the failure of the appointment procedure, or where those powers are imported into s.16 to compensate for a deficient or non-existent appointment procedure, (s.16(7)).

Points

[19C]   *Effect*

The object of the section is, as far as possible, to respect the parties' agreement as to the required qualifications of the tribunal, notwithstanding an intervention, or potential intervention, by the court. As we interpret the section, the court need not be bound by the parties' agreement, but it is clear that it should normally follow it. Obvious examples are charterparty clauses which require as arbitrators 'members of the Baltic Exchange' and/or someone 'engaged in the shipping or grain trades'.

[19D]   *Non-mandatory status*

In the first edition we said that the wording of the section suggests that it is not one that could be directly excluded by agreement of the parties.

On reflection we consider that if an arbitration agreement requires particular qualifications of arbitrators but also expressly provides that this section should not apply to it, and the court is involved in an appointment, it could simply choose whether or not to have regard to the specified qualifications.

[19E]   *Model Law*

Art.11 (5) of the Model Law provides (in part):

> 'The court or other authority, in appointing an arbitrator, shall have due regard to any qualifications required of the arbitrator by the agreement of the parties...'

Section 20 – Chairman

20.–(1) Where the parties have agreed that there is to be a chairman, they are free to agree what the functions of the chairman are to be in relation to the making of decisions, orders and awards.

(2) If or to the extent that there is no such agreement, the following provisions apply.

(3) Decisions, orders and awards shall be made by all or a majority of the arbitrators (including the chairman).

(4) The view of the chairman shall prevail in relation to a decision, order or award in respect of which there is neither unanimity nor a majority under subsection (3).

Definitions

> 'agreement', 'agree', 'agreed': s.5(1).
> 'arbitrator': s.82(1).
> 'party': ss.82(2), 106(4).

[20A] Status

This section derives from the Model Law, Art.29, and the Arbitration Act 1950, s.9. It reiterates the law prior to this Act with one change. It is *non-mandatory*, (s.4(2)).

[20B] Summary

This section defines the functions of the chairman where one is appointed. Legislation has not previously acknowledged the role of chairman. That of third arbitrator, although commonplace today, was, as a result of the wording of s.9 of the 1950 Act, less familiar to the English jurisdiction than that of umpire (whose functions are defined in s.21). The amendment of s.9 effected by s.6(2) of the 1979 Act meant that the third arbitrator appeared far more often, and in practice third arbitrators have normally (although not invariably) chaired tribunals. The particular functions of the chairman are here introduced.

Points

[20C] *Agreement – Subs.(1)*

The parties may specifically agree that a chairman should be appointed, and in that case, they may also agree his functions. As we suggest below, it may be sensible, for instance, to permit the chairman to act on

121

his own in respect of certain (or all) interlocutory matters. Any agreement must be in writing to accord with s.5.

[20D]   *Functions in default of agreement – Subss.(2) to (4)*
To the extent that the parties do not agree, or in cases where the chairman is a required addition to the tribunal because the parties' agreement provided for the appointment of an even number of arbitrators (see s.15(2)), the default provisions of the section apply.

Decisions, orders and awards are to be made unanimously or by a majority, in which case the chairman's view has no more weight than that of any other arbitrator. However in cases where there is neither unanimity nor a majority then the chairman's view will prevail, thus preventing a deadlock.

This situation might arise where there is an even number of arbitrators – perhaps four – or where in an odd-numbered tribunal, there is an abstention. Even in a three-arbitrator tribunal it is possible to envisage situations where there is no majority, when the chairman's view would prevail. For instance, one arbitrator may hold that there is no liability; another may hold that there is and that specific performance should be ordered; whilst the third may agree on liability but wish to award damages. Or each may disagree on the quantum of damages (although there is then an argument that there is a majority in favour of at least the middle figure).

Subs.(3) appears to contemplate that all the arbitrators are to be involved in the decision-making process for all 'decisions, orders and awards'. This may be varied by agreement, and it is often sensible for parties to agree that certain, or all, interlocutory matters be dealt with by the chairman alone. Such agreements save time and cost.

[20E]   *Model Law*
Art.29 provides:

> 'In arbitral proceedings with more than one arbitrator, any decision of the arbitral tribunal shall be made, unless otherwise agreed by the parties, by a majority of all its members. However, questions of procedure may be decided by a presiding arbitrator, if so authorised by the parties or all members of the arbitral tribunal.'

Unlike under the Act, there is no fall back for when a majority cannot be obtained, and the arbitrators will, in practice, have to persuade each other, until there is at least a majority view.

[20F]   *Rules*
Domestic rules normally provide for a single arbitrator (though the CIArb Rules reflect the possibility of the parties agreeing otherwise). The LCIA Rules provide for a chairman (who is not a party nominated arbitrator) where there are three arbitrators. By Art.14.3, with the prior consent of the other two arbitrators, he may make procedural rulings alone. By Art.26.3, where there is no majority decision on an issue, the

chairman decides it. (Art.25 of the ICC Rules has similar effect to Art.26.3 of the LCIA Rules). The UNCITRAL Rules provide for a 'presiding arbitrator', which we suggest amounts to a chairman (see paragraph 15D). By Art.31 awards or decisions are made by a majority. In the case of questions of procedure, when there is no majority or when the arbitral tribunal so authorises, the presiding arbitrator may decide on his own, subject to revision, if any, by the arbitral tribunal. The UNCITRAL Rules do not, therefore, deal with the situation contemplated by Art.26.3 of the LCIA Rules and, should it arise, s.20(4) would remain applicable to produce the same result.

Section 21 – Umpire

21.-(1) Where the parties have agreed that there is to be an umpire, they are free to agree what the functions of the umpire are to be, and in particular–

(a)   whether he is to attend the proceedings, and
(b)   when he is to replace the other arbitrators as the tribunal with power to make decisions, orders and awards.

(2) If or to the extent that there is no such agreement, the following provisions apply.

(3) The umpire shall attend the proceedings and be supplied with the same documents and other materials as are supplied to the other arbitrators.

(4) Decisions, orders and awards shall be made by the other arbitrators unless and until they cannot agree on a matter relating to the arbitration.
  In that event they shall forthwith give notice in writing to the parties and the umpire, whereupon the umpire shall replace them as the tribunal with power to make decisions, orders and awards as if he were sole arbitrator.

(5) If the arbitrators cannot agree but fail to give notice of that fact, or if any of them fails to join in the giving of notice, any party to the arbitral proceedings may (upon notice to the other parties and to the tribunal) apply to the court which may order that the umpire shall replace the other arbitrators as the tribunal with power to make decisions, orders and awards as if he were sole arbitrator.

(6) The leave of the court is required for any appeal from a decision of the court under this section.

Definitions

'agreement', 'agree', 'agreed': s.5(1).   'party': ss.82(2), 106(4).
'arbitrator': s.82(1).   'upon notice' (to the parties or the
'the court': s.105.   tribunal): s.80.
'notice (or other document)': s.76(6).   'in writing': s.5(6).

[21A]   **Status**

This section derives in part from the Arbitration Act 1950, s.8(2) and (3). It is *non-mandatory*, (s.4(2)).

[21B]   **Summary**

The functions of the umpire are here defined. The parties are at liberty to agree the functions of the umpire, with particular regard to his attendance at the proceedings, and the point at which he replaces the other arbitrators as the deciding tribunal. Again, any agreement must be in writing to accord with s.5.

To the extent that the parties do not so agree, the section sets out provisions that represent something of a compromise between different possible interpretations of the umpire's function.

Points

[21C]   *Functions in default of agreement – Subss.(3), (4)*
Where there is no agreement covering the matter, the umpire is required to attend the proceedings and to be supplied with the same documents and materials as the other arbitrators. He does not, initially, take part, but merely observes the arbitration proceedings, unless and until a point is reached where the arbitrators fail to agree. In that event, the arbitrators are required to notify the umpire and the parties, and the umpire then replaces them as the tribunal, operating as a sole arbitrator. He is, of course, in a position to assume that role immediately because of his attendance throughout the proceedings. As noted in paragraph 16C, we think that 'the proceedings' may amount to more than merely the 'substantive hearing' referred to in s.16(6)(b).

It has long been the practice for an umpire to attend the proceedings in any event, where one was likely to be appointed, and these subsections simply have the effect that attendance by the prospective umpire will now be the ordinary expectation. In any event, note *Fletamentos Maritimos* v. *Effjohn* [1995] 1 Lloyd's Rep. 311, in which it was held to be implied that such a potential umpire might attend the proceedings.

It was at one stage suggested that an umpire might be expressly authorised to deliberate with arbitrators prior to disagreement, but the DAC ultimately rejected that approach. However, it is open to the parties to agree that such deliberation might take place. In any event it is usual and useful for the umpire to hear the arbitrators' deliberations, if only so that he knows precisely why any eventual disagreement has arisen.

[21D]   *Order to replace arbitrators – Subss.(5), (6)*
Failure by the arbitrators to give notice of disagreement and thus effectively to vest power in the umpire at the appropriate time will permit any party to the arbitral proceedings to apply to the court for an appropriate order.

[21E]   *The application*
The application under s.21(5) is an arbitration application falling within the terms of the Practice Direction on Arbitrations under CPR Part 49. We deal generally with arbitration applications in Part 3, Materials, Section H. Specifically in respect of an application under s.21:

(i)   the application must be on notice to the other parties and to the tribunal;

(ii)   the requirement for giving notice to the other parties is to be met by making those parties respondents to the application and serving on them the arbitration claim form and evidence in support;

(iii)   the requirement for giving notice to the tribunal is met by sending copies of the arbitration claim form to the arbitrators for their information at their last known addresses together with copies of the evidence in support;

(iv)   the arbitration claim form must show that the arbitrators cannot agree and have failed to give notice of that fact or that one of them has failed to join in giving such notice; and

(v)   unless otherwise ordered, the hearing will be in private.

[21F]   ***The future of the umpire***
The umpire is a very English concept and, at least prior to the Arbitration Act 1979, umpires were commonplace in those areas of arbitral activity (particularly, for example, shipping) where it has been the norm to have party-appointed arbitrators rather than a sole arbitrator. It is now probably more common, where the parties are prepared to go to the expense of having a three-man tribunal, for there to be three arbitrators, one being appointed as a chairman. The Act reflects this.

The DAC contemplated sweeping away the concept of the umpire altogether. However it decided against so doing, for the benefit of those who still wish to use this form of dispute resolution. We doubt that it will be used much, apart from where standard contract forms containing 'umpire' arbitration clauses continue in service. But clearly its retention is consonant with the aim of encouraging the widest possible means of resolving disputes.

[21G]   ***USA***
It should be noted that the term 'umpire' is used in America. It is, however, used to describe what we would call the 'chairman'.

[21H]   ***Model Law***
There is no similar provision, the umpire being unknown in most other jurisdictions.

Section 22 – Decision-making where no Chairman or Umpire

22.–(1) Where the parties agree that there shall be two or more arbitrators with no chairman or umpire, the parties are free to agree how the tribunal is to make decisions, orders and awards.

(2) If there is no such agreement, decisions, orders and awards shall be made by all or a majority of the arbitrators.

Definitions

'agreement', 'agree': s.5(1).
'arbitrator': s.82(1).
'party': ss.82(2), 106(4).

[22A] ## Status

This section derives in part from the Model Law, Art.29. It is *non-mandatory*, (s.4(2)).

[22B] ## Summary

This section provides for decision-making where there is more than one arbitrator, but no chairman or umpire. It therefore closes a gap left by s.20(3) and (4), which only applies to decision-making where there is a chairman, and by s.21(4), which only applies where there is an umpire.

Points

[22C] *Effect*
The parties are at liberty to agree how decisions should be made in such a situation. The agreement must be in writing, in accordance with s.5. In default of agreement, the section provides for unanimous or majority decisions.

It is apparent, however, that if a tribunal comprising an even number of arbitrators is evenly divided in respect of a matter, there is no statutory provision enabling that deadlock to be broken. For a statute to leave this position open is not uncommon. Art.29 of the Model Law (which we have set out at paragraph 20E) does the same. The practice is, we believe, to rely on persuasion and common sense to avoid failure of the arbitration process.

In any event, we believe that parties are very unlikely ever to agree the sort of tribunal contemplated by this section. The parties have nothing to gain from agreeing a panel with an equal number of arbitrators and equally nothing to gain from not providing for a chairman or umpire unless, of course, they expressly want to anticipate the possibility of a 'tie' and the consequent failure of the arbitral process. Whether they would then be free to go to court may be questionable. While the section closes a gap in the scheme of the Act, in practice it was barely necessary to provide it.

[22D]   *Model Law*
We have set out Art.29 at paragraph 20E above.

Section 23 – Revocation of Arbitrator's Authority

23.–(1) The parties are free to agree in what circumstances the authority of an arbitrator may be revoked.

(2) If or to the extent that there is no such agreement the following provisions apply.

(3) The authority of an arbitrator may not be revoked except–

(a) by the parties acting jointly, or
(b) by an arbitral or other institution or person vested by the parties with powers in that regard.

(4) Revocation of the authority of an arbitrator by the parties acting jointly must be agreed in writing unless the parties also agree (whether or not in writing) to terminate the arbitration agreement.

(5) Nothing in this section affects the power of the court–

(a) to revoke an appointment under section 18 (powers exercisable in case of failure of appointment procedure), or
(b) to remove an arbitrator on the grounds specified in section 24.

Definitions

'agreement', 'agree': s.5(1).
'arbitration agreement': ss.6, 5(1).
'arbitrator': s.82(1).

'the court': s.105.
'party': ss.82(2), 106(4).
'in writing': s.5(6).

[23A] Status

This section derives from the Arbitration Act 1950, s.1, and may be compared with the Model Law, Art.14. It is *non-mandatory*, s.4(2).

[23B] Summary

This section deals with the revocation of the arbitrator's authority. It is one of a number of sections concerning the circumstances in which an arbitrator may 'cease to hold office', and the consequences of that cessation. This section provides for the termination of the arbitrator's appointment by the concerted action of the parties, or by an arbitral or other relevant institution or person.

Note that where only one of the parties seeks the termination of the arbitrator's appointment, they may do so by applying to the court for his removal, see s.24. Where it is the arbitrator, rather than the parties, who effects the termination of his own appointment by resigning, s.25 provides for the consequences of that resignation. S.26 deals with the effect of the death of the arbitrator. S.27 deals with a number of other matters that arise when an arbitrator ceases to hold office.

Points

[23C] *Effect*

Essentially, the parties are at liberty to agree the circumstances in which the arbitrator's authority may be revoked and his appointment terminated. To the extent that there is no such agreement, the arbitrator's authority is irrevocable except at the instance of the parties acting jointly, or at the instance of a third party (which may be an arbitral institution) in whom the parties have vested the relevant powers. This is a change from the law prior to the Act under which, in the absence of a contrary intention in the arbitration agreement, revocation was only possible with leave of the High Court.

The court exercising a power either under s.18 (revoking an appointment already made in the context of establishing the tribunal) or under s.24 (removal for partiality, incapacity, etc) will also effectively revoke the arbitrator's authority, and such revocation expressly takes effect notwithstanding the fundamental principle of irrevocability, (subs.(5)).

It may be suggested that the different terms used in the Act, such as 'removal' and 'revocation of authority' have different legal significance. Paragraph 98 of the DAC Report (to which reference may be made for some purposes of interpretation) has sought to give the lie to any such argument.

[23D] ***Freedom to agree circumstances of revocation – Subs.(1)***

In our view such agreements are likely to be complex and, in practice, will mostly be imported from institutional rules. To take one example, the UNCITRAL Arbitration Rules, Arts.10-12, provide a regime for the challenge to an arbitrator where there may be doubts as to his impartiality or independence. (Independence is not a ground for removal under s.24, but might be added by adopting these Rules). Such challenges may occur well into a reference, provided the challenge is notified within 15 days of the circumstances giving rise to the doubt surfacing. Unless the challenge is resolved by concession, other authorities become involved. Art.12.1 provides:

> 'If the other party does not agree to the challenge and the challenged arbitrator does not withdraw, the decision on the challenge will be made:
>
> (a) when the initial appointment was made by an appointing authority, by that authority;
> (b) when the initial appointment was not made by an appointing authority but an appointing authority has been previously designated, by that authority;

(c) in all other cases, by the appointing authority to be designated in accordance with the procedure for designating an appointing authority as provided for [elsewhere in the Rules].'

Further the agreement needs to provide for what is to happen where there has been a revocation. Art.12.2 of the UNCITRAL Rules, for instance, goes on to provide for the means of appointment of the replacement arbitrator.

The LCIA Rules, in Arts.10 and 11, and the ICC Rules, in Arts.11 and 12, both deal at length with challenges to arbitrators, revocation of their appointments and replacement procedures.

[23E]   *Formalities of revocation by parties – Subs.(4)*
In keeping with the general requirement (set out in s.5(1)) for agreements between the parties to be in writing in order to be effective for the purposes of this Part, an agreement to revoke an arbitrator's authority made by the parties acting jointly must be in writing.

By contrast, and as an exception to the general requirement, an agreement to terminate an arbitration need not be in writing, due to the impracticability of imposing such a requirement in certain of the circumstances in which an arbitration may be mutually allowed by the parties to lapse. Accordingly, where the agreement to revoke the arbitrator's authority is made in the context of an agreement to terminate the arbitration, then neither need be in writing.

[23F]   *Model Law*
Art.14 provides:

> '(1) If an arbitrator becomes *de jure* or *de facto* unable to perform his functions or for other reasons fails to act without undue delay, his mandate terminates if he withdraws from his office or if the parties agree on the termination. Otherwise, if a controversy remains concerning any of these grounds, any party may request the court or other authority specified ... to decide on the termination of the mandate, which decision shall be subject to no appeal.'

[23G]   *Rules*
CIMAR and the CIArb Rules do not provide for revocation so that s.23 applies. The ICE Procedure contains a provision at Rule 24.1 that revokes the arbitrator's appointment within 14 days of the appointment if the arbitrator stipulates at the time of his appointment that he requires the ICE Arbitration Procedure to apply and the parties agree otherwise. We have dealt with the international rules above.

Section 24 – Power of Court to Remove Arbitrator

24.-(1) A party to arbitral proceedings may (upon notice to the other parties, to the arbitrator concerned and to any other arbitrator) apply to the court to remove an arbitrator on any of the following grounds–

(a) that circumstances exist that give rise to justifiable doubts as to his impartiality;
(b) that he does not possess the qualifications required by the arbitration agreement;
(c) that he is physically or mentally incapable of conducting the proceedings or there are justifiable doubts as to his capacity to do so;
(d) that he has refused or failed–

  (i) properly to conduct the proceedings, or
  (ii) to use all reasonable despatch in conducting the proceedings or making an award,

  and that substantial injustice has been or will be caused to the applicant.

(2) If there is an arbitral or other institution or person vested by the parties with power to remove an arbitrator, the court shall not exercise its power of removal unless satisfied that the applicant has first exhausted any available recourse to that institution or person.

(3) The arbitral tribunal may continue the arbitral proceedings and make an award while an application to the court under this section is pending.

(4) Where the court removes an arbitrator, it may make such order as it thinks fit with respect to his entitlement (if any) to fees or expenses, or the repayment of any fees or expenses already paid.

(5) The arbitrator concerned is entitled to appear and be heard by the court before it makes any order under this section.

(6) The leave of the court is required for any appeal from a decision of the court under this section.

Definitions

'arbitration agreement': ss.6, 5(1).
'arbitrator': s.82(1).
'the court': s.105.
'notice (or other document)': s.76(6).

'party': ss.82(2), 106(4).
'upon notice' (to the parties or the tribunal): s.80.

[24A] ## Status

This section derives from the Arbitration Act 1950, ss.13(3), 23(1) and 24(1), and from the Model Law, Arts.12, 13 and 14. It is *mandatory*, (s.4(1) and Sched.1).

[24B] ## Summary

This section covers the power of the court to remove an arbitrator. It is one of a number of sections concerning the circumstances in which an arbitrator may 'cease to hold office' and the consequences of that cessation. It applies where only one of the parties seeks the termination of the arbitrator's appointment, and provides that they may do so by applying to the court upon notice (s.80) for his removal. For other sections concerning the arbitrator's cessation of office, see our commentary at paragraph 23B.

By contrast with the Arbitration Act 1950 which (apart from specifying unreasonable delay) required a party to show that the arbitrator had 'misconducted himself or the proceedings' as the basis for his removal, this section sets out separately and in greater detail the grounds on which the court may exercise its power. Arbitrators welcome the abandonment of the word and notion 'misconduct'. On the other hand, as has been pointed out, it would seem to be no less a chastisement to be removed for any of the reasons defined in this section as to be removed for misconduct.

Points

[24C] *Doubts as to impartiality – Subs.(1)(a)*
By contrast with the Model Law, Art.12 (see paragraph 24I below), the first ground refers only to justifiable doubts about the arbitrator's *'impartiality'*, without making any mention of his *'independence'*. This, the DAC Report tells us, is in essence because it is possible to be impartial without being wholly independent, and anyone who is affected by lack of independence will demonstrate partiality. The further ground is, therefore, unnecessary.

The DAC Report gives a particular example of the English Bar (whose members participate in arbitrations both as advocates and as arbitrators). Since barristers are self-employed, members of the same chambers may be (and not infrequently are) instructed to appear in front of their colleagues while the latter are sitting as arbitrators. In such a case, if 'independence' were an available ground, the strict independence of the arbitrator could be called into question and used as the basis of a challenge, whereas his ability to act impartially was never in issue.

In fact there has been a challenge to a barrister arbitrator on the grounds that he belonged to the same set of chambers as the barrister acting for one of the parties. The challenge failed, it being held that the test for deciding whether there were justifiable doubts as to his impartiality was an objective one. The judge commented that it has long been an everyday occurrence for a barrister to appear against a member of his own chambers, and for other members of his chambers to appear before him when he is acting as an arbitrator or deputy judge or recorder: *Laker Airways Inc.* v. *FLS Aerospace Ltd* [1999] 2 Lloyd's Rep. 45.

Plainly, however, lack of independence in the arbitrator may be relevant if it is such as to give rise to justifiable doubts about his impartiality. For commentary on 'independence' in the international context, see paragraph 24I below.

In another case, where a construction arbitrator wrote ill-judged letters indicating irritation at one party's refusal to agree his terms, and in particular a minimum fee, and where he made an unusual order as to costs, the Court of Appeal held, with some apparent hesitation, that in all the circumstances there was no real danger of bias such as to justify the arbitrator's removal: *Andrews* v. *Bradshaw and Anor* (The Times, 11 October 1999).

The approach to bias is that laid down in *R.* v. *Gough* [1993] AC 646: see *Laker Airways Inc.* v. *FLS Aerospace* (above) and *Rustal Trading Ltd* v. *Gill & Duffus SA* [2000] 1 Lloyd's Rep. 14 (in which the particular situation of trade arbitrators is helpfully dealt with).

[24D]   *Failure to conduct the proceedings properly – Subs.(1)(d)*
The fourth ground of challenge refers to the arbitrator's proper conduct of the proceedings and to his use of reasonable despatch. Readers should refer to s.33, where positive provision is made as to the duties of the tribunal in relation to the proceedings. In short, these duties are that in its conduct of the arbitral proceedings, in its decisions and in the exercise of all other powers conferred on it the tribunal must act fairly and impartially between the parties, giving each a reasonable opportunity of putting its case, and must adopt suitable procedures, avoiding unnecessary delay or expense, so as to provide a fair means for the resolution of the issues.

In addition to the relevant failures in the arbitrator's conduct of the case, the applicant must also show substantial injustice to himself in order to justify removing the arbitrator. What constitutes 'substantial injustice' will have to be worked out by the courts on a case-by-case basis. These provisions are not intended 'to allow the court to substitute its own view as to how the arbitral proceedings should be conducted', (see the DAC Report, paras.105 and 106). Their purpose is to catch those rare cases where it can be said that the procedure adopted by the tribunal is frustrating, rather than furthering, the object of arbitration, as defined in s.1, namely the procurement of 'the fair resolution of disputes by an impartial tribunal without unnecessary delay or expense'.

[24E]   *Exhausting other recourses – Subss.(2), (3)*
There are provisions intended to prevent a party from making an application for removal either prematurely or belatedly. Thus subs.(2) requires the applicant first to have recourse to any arbitral institution,

where appropriate. However note that, pursuant to s.73, any objection as to the conduct of the arbitration must be made promptly, since knowing participation in the proceedings thereafter will lose the party his right to object.

A tactical challenge on spurious grounds, with a view to holding up the arbitration, is discouraged by subs.(3), which provides that the tribunal is enabled to continue the proceedings and make an award pending an application under this section. The award could nonetheless still be set aside (or a declaration made that it had no effect) under s.68 (most likely under subs.(2)(a)). If a serious challenge is made under this section arbitrators should consider very carefully whether to interrupt the arbitration process.

[24F]   *Adjustment to arbitrator's remuneration – Subs.(4)*
Where an arbitrator is removed by the court, it has the power to adjust his remuneration, whether already paid or still expected. The purpose of this provision is evidently to enable the court to sanction his conduct, where appropriate, by depriving him of some, or all of his fees and expenses. (We decline to comment on the VAT consequences of the court ordering a repayment, but there are obvious complexities!)

[24G]   *Arbitrator's right to appear – Subss.(5) and (6)*
The arbitrator concerned is entitled to appear and be heard on the application to remove him. Under the procedure now followed (see paragraph 24H below) such an arbitrator will be made a respondent to the proceedings. In the previous edition we suggested that the arbitrator would rarely be well advised to appear. Under the new procedure, as a respondent he is more or less forced to take some part in the proceedings unless he is prepared to risk removal and an order for costs against him. We expect, therefore, most arbitrators to acknowledge service and then consider the extent to which they wish to put in evidence and participate at the hearing. Many will still, no doubt, substantially leave it to the party seeking to uphold their conduct to fight their corner.

Leave of the court is required for any appeal from its decision.

[24H]   *The application*
The application under s.24(1) is an arbitration application falling within the terms of the Practice Direction on Arbitration under CPR Part 49. We deal generally with arbitration applications in Part 3, Materials, Section H. Specifically in respect of an application under s.24:

(i)   the application must be on notice to the other parties, to the arbitrator concerned and to any other arbitrator;
(ii)   the requirement for giving notice to the other parties is to be met by making those parties respondents to the application and

serving on them the arbitration claim form and evidence in support;

(iii) the requirement for giving notice to the arbitrator concerned is to be met by making him a respondent to the application and serving on him the arbitration claim form and evidence in support;

(iv) the requirement for giving notice to the balance of the tribunal is met by sending copies of the arbitration claim form to the arbitrators for their information at their last known addresses together with copies of the evidence in support;

(v) the arbitration claim form must show that the applicant has exhausted any available recourse to any institution or person vested by the parties with power to remove the arbitrator;

(vi) unless otherwise ordered, the hearing will be in private.

[24I] *Model Law*

Art.12 provides:

> '(1) When a person is approached in connection with his possible appointment as an arbitrator, he shall disclose any circumstances likely to give rise to justifiable doubts as to his impartiality or independence. An arbitrator, from the time of his appointment and throughout the arbitral proceedings, shall without delay disclose any such circumstances to the parties unless they have already been informed of them by him.
>
> (2) An arbitrator may be challenged only if circumstances exist that give rise to justifiable doubts as to his impartiality or independence, or if he does not possess qualifications agreed to by the parties. A party may challenge an arbitrator appointed by him, or in whose appointment he has participated, only for reasons of which he becomes aware after the appointment has been made.'

We have already observed that the Model Law includes lack of independence as a ground for removal. The DAC were not convinced that more than impartiality was required, since 'culpable' lack of independence gives rise to lack of impartiality. In our view, the generally and internationally accepted requirement of independence exists more to ensure that fairness is being seen to be achieved, rather than actually to achieve it, and in that sense, the DAC were correct to find independence as a separate requirement unneccesary.

This view was supported, for instance, by *Redfern and Hunter* (2nd edition) page 223: 'Independence and impartiality are distinct but related concepts. The rationale of a requirement of independence must be to ensure that a connection to one of the parties or a witness is not so close as to give rise to a danger that the arbitrator will lack impartiality.'

Where we sense a difficulty over the DAC's decision, is that many institutional rules will import the requirement. Thus Art.10.1 of the UNCITRAL Rules:

> 'Any arbitrator may be challenged if circumstances exist that give rise to justifiable doubts as to the arbitrator's impartiality or independence.'

and Art.7.1 of the ICC Rules:

> 'Every arbitrator must be and remain independent of the parties involved in the arbitration.'

and the LCIA Rules in Art.5.2:

> 'All arbitrators conducting an arbitration under these Rules shall be and remain at all times impartial and independent of the parties; and none shall act in the arbitration as advocates for any party. No arbitrator, whether before or after appointment, shall appraise any party on the merits or outcome of the dispute.'

In international arbitration at least, therefore, the difference between the Act and the Model Law will make little impact.

(We note, in passing, that in our view Art.1(2) of the UNCITRAL Rules would not have the effect of rendering Art.10(1) ineffective as being in conflict with a provision of English law. S.23 of the Act entitles the parties to agree the circumstances in which the authority of an arbitrator is revoked, and his lack of independence may be the subject of such an agreement).

[24J] *Other points arising from the Model Law*
We note that the Act has no continuing duty of disclosure expressed in it, as does the Model Law. But we think it implicit.

The Model Law ground that the arbitrator does not possess the qualifications required is imported into the Act directly at s.24(1)(b). This, too, might emerge in the course of the arbitration proceedings.

The limitation that the party who has appointed an arbitrator can only challenge him for reasons of which he becomes aware after the appointment has been made is not expressed in the Act. But it is, in effect, present in s.73. A party cannot raise objection after he has made the appointment and taken further part in the proceedings if he has failed 'forthwith' to object, unless 'he did not know and could not with reasonable diligence have discovered the grounds for the objection'. We deal in more detail with s.73, which is more complex than we intimate here, below.

[24K] *Rules*
We have covered the international rules above. As might be expected given the mandatory nature of the provision, the domestic rules do not cover the matter, save that Rule 1.1 of the ICE Arbitration Procedure notes as its objective the resolution of the dispute by an 'impartial' arbitrator.

Section 25 – Resignation of Arbitrator

25.-(1) The parties are free to agree with an arbitrator as to the consequences of his resignation as regards–

(a)   his entitlement (if any) to fees or expenses, and
(b)   any liability thereby incurred by him.

(2) If or to the extent that there is no such agreement the following provisions apply.

(3) An arbitrator who resigns his appointment may (upon notice to the parties) apply to the court–

(a)   to grant him relief from any liability thereby incurred by him, and
(b)   to make such order as it thinks fit with respect to his entitlement (if any) to fees or expenses or the repayment of any fees or expenses already paid.

(4) If the court is satisfied that in all the circumstances it was reasonable for the arbitrator to resign, it may grant such relief as is mentioned in subsection (3)(a) on such terms as it thinks fit.

(5) The leave of the court is required for any appeal from a decision of the court under this section.

Definitions

'agreement', 'agree': s.5(1).   'party': ss.82(2), 106(4).
'arbitrator': s.82(1).   'upon notice' (to the parties or the
'the court': s.105.   tribunal): s.80.
'notice (or other document)': s.76(6).

[25A]   ## Status

This is a new section, though it is closely related to Art.14 of the Model Law. It is *non-mandatory*, (s.4(2)).

[25B]   ## Summary

This section covers the consequences of an arbitrator's resignation. It assumes a resignation; it does not deal with the right to resign, or give one. It proceeds on the basis that a resignation may occur, and that such resignation is likely to be a breach of the agreement to act.

  As we have said above, this is one of a number of sections concerning the circumstances in which an arbitrator may 'cease to hold office' and the consequences of that cessation. Where it is the arbitrator, rather than the parties, who effects the termination of his own appointment by resigning, this section provides for the consequences of that resignation. For other sections concerning the arbitrator's cessation of office, see our commentary at paragraph 23B.

Points

[25C]   *Agreement – Subs.(1)*
This section essentially provides that the parties are at liberty to agree with an arbitrator as to the consequences of resignation on his entitlement to fees or expenses and any liability incurred by him, the latter resulting from breach of his agreement to act as arbitrator. The liability might include such loss to the parties as their own expenses thrown away in connection with an arbitration which has to be restarted.

The agreement contemplated may be made before or after the resignation. Where it is before, it is likely to be found in institutional rules. We include an example of an agreement made with an arbitrator on resignation in Materials section E earlier in this book.

We do not think that an agreement between one party only and, for example, its own appointee would fall within this section. Whilst it might bind that party and the arbitrator, the consequences of resignation as between the parties would have to be the subject of an agreement between them, or be determined by the court.

[25D]   *Provisions in default of agreement – Subss.(2) to (4)*
To the extent that there is no such agreement, subss.2 to 5 provide a route for an arbitrator who resigns for good cause to be relieved of any potential liability and to obtain part or all of his fees. The subsections do not provide for the arbitrator who resigns without good cause (who is presumably left to sue, if he so wishes).

The section to some extent (although it has wider scope) reflects the need for a remedy to resolve the conflict that may arise consequent upon the duties of fairness and expedition placed on the arbitrator by s.33 and, on the other hand, the autonomy given to the parties to agree their own procedure by s.34. The DAC saw no conflict between ss.33 and 34, but were concerned that an arbitrator might have imposed upon him a procedure he did not want, or did not think compatible with his duties under s.33. Resignation might be his only option. In those circumstances, it was felt he should have access to a remedy of the sort envisaged in subs.(3), namely to be relieved of any liability he might incur by being in breach of his duty to act, and to receive (or not to have to repay) any fees and expenses earned or incurred.

There will, of course, be other circumstances in which it may be reasonable for the arbitrator to resign, such as when the parties are so slow to act that the arbitration has become an unfair burden, (see the DAC Report, para.115).

In granting relief, the court may impose terms.

[25E]   *Relationship to immunity*
It should be noted that s.29(3) expressly excludes from the scope of the immunity that otherwise applies to the arbitrator any liability incurred by him by reason of his resigning. Such immunity can only be obtained upon application under subs.(3)(a).

[25F]   *Appeal – Subs.(5)*
Leave is required for any appeal from the decision of the court.

[25G]   **The application**
The application under s.25(3) is an arbitration application falling within the terms of the Practice Direction on Arbitrations under CPR Part 49. We deal generally with arbitration applications in Part 3, Materials, Section H. Specifically in respect of an application under s.25:

(i)   the application by the arbitrator must be on notice to the parties;
(ii)   the requirement for giving notice to the parties is to be met by making the parties respondents to the application and serving on them the arbitration claim form and evidence in support;
(iii)   unless otherwise ordered, the hearing will be in private.

[25H]   *Model Law*
Like the Act, the Model Law does not give the arbitrator a specific right to withdraw (although it comes close to doing so in Art.14) but assumes that he will or may do so in certain circumstances. It does not deal with fees and liability as this section does:

> Art.13 (Challenge Procedure)
> '(2) ... Unless the challenged arbitrator withdraws from his office ...'
> Art.14 (Failure or impossibility to act)
> '(1) If an arbitrator becomes *de jure* or *de facto* unable to perform his functions or for other reasons fails to act without undue delay, his mandate terminates if he withdraws from his office ...'

[25I]   *Rules*
Of those domestic rules we have considered, we are not aware of any that provide for a regime different from that provided for under the default provisions of s.25. Note, in addition, that Rule 24.4 of the ICE Arbitration Procedure does give the arbitrator a specific right to resign if agreement is reached between the parties inconsistent with the procedure. He is then entitled to his reasonable fees and expenses and could presumably sue for them independently of this provision.

Of the international rules, Art.10.1 of the LCIA Rules provides:

> 'If ... (a) any arbitrator gives written notice of his desire to resign as arbitrator to the LCIA Court, to be copied to the parties and the other arbitrators (if any) ... the LCIA Court may revoke that arbitrator's appointment and appoint another arbitrator. The LCIA Court shall decide upon the amount of fees and expenses to be paid for the former arbitrator's services (if any) as it may consider appropriate in all the circumstances.'

The provision covers, at least, the ground contemplated by s.25(1)(a). Art.12 of the ICC Rules and Art.13 of the UNCITRAL Arbitration Rules provide that a substitute arbitrator shall be appointed, but do not deal with fees or liabilities incurred.

Section 26 – Death of Arbitrator or Person Appointing Him

26.–(1) The authority of an arbitrator is personal and ceases on his death.

(2) Unless otherwise agreed by the parties, the death of the person by whom an arbitrator was appointed does not revoke the arbitrator's authority.

Definitions

'agreed': s.5(1).
'arbitrator': s.82(1).
'party': ss.82(2), 106(4).

Status

[26A]

Subs.(1) is a new provision. It is *mandatory*, (s.4(1) and Sched.1). Subs.(2) widens what was previously the Arbitration Act 1950, s.2(2). It is *non-mandatory*, (s.4 (2)).

Summary

[26B]

This section covers the consequences on the arbitrator's authority of the death of that arbitrator, or of a person appointing him. It complements s.8 (1), which is non-mandatory, and which covers the consequences on the arbitration agreement of the death of a party to the arbitration.

As we have said above, this is one of a number of sections concerning the circumstances in which an arbitrator may 'cease to hold office' and the consequences of that cessation. For other sections concerning the arbitrator's cessation of office, see our commentary at paragraph 23B.

Points

Death of the arbitrator – Subs.(1)

[26C]

This subsection provides that the authority of an arbitrator is personal and so ceases on his death. This was implicit in the Arbitration Act 1950, ss.7 and 10 of which provided (in different circumstances) for the appointment of a new arbitrator where death had occurred.

Death of person appointing the arbitrator – Subs.(2)

[26D]

The Arbitration Act 1950, s.2(2), provided that the authority of the arbitrator should not be revoked by the death of any party by whom he was appointed. This subsection uses the word 'person' in place of 'party'. Thus the ambit of the subsection is widened to make it clear that, subject to contrary agreement, an appointment by a third party will not terminate on the death of that third party.

The third party is quite likely to be the president of an institution, but it may equally be another arbitrator. In the latter event, the subsection appears to overlap with subs.27(5), which has a similar effect.

[26E]   *Model Law*
There is no comparable provision in the Model Law.

[26F]   *Rules*
We are not aware of any rules which conflict with the mandatory provision of s.26(1) or the default position under s.26(2). It is common for rules to provide for what happens in the event of death of an arbitrator – see s.27 and commentary following.

Section 27 – Filling of Vacancy, etc.

27.–(1) Where an arbitrator ceases to hold office, the parties are free to agree–

(a) whether and if so how the vacancy is to be filled,
(b) whether and if so to what extent the previous proceedings should stand, and
(c) what effect (if any) his ceasing to hold office has on any appointment made by him (alone or jointly).

(2) If or to the extent that there is no such agreement, the following provisions apply.

(3) The provisions of sections 16 (procedure for appointment of arbitrators) and 18 (failure of appointment procedure) apply in relation to the filling of the vacancy as in relation to an original appointment.

(4) The tribunal (when reconstituted) shall determine whether and if so to what extent the previous proceedings should stand.

This does not affect any right of a party to challenge those proceedings on any ground which had arisen before the arbitrator ceased to hold office.

(5) His ceasing to hold office does not affect any appointment by him (alone or jointly) of another arbitrator, in particular any appointment of a chairman of umpire.

Definitions

'agreement', 'agree': s.5(1).
'arbitrator': s.82(1).
'party': ss.82(2), 106(4).

[27A] Status

This section derives in part from the Model Law, Art.15, and in part from the Arbitration Act 1950, s.25. It is *non-mandatory*, (s.4(2)).

[27B] Summary

The section concerns a number of matters that arise when an arbitrator ceases to hold office. As we have said above, it is one of a number of sections concerning the circumstances in which an arbitrator may 'cease to hold office' and the consequences of that cessation. For other sections concerning the arbitrator's cessation of office, see our commentary at paragraph 23B.

Points

[27C]   *Effect – Subs.(1)*
Essentially, the parties are at liberty to agree as to certain important matters that arise following an arbitrator ceasing to hold office, for whatever reason. These matters are the filling of the vacancy that has been created, the status of the proceedings up to that point and the effect of the arbitrator's departure on any appointment he may have made or in which he may have participated.

[27D]   *Filling of vacancy – Subs.(3)*
To the extent that the parties do not agree as to filling vacancies, a vacancy is to be dealt with by referring the parties back to the appointment procedures that would apply to an original appointment. Thus s.16, which provides for the procedure for the appointment of arbitrators, and s.18, which gives the court powers in relation to the appointment procedure and on its failure, both come into operation, to whatever extent necessary.

In a case where an arbitration clause provided for a respondent to appoint an arbitrator within 30 days of receiving notice of an appointment by the claimant, it has been held that the respondent is entitled to the full 30 day period in the case of a replacement arbitrator having to be appointed, and is not limited to the 14 day period referred to in s.16(4). The court also held, *obiter*, that when read pursuant to s.27(3), s.16(1) is to be construed as if it read:

> 'The parties are free to agree on the procedure for appointing the arbitrator or arbitrators, including the procedure for appointing any chairman or umpire, and any such agreement shall apply in relation to the filling of the vacancy as in relation to an original appointment.'
> (*Federal Insurance Co.* v. *Transamerica Occidental Life Insurance Co.*
> [1999] 2 All ER (Comm.) 138)

It is noteworthy that the court has not retained the power it formerly had under s.25 of the Arbitration Act 1950 to make a direct appointment to fill a vacancy. It will now only become involved if, in the course of the parties' appointment procedure, it is required to do so. The DAC Report indicated that the reason for this decision was the need to maintain fairness between the parties: it reduces the risk that one party might have its own nominee, whilst the second arbitrator is a court appointment.

[27E]   *Previous proceedings – Subs.(4)*
To the extent that the parties do not agree as to the status of previous proceedings, the reconstituted tribunal is empowered to make a determination. However that determination will not deprive a party of the right to challenge the proceedings on a ground that arose before the vacancy occurred.

We anticipate that it will not be easy, in practice, for the new arbitrator or arbitrators to determine to what extent previous proceedings should stand. In particular, the situation is likely to differ according to the number of arbitrators. Thus where there is a sole arbitrator, repetition of previous phases of the arbitration is more likely to occur than where there are three arbitrators, although a change in a party-appointed second arbitrator may lead to that new arbitrator insisting on repetition on the ground of fairness alone.

The determination under subs.(4) is, presumably, a decision within the meaning of ss.20 and 22, so it will be made by majority, in the absence of unanimity, or by the chairman (if there is one), in the absence of both unanimity and a majority. (As we noted in paragraph 22C above, there is a lacuna in the scheme of the Act concerning decision-making where there is no chairman. However, we think this is unimportant in reality.)

In practice it is likely that rules adopted by the parties will assist. Thus, for instance, UNCITRAL Rules, Art.14, provide:

> 'If ... the sole or presiding arbitrator is replaced, any hearings held previously shall be repeated; if any other arbitrator is replaced, such prior hearings may be repeated at the discretion of the arbitral tribunal.'

Article 12.4 of the ICC Rules provides, in part:

> 'Once reconstituted, and after having invited the parties to comment, the Arbitral Tribunal shall determine if and to what extent prior proceedings shall be repeated before the reconstituted Arbitral Tribunal.'

[27F]   ***Appointments made by departing arbitrator – Subs.(5)***
To the extent that there is no agreement between the parties as to the effect upon appointments – for example of a chairman or umpire – which the departing arbitrator had made, or in which he had participated, these will survive the departure unaffected by it.

At paragraph 26D above, we have noted that this subsection overlaps with s.26(2), where the reason for ceasing to hold office is death.

[27G]   *Model Law*
This section is intended to reflect Art.15 of the Model Law. That provides:

> 'Where the mandate of an arbitrator terminates under article 13 or 14 or because of his withdrawal from office for any other reason or because of the revocation of his mandate by agreement of the parties or in any other case of termination of his mandate, a substitute arbitrator shall be appointed according to the rules that were applicable to the appointment of the arbitrator being replaced.'

Art.13 concerns challenges; Art.14 concerns termination of the mandate by withdrawal, agreement of the parties or order of the court or other authority where the arbitrator is *de jure* or *de facto* unable to perform his functions, or is guilty of undue delay.

We note that the appointment of a substitute arbitrator is mandatory under Art.15, whereas it is not so under s.27, which specifically permits a decision not to fill the vacancy. We also note that s.27 is wider than the corresponding Model Law provision, since it also deals with the effect of a vacancy on previous proceedings and appointments.

[27H]   *Rules*

We have noted some relevant rules above in paragraph 27E. Additionally, CIMAR and the ICE Arbitration Procedure rely on the default position, whilst the CIArb Rules provide, in Art.4.4, that if any arbitrator dies, or is unable, or refuses, to act, the appointing authority (i.e. that agreed by the parties, failing which the President or a Vice President for the time being of the Institute) will appoint a substitute upon the application of a party or any remaining arbitrators. The LCIA Rules cover the ground of s.27(1)(a) in Arts.10 and 11, and the UNCITRAL Rules cover both s.27(1)(a) and (b) through Arts.11.3, 12.2, 13 and 14.

Section 28 – Joint and Several Liability of Parties to Arbitrators for Fees and Expenses

28.–(1) The parties are jointly and severally liable to pay to the arbitrators such reasonable fees and expenses (if any) as are appropriate in the circumstances.

(2) Any party may apply to the court (upon notice to the other parties and to the arbitrators) which may order that the amount of the arbitrators' fees and expenses shall be considered and adjusted by such means and upon such terms as it may direct.

(3) If the application is made after any amount has been paid to the arbitrators by way of fees or expenses, the court may order the repayment of such amount (if any) as is shown to be excessive, but shall not do so unless it is shown that it is reasonable in the circumstances to order repayment.

(4) The above provisions have effect subject to any order of the court under section 24(4) or 25(3)(b) (order as to entitlement to fees or expenses in case of removal or resignation of arbitrator).

(5) Nothing in this section affects any liability of a party to any other party to pay all or any of the costs of the arbitration (see sections 59 to 65) or any contractual right of an arbitrator to payment of his fees and expenses.

(6) In this section references to arbitrators include an arbitrator who has ceased to act and an umpire who has not replaced the other arbitrators.

Definitions

'arbitrator': s.82(1).
'costs of the arbitration': s.59.
'the court': s.105.
'notice (or other document)': s.76(6).

'party': ss.82(2), 106(4).
'upon notice' (to the parties or the tribunal): s.80.

[28A] Status

This is a new section, although so far as it concerns joint and several liability, it reflects what we understood to be the law prior to the Act. It is *mandatory*, (s.4(1) and Sched.1).

[28B] Summary

This section covers the liability of the parties to arbitrators for their fees and expenses. Two circumstances may be postulated. There may be a contractual arrangement between one or more parties and an arbitrator or the arbitrators; or there may be no such arrangement. The Act deals with both situations.

In particular, the contractual right of an arbitrator to his fees agreed with a party is preserved, see subs.(5). Otherwise, the parties are liable for such reasonable fees and expenses (if any) as are appropriate in the circumstances.

The section is not at all concerned with the question of where, as between the parties, the burden of the costs of the arbitration should fall, (subs. (5)). The liability of one party to another for costs (whether that party's or the arbitrators') is dealt with in ss.59 to 65.

Points

[28C]   *Joint and several liability – Subs.(1)*
Under this subsection the parties are jointly and severally liable to pay to the arbitrators such reasonable fees and expenses (if any) as are appropriate in the circumstances. 'Expenses' include the fees and expenses of an expert, legal adviser or assessor appointed by the tribunal for which the arbitrators are liable, (s.37(2)).

The subsection thus prescribes the minimum to which the arbitrator or arbitrators may be entitled (subject to orders of the court under ss.24(4) or 25(3)(b), in cases of removal or resignation). It interacts, however, with the other subsections, as follows:

(a) where both parties have agreed fees with the arbitrator or arbitrators which are in excess of what is reasonable, they will nonetheless be bound to pay the excess fees as a matter of contract law, subject to any other defences they may have, (see subs.(5));
(b) where one party only has agreed fees with the arbitrator or arbitrators which are in excess of what is reasonable, the party not making the agreement will be jointly and severally liable with the party making the agreement up to such amount as is reasonable, whilst the party making the agreement is bound as a matter of contract to pay the excess. This is the combined effect of subss.(1) and (5);
(c) where there is no agreement as to the fees payable by either party, the parties are simply jointly and severally liable to the arbitrators in respect of their reasonable fees and expenses. Amongst other things, this would mean that the parties were not bound to reimburse arbitrators in respect of any unreasonable element of fees and expenses paid to experts, legal advisers or assessors under s.37.

Because of the mandatory nature of the section it would seem that the parties could not make an agreement with the arbitrators under which they were not jointly and severally liable; although, presumably, they could make an agreement to pay fees which were less than those which might (from a commercial standpoint) be regarded as reasonable, since the 'circumstance' of that agreement would itself have to be taken into account.

Equally, where there is a scheme under which the arbitrators are reimbursed by a third party, such as the Lloyd's Arbitration Scheme, 'reasonable fees' would, we expect, be nil, the relevant 'circumstance' being that the scheme specifically contemplates payment by Lloyd's itself. It might well be, however, where payment by a third party was contemplated and that third party failed to pay anything, that the arbitrator could claim at least reasonable fees from the parties.

[28D] *Adjustment by the court – Subss.(2), (3)*

If a question arises as to the reasonableness of an arbitrator's fees (where neither party has agreed fees, or only one party has and the other does not accept them) an application may be made to the court for an appropriate adjustment to those fees and expenses.

The application may be made after, as well as before, the fees have been paid, and the repayment of any excessive amount may be ordered, but only if the court considers it reasonable in the circumstances to order repayment. For example, as paragraph 124 of the DAC Report pointed out, an applicant who delays in making the application is likely to receive short shrift from the court – the arbitrators might already have spent the money!

Note that the amount of fees and expenses that a party is liable to pay under this section is also the amount that is properly payable by an applicant (in the absence of an agreement relating to payment with the arbitrators) on an application to the court to secure the release of an award by means of a payment into court, with an adjustment of fees, (s.56(3)). The fact that payment has already been made to the arbitrators to obtain the release of the award does not prevent an application under this section, (s.56(8)).

[28E] *Ss. 28, 64 and the arbitrator's award*

In paragraph 28D above we have noted that a party may apply to the court for adjustment of the arbitrator's fees at any time. The application may be made after an award has been made in the arbitration, and, indeed, after money has been paid to obtain the release of the award. The question arises as to what the arbitrator should include in the award when he allocates costs in respect of his own fees.

In our view, he should now make a very simple statement. It is clearly no longer appropriate that he 'tax and settle'. We suggest the use of a formula such as:

> 'The respondent shall pay my fees and expenses in the sum of £[   ] plus VAT; provided that if, in the first instance, the claimant shall have paid any amount in respect of my fees and expenses, he shall be entitled to an immediate refund of the sum so paid from the respondent.'

See also paragraphs 49F and 61E.

Insofar as such fees are irrecoverable as between the parties, they may be challenged under s.64(2). Only 'reasonable' fees are recoverable. Equally, arbitrators are only entitled to their 'reasonable' fees as between themselves and the parties. Under both types of application the same test will therefore be applied.

Suppose, for example, an arbitrator who has no agreement as to his fees charges £1,000, and the claimant pays that sum to obtain the award. In the award, costs are allocated against the respondent. An application to the court under s.64(2) might produce the result that, say, £800 represented reasonable fees. That would be the figure which, as recoverable costs, the respondent would actually pay the claimant. An application under s.28 should similarly produce the result that £800 was appropriate, and the excess of £200 may be ordered to be repaid to the claimant by the arbitrator, if it were reasonable so to order, (s.28(3)). See also paragraph 64F.

[28F] *Subject to other court orders – Subs.(4)*
Where an arbitrator has been removed by the court or resigned, the court has power to make orders in relation to his entitlement to fees, (ss.24(4) and 25(3)(b)). Subs.(4) provides that this section takes effect subject to any such order.

[28G] *Judge-arbitrators*
Note that if the tribunal is a judge-arbitrator, he may exercise the power of the court to order the consideration and adjustment of fees under subs.(2). His power will still be subject to any order made under ss.24(4) or 25(3)(b), but any such order (in relation to his removal or resignation) would be made by the Court of Appeal, see s.93 and Sched.2, para.3.

[28H] *Former and potential arbitrators – Subs.(6)*
Subs.(6) expressly extends the section to include an umpire who has not yet replaced the other arbitrators; and to arbitrators who have ceased to act, either because they have themselves been replaced by an umpire or because they have 'ceased to hold office' by reason of revocation of authority (s.23), removal by the court (s.24), resignation (s.25) or death (s.26).

[28I] *'Party'*
Questions may arise as to who is a 'party' for the purpose of attracting joint and several liability. Persons alleged to be parties but who take no part in the proceedings (and in respect of whom it is determined that the tribunal does not have jurisdiction) would presumably not be included.

Persons alleged to be parties who participate in the arbitral proceedings to the extent of successfully challenging the jurisdiction of the tribunal would probably also not be included. We tend to the view that a successful challenger has no liability for fees because he is a party to neither the arbitration agreement nor the arbitration. Justice would also require that liability rest with the party who wrongfully asserted an agreement to which the other was not in fact a party. However the position is not certain.

   Note that a challenge to the arbitrators' fees under s.28 may be made by *any* party, so the gap in the procedure prior to this Act identified by Mr Justice Donaldson in *Rolimpex* v. *Hadji E. Dossa* [1971] 1 Lloyd's Rep. 380 – namely that a party other than the one taking up the award had no route to challenge the tribunal's fees where they were taxed in the award – is now filled.

[28J]   *The application*
The application under s.28(2) is an arbitration application falling within the terms of the Practice Direction on Arbitrations under CPR Part 49. We deal generally with arbitration applications in Part 3, Materials, Section H. Specifically in respect of an application under s.28:

(i)   the application must be on notice to the other parties and to the arbitrators;
(ii)   the requirement for giving notice to *both* the other parties and the arbitrators is to be met by making them all respondents to the application and serving on them the arbitration claim form and evidence in support;
(iii)   unless otherwise ordered, the hearing will be in private.

[28K]   *The arbitrator(s) and the application*
The arbitrator (or arbitrators) will be made respondents to the proceedings, and it is likely that the applicant will seek costs in his claim form against an arbitrator whom he believes to be seeking (or to have been paid) excessive fees. The arbitrator is, therefore, likely to acknowledge service and, thereafter, consider the extent to which he wishes to put in evidence and participate at the hearing. An arbitrator who 'loses' a hearing under s.28 is vulnerable to an order for the costs of the application against him, so that it will be sensible for him to take advice on the possible making of an offer that is expressed to be 'without prejudice except as to costs' to protect his position.

[28L]   *Appeals to Court of Appeal*
The Court of Appeal has jurisdiction to hear appeals from a judge or a court under this section, and a party may apply to the judge or to the Court of Appeal for permission to appeal and such judge and the court have power to grant that permission: *Inco Europe Ltd* v. *First Choice Distribution* [1999] 1 WLR 270.

[28M]   *Model Law*
The Model Law contains no provision for the fees of the arbitrators.

[28N]   *Rules*
This is a mandatory provision and accordingly embodied in every arbitration and arbitration agreement governed by the Act. Domestic

arbitration rules generally add nothing. Of the international rules we have considered, the LCIA rules provide that the fees are determined by the LCIA Court in accordance with the schedule of costs, the parties being jointly and severally liable to the arbitral tribunal and the LCIA. Similarly, the ICC Rules provide, in the normal case, for the institutional court to fix the fees in accordance with scales.

UNCITRAL Arbitration Rules, Art.39, is in these terms:

> '1 The fees of the arbitral tribunal shall be reasonable in amount, taking into account the amount in dispute, the complexity of the subject matter, the time spent by the arbitrators and any other relevant circumstances of the case.'

The balance of the Article makes provision for account to be taken of the fee schedule, if any, of the appointing authority, and for consultation between the arbitrators and that appointing authority. No doubt the matters set out here under Art.39 would fall for consideration under subs.(1).

Section 29 – Immunity of Arbitrator

29.–(1) An arbitrator is not liable for anything done or omitted in the discharge or purported discharge of his functions as arbitrator unless the act or omission is shown to have been in bad faith.

(2) Subsection (1) applies to an employee or agent of an arbitrator as it applies to the arbitrator himself.

(3) This section does not affect any liability incurred by an arbitrator by reason of his resigning (but see section 25).

Definitions

'arbitrator': s.82(1).

[29A]   Status

This is a new provision. It is *mandatory*, (s.4(1) and Sched.1).

[29B]   Summary

This section concerns the immunity of arbitrators. It also applies to umpires by virtue of the Act's definition of 'arbitrator', (s.82(1)), and to employees and agents of the arbitrator or umpire, (subs.(2)).

Points

[29C]   *Effect – Subs.(1)*
The section resolves what has hitherto been an area of uncertainty by providing that arbitrators are immune from suit for acts or omissions in the course of their functions as arbitrators. It follows, for instance, that a party losing an arbitration will not be able effectively to re-open the issues by alleging negligence on the part of the arbitrator and bringing proceedings against him.

If arbitrators were not immune from such actions and they were exposed to an open-ended liability to the parties, considerable harm would be done to the finality of the arbitral process, and it might be difficult to find arbitrators willing to serve at all. If nothing else, insurance would become essential, and the cost might well prove prohibitive.

However the arbitrator's immunity does not extend to acts or omissions that are shown to have been in bad faith. The term 'bad faith' is not further defined, and may have a variety of meanings in different contexts. See, for example, the views of Mr Justice Lightman in *Melton*

153

Medes v. *Securities and Investments Board* [1995] 3 All ER 880, at 889j to 890b. It remains to be seen whether, in the context of the Act, the court will decide that bad faith has a moral ingredient, and connotes, for example, malice or dishonesty (in the sense of knowing of the absence of a basis for making a particular decision), or whether it will bear a wider interpretation.

The section has mandatory status, so that the parties are not able, by agreement between themselves, to deprive the arbitrator of this protection.

[29D] *No immunity on resignation – Subs.(3)*

This subsection expressly excludes any liability incurred by an arbitrator by reason of his resigning from the scope of the immunity that otherwise applies to him. In short, the arbitrator is not protected from a claim that he has resigned in breach of contract. However he may, in those circumstances, apply to the court for relief from liability together with an order with respect to his entitlement to fees and expenses under the procedure provided by s.25(3).

We note that the DAC suggested (paragraph 361 of the Report) that some comparable provision should be made to enable the court to make orders with respect to an arbitrator's immunity where it removed him. The DAC reasoned that it was anomalous that on removal the arbitrator should retain his immunity (although the court can adjust his fees under s.24(4)). However, for reasons best known to government, the DAC's recommendation was not accepted.

[29E] *Arbitral institutions*

Note that s.74 makes complementary provision for the immunity of arbitral institutions.

[29F] *Model Law*

There is no comparable provision in the Model Law. Any country adopting it will have to decide whether to add such a provision, unless immunity or lack of it (if that is what is sought) is clear from other sources of law in that country. In practice most countries seek to confer at least a degree of immunity.

[29G] *Rules*

Since this is a mandatory section, it is not surprising that the domestic rules we have considered do not cover this ground. Art.31 of the LCIA Rules provides an exclusion which is effective 'save where the act or omission is shown ... to constitute conscious and deliberate wrong-doing', Art.34 of the ICC Rules provides an exclusion for 'any act or omission in connection with the arbitration' and the UNCITRAL Arbitration Rules are silent. An arbitration proceeding under UNCITRAL Arbitration Rules is therefore simply subject to s.28. Art.31 of the LCIA Rules must be qualified as regards arbitrations where the seat is England, Wales and Northern Ireland by saying that if the exclusion is narrower than 'bad faith' as interpreted by the courts, it will be broadened by the section and Art.34 of the ICC Rules must be qualified by the exclusion in respect of bad faith.

Jurisdiction of the Arbitral Tribunal

Section 30 – Competence of Tribunal to Rule on its Own Jurisdiction

30.–(1) Unless otherwise agreed by the parties, the arbitral tribunal may rule on its own substantive jurisdiction, that is, as to–

(a)   whether there is a valid arbitration agreement,
(b)   whether the tribunal is properly constituted, and
(c)   what matters have been submitted to arbitration in accordance with the arbitration agreement.

(2) Any such ruling may be challenged by any available arbitral process of appeal or review or in accordance with the provisions of this Part.

Definitions

'agreement', 'agreed': s.5(1).
'arbitration agreement': ss.6, 5(1).
'available arbitral process': s.82(1).
'party': ss.82(2), 106(4).

[30A]   ## Status

This section derives from the Model Law, Art.16(1). Unlike Art.16, it is *non-mandatory*, (s.4(2)).

[30B]   ## Summary

This section allows the tribunal to rule on its own jurisdiction, subject to the contrary agreement (in writing) of the parties.

Points

[30C]   *Effect – Subs.(1)*
The section enacts the internationally recognised doctrine of 'Kompetenz-Kompetenz', pursuant to which the tribunal may decide issues concerning its own jurisdiction in relation to the arbitral proceedings. The scope of the tribunal's substantive jurisdiction is here set out. The tribunal is given only a power to 'rule'. It is not given a power of final

decision, although its decision may in practice be final in certain circumstances. However, as we observe in paragraph 67E below, on hearing an appeal against a ruling made under this section, the court may decide to hear the whole matter – evidence included – again. Accordingly, in cases involving substantial issues of fact, including one as to whether a party had ever been a party to a contract, it might be preferable to go straight to court, either by agreement between the parties or upon application and with the tribunal's consent, rather than have the tribunal determine its own jurisdiction: *Azov Shipping Co.* v. *Baltic Shipping Co.* [1999] 1 Lloyd's Rep. 68 and [1999] 2 Lloyd's Rep. 159.

Since the section is non-mandatory, the parties may agree that the tribunal should not be able to exercise this power. Such an agreement might simply be effected by the choice of a foreign governing law which did not provide for such a power, (s.4(5)), or by rules which did not permit it.

[30D] *Statutory arbitrations*
S.96(2) adapts the section for the purpose of statutory arbitrations by providing that where subs.(1)(a) refers to 'whether there is a valid arbitration agreement', that reference should be construed as a reference to 'whether the enactment applies to the dispute or difference in question'.

[30E] *Related provisions – Subs.(2)*
For the purposes of challenge, and otherwise, the section links with a number of other provisions of this Part of the Act, as follows.

S.31 sets out in detail the procedure to be followed where there is an objection to the substantive jurisdiction of the tribunal; s.32 sets out the circumstances in which the court may determine a question as to substantive jurisdiction as a preliminary point; s.66(3) specifically provides that lack of substantive jurisdiction may be raised as a defence to the enforcement of an award; s.67 deals with challenging the award on the basis of lack of substantive jurisdiction; and s.73 covers the loss of the right to object, amongst other things, to a lack of substantive jurisdiction.

[30F] *Relationship with s.9*
On an application under s.9 to stay proceedings in favour of arbitration where there was an issue as to whether a contract had been entered into incorporating an arbitration agreement, it was treated as common ground that the court could proceed on the basis of various procedural options:

(1) that there was an arbitration agreement between the parties;
(2) that it should stay the proceedings pursuant to s.30 of the Act;
(3) that it should not decide the issue immediately but should order an issue to be tried under the then Order 73, rule 6(2), or

(4)   that it could decide that there was no arbitration agreement and dismiss the application to stay.

The court held that all the circumstances should be considered before weighing the various options, the dominant factors being the interests of the parties and the avoidance of unnecessary delay or expense.

It held that whilst an arbitral tribunal can decide questions of jurisdiction under s.30, the court does not have always to refer a dispute about whether or not an arbitration agreement exists to the tribunal whose competence is itself disputed. In some cases it would be better to act under the then Order 73, rule 6 (now CPR Part 49) and in others it would be appropriate to leave the matter to be decided by an arbitration tribunal, though that course would be likely to be adopted only where the court considered that it was virtually certain that there was an arbitration agreement, or if there was only a dispute about the ambivalent scope of the arbitration agreement. The court should consider the likelihood of a challenge to an award on jurisdiction on some important point of law connected to the existence of the agreement since it could not be in the interests of parties to have to return to the court to get a definitive answer to a question which could and should have been decided by the court before the arbitration got under way seriously.

In the particular matter the court found the facts to be clear but in any case would have acceded to an application under Order 73 rule 6 since the relevant issue could have been tried promptly; and in any event it was highly desirable that an issue such as the formation of a contract incorporating an arbitration agreement should be determined by the court before the arbitration got under way and before time and money was expended on an assumption which might turn out to be invalid: *Birse Construction Limited* v. *St David Ltd* [1999] BLR 194.

The correctness of the approach adopted in this case may be open to question. The DAC Report, in para.141(iii), anticipated that there might be cases where it could be cheaper and quicker for the party wishing to arbitrate to go directly to the court to seek a favourable ruling on jurisdiction, but said that such an approach would be very much the exception, and for that reason s.32 (see below) was drawn narrowly. The provisions of s.32(2) support the view that, in general, matters of substantive jurisdiction should in the first instance be left to the arbitration tribunal.

[30G]   *Model Law*

Art.16(1) of the Model Law provides:

> 'The arbitral tribunal may rule on its own jurisdiction, including any objections with respect to the existence or validity of the arbitration agreement. For that purpose, an arbitration clause which forms part of a contract shall be treated as an agreement independent of the other terms of the contract. A decision by the arbitral tribunal that the contract is null and void shall not entail *ipso jure* the invalidity of the arbitration clause.'

We comment that, as noted in paragraph 7E above, the Model Law deals with the tribunal ruling on its own jurisdiction and separability of the arbitration clause in the same Article. The DAC considered these as different matters, and separability is dealt with in s.7 of the Act.

The Act contains a more comprehensive definition of substantive jurisdiction, including whether the tribunal is properly constituted, and what matters have been submitted to it. Such matters, although not spelt out, would probably fall for consideration under Art.16.

Insofar as there have been cases under this Article of the Model Law, they have mostly concerned the inter-relationship between it and Art.8 (stay of legal proceedings). Art.8 provides, as does s.9, that there should be a stay unless the court finds the arbitration agreement to be null and void, inoperative, or incapable of being performed. Thus a claimant who maintains such ineffectiveness may force an earlier decision as to the jurisdiction of the tribunal by commencing proceedings in court.

[30H]   *Rules*

Of the domestic rules, CIMAR confirms that the arbitrator has the power set out in s.30 (rule 4.1), and rule 7 of the ICE Arbitration Procedure both gives the tribunal power to rule on its own jurisdiction and defines the matters it may rule on in wider terms than the Act. Apart from giving the arbitrator all the powers given by the Act in Art.7.1, the CIArb Rules are silent. Art.23.1 of the LCIA rules, Art.6.2 of the ICC Rules and Art.21 of the UNCITRAL Arbitration Rules give the tribunal power to rule on its own jurisdiction. Although the international rules do not refer to each of heads (a), (b) and (c) under s.30(1), it is our view that the wording 'rule on own jurisdiction' or similar is sufficient to import each of these heads into the arbitrator's jurisdiction.

For a case in which a party was held not to have taken part in arbitral proceedings by advising the ICC that it was not a party to the contract and thus not a party to the arbitration agreement, see *Caparo Group Ltd v. Fagor Arrasate Soc. Cooperativa* (unreported, Clarke J, 7 August 1998).

Section 31 – Objection to
Substantive Jurisdiction of Tribunal

31.–(1) An objection that the arbitral tribunal lacks substantive jurisdiction at the outset of the proceedings must be raised by a party not later than the time he takes the first step in the proceedings to contest the merits of any matter in relation to which he challenges the tribunal's jurisdiction.

A party is not precluded from raising such an objection by the fact that he has appointed or participated in the appointment of an arbitrator.

(2) Any objection during the course of the arbitral proceedings that the arbitral tribunal is exceeding its substantive jurisdiction must be made as soon as possible after the matter alleged to be beyond its jurisdiction is raised.

(3) The arbitral tribunal may admit an objection later than the time specified in subsection (1) or (2) if it considers the delay justified.

(4) Where an objection is duly taken to the tribunal's substantive jurisdiction and the tribunal has power to rule on its own jurisdiction, it may–

(a)   rule on the matter in an award as to jurisdiction, or
(b)   deal with the objection in its award on the merits.

If the parties agree which of these courses the tribunal should take, the tribunal shall proceed accordingly.

(5) The tribunal may in any case, and shall if the parties so agree, stay proceedings whilst an application is made to the court under section 32 (determination of preliminary point of jurisdiction).

Definitions

> 'agree': s.5(1).
> 'the court': s.105.
> 'party': ss.82(2), 106(4).
> 'substantive jurisdiction' (in relation to an arbitral tribunal):
>   s.82(1), and see s.30(1)(a) to (c).

[31A]   Status

This section derives from the Model Law, Art.16(2) and (3). It is *mandatory*, (s.4(1) and sched.1), but plainly subs.(4) is irrelevant if the parties have agreed that the tribunal does not have power to rule on its own jurisdiction, (s.30). It would seem, in such a case, perhaps somewhat curiously, that under subs.(3) a tribunal may admit a late objection even though only the court will be able to rule on it.

[31B]   Summary

This section is the first of two which deal with the procedures available where there is a challenge to the substantive jurisdiction of the tribunal.

Points

[31C] *Alternative procedures*

Ss.31 and 32 provide for the ways in which a challenge to the tribunal's jurisdiction may proceed, depending, to some extent, upon whether the tribunal has power to rule on its own jurisdiction under s.30 (which is a non-mandatory provision).

If the tribunal has that power, then, pursuant to s.31, it may be called on to exercise it either by making a specific award as to jurisdiction, or by dealing with the jurisdiction challenge in its eventual award on the merits. This choice is primarily one for the tribunal's discretion; however, the parties are able, by agreement, to require the tribunal to adopt one or other of these courses, (subs.(4)). In either case, the tribunal's award would in turn be subject to challenge on the jurisdiction aspect on an application to the court by a party pursuant to s.67.

Even in a case where the parties have agreed that their awards should not be reasoned (e.g. under the LMAA Terms: see paragraph 52D below) it would seem advisable that an award deciding a tribunal's substantive jurisdiction should contain the reasons for the conclusion on that topic, for s.67 is mandatory and thus there is an automatic right of challenge; and under s.70(4) if an award does not contain reasons the court may order the tribunal to state those reasons sufficiently to enable it to consider the application. It is, of course, most undesirable that a case should bounce backwards and forwards between the arbitration tribunal and the court in this way.

As an alternative course, an application may be made to the court, pursuant to s.32, for the determination of a preliminary point of jurisdiction before any award is made. If the tribunal does not have power to rule on its own jurisdiction, then this is the only mode of challenge available to a party. In such a case the tribunal has a discretion either to stay the arbitral proceedings whilst an application is pending, or to continue those proceedings and make an award. However, the parties may, by agreement, require the tribunal to stay the arbitral proceedings, (subs.(5) and s.32(4)).

[31D] *Requirements – Subss.(1) to (3)*

Whichever course the challenge takes, it must be raised in accordance with the requirements of these subsections. That follows from the mandatory status of the section.

Thus, subject to the tribunal's discretion to admit challenges made late where the delay is justified (subs.(3)), a challenge arising at the outset of the proceedings must be made before the first step in the proceedings contesting the merits of any matter affected by it (subs.(1)); and a challenge arising during the course of the proceedings must be made as soon as possible (subs.(2)). In respect of a challenge at the outset, mere appointment or participation in the appointment of an arbitrator is to be disregarded.

Subs.(1) has been carefully drafted to avoid specific reference to an event which, in the circumstances of any particular arbitration, might not happen (compare, for instance, the Model Law's 'the submission of the statement of defence'). Nonetheless we think it may give rise to some debate. The use of the words 'proceedings' and 'arbitral tribunal' clearly indicates that the arbitration must have commenced and the arbitrators must have been appointed, so matters prior to that are irrelevant. A subsequent letter from one party to another contesting the other's position is, we think, unlikely to be a 'step in the proceedings to contest the merits'. We prefer the view, and think it intended, that the step has to be a formal one, possibly ordered by the tribunal. Thus, for instance, attendance at a preliminary meeting would not of itself amount to a contest on the merits, but submission of a later formal denial of the claim would, whether it took the form of a defence document or some other form. We suppose it might be argued that an oral statement at a preliminary meeting that a claim was denied, perhaps accompanied by some explanation of the reasons for that denial, could amount to 'the first step in the proceedings to contest the merits'; but we would hope that the courts would look for some greater degree of formality. In the long run it is likely that problems arising in this area will have to be resolved on a case-by-case basis. In paragraph 31I below, we venture to suggest that where arbitration rules (such as LCIA or UNCITRAL) cite the statement of defence as the latest point for challenge, it is unlikely that the court will wish to reach any conflicting decision.

In respect of a challenge during the course of the proceedings, again we can see considerable scope for argument over the meaning of 'as soon as possible'. We think there must be an element of reasonableness implied. If subs.(2) were applied literally it could needlessly prevent the tribunal from acting under s.31(4) and ruling on its own jurisdiction for wholly unmeritorious reasons. (The literal effect is, of course, mitigated by subs.(3), but the onus here is upon the challenging party to demonstrate that their delay was justified).

It will be noted that the right to object may also be lost under s.73.

[31E]   *Tribunal ruling on its own jurisdiction – Subs.(4)*
Even though the tribunal has power to rule on its own jurisdiction, this subsection will not be effective if the parties proceed under s.32, so that the matter is determined in court, or if the objection is taken too late. In the latter case, the right to challenge at all may be lost.

As we have noted above, the parties may require the arbitrators to proceed in accordance with their wishes in relation to whether the ruling is given in an interim or final award. This is one of those instances where the parties' wishes could come into conflict with the general duty of the tribunal under s.33, possibly leading to the resignation of the arbitrator in circumstances in which he may be entitled to relief from the consequences of breach of his agreement to act (s.25).

As to what constitutes an award for these purposes, it has been held that a letter from an arbitrator finding and declaring that there was a contract between parties which included a written arbitration agreement, and that he had been correctly appointed under that agreement and that all disputes under the contract had been referred to arbitration, was an award as to the arbitrator's substantive jurisdiction even though it did not comply with the requirements of s.52 in that it did not state the seat of the arbitration or the arbitrator's reasons: *Ranko Group v. Antarctic Maritime SA* (1998) LMLN 492.

[31F]   *Stay of proceedings pending application to court – Subs.(5)*
We make a similar observation as under subs.(4). Since this is a subsection which may require arbitrators to proceed in accordance with the parties' wishes, a resignation might result, particularly if there were long delays due to the application to court. A resigning arbitrator might then obtain relief from the consequences under s.25.

[31G]   *Position of party who takes no part*
It should be noted that a person alleged to be a party to arbitral proceedings, but who takes no part in them, may question the substantive jurisdiction of the tribunal by court proceedings, and may also challenge an award on the ground of lack of jurisdiction by an application to court under s.67. See s.72 for the special considerations that apply in such a case.

[31H]   *Model Law*
Art.16 of the Model Law provides:

> '(2) A plea that the arbitral tribunal does not have jurisdiction shall be raised not later than the submission of the statement of defence. A party is not precluded from raising such a plea by the fact that he has appointed, or participated in the appointment of, an arbitrator. A plea that the arbitral tribunal is exceeding the scope of its authority shall be raised as soon as the matter alleged to be beyond the scope of its authority is raised during the arbitral proceedings. The arbitral tribunal may, in either case, admit a later plea if it considers the delay justified.
> (3) The arbitral tribunal may rule on a plea referred to in paragraph (2) of this article either as a preliminary question or in an award on the merits. If the arbitral tribunal rules as a preliminary question that it has jurisdiction, any party may request, within thirty days after having received notice of that ruling, the court specified in article 6 to decide the matter, which decision shall be subject to no appeal; while such a request is pending, the arbitral tribunal may continue the arbitral proceedings and make an award.'

We have already observed that Art.16 refers specifically to the statement of defence. The DAC avoided the use of this expression because the flexibility inherent in the arbitration process contemplated by s.34 (below) means that there may not inevitably be a defence document of any conventional type in every case. Art.23 does, however, so contemplate.

The balance of Art.16(2) is virtually repeated in the Act save that the Act defines more closely what an objection to the tribunal's 'authority' may be, (s.30(1)). The opportunity for the parties to dictate the choice between ruling at once or in the award on merits is not present in the Model Law; and there is no opportunity for immediate referral to the court. The effect of Art.16 is that the tribunal must make a ruling. It is final unless challenged within 30 days. (Under the Act appeals must be brought within 28 days, see ss.67(1) and 70(3).)

[311] *Rules*

As a mandatory section, any rules in conflict with it are, by s.4(1), ineffective. Thus CIMAR and the CIArb rules do not cover this ground; it is unnecessary to do so. Nor does the ICE Arbitration Procedure except (at rule 7.2) to include a provision that should a party refer a decision in respect of jurisdiction to the court the arbitrator shall direct whether or not the arbitration should continue pending the outcome. Presumably such a discretion would be overriden by the joint agreement of the parties.

Of the international rules, as we would expect, the LCIA Rules reflect s.31 in Art.23, but we note that Art.23.3 permits the tribunal to decide its course of action 'as it considers appropriate in the circumstances'. Compliance with s.31, we think, is nonetheless achieved via Art.14 which permits the parties to impose their agreement on the tribunal. The ICC Rules are silent on the ground covered by s.31, except that Art.33 may give rise to a waiver in some circumstances. Art.21 of the UNCITRAL Arbitration Rules is consistent with s.31, but not so complete. S.31, in effect therefore, supplies additional rules. Both the LCIA Rules and the UNCITRAL Arbitration Rules refer to the statement of defence as the latest point for noting a challenge, and we doubt that any court would wish to reach a conflicting conclusion.

Section 32 – Determination of Preliminary Point of Jurisdiction

32.-(1) The court may, on the application of a party to arbitral proceedings (upon notice to the other parties), determine any question as to the substantive jurisdiction of the tribunal.

A party may lose the right to object (see section 73).

(2) An application under this section shall not be considered unless–

(a) it is made with the agreement in writing of all the other parties to the proceedings, or

(b) it is made with the permission of the tribunal and the court is satisfied–

    (i) that the determination of the question is likely to produce substantial savings in costs,

    (ii) that the application was made without delay, and

    (iii) that there is good reason why the matter should be decided by the court.

(3) An application under this section, unless made with the agreement of all the other parties to the proceedings, shall state the grounds on which it is said that the matter should be decided by the court.

(4) Unless otherwise agreed by the parties, the arbitral tribunal may continue the arbitral proceedings and make an award while an application to the court under this section is pending.

(5) Unless the court gives leave, no appeal lies from a decision of the court whether the conditions specified in subsection (2) are met.

(6) The decision of the court on the question of jurisdiction shall be treated as a judgment of the court for the purposes of an appeal.

But no appeal lies without the leave of the court which shall not be given unless the court considers that the question involves a point of law which is one of general importance or is one which for some other special reason should be considered by the Court of Appeal.

Definitions

'agreement', 'agree': s.5(1).
'agreement in writing': s.5(2) to (5).
'the court': s.105.
'notice (or other document)': s.76(6).
'party': ss.82(2), 106(4).

'substantive jurisdiction' (in relation to an arbitral tribunal: s.82(1), and see s.30(1)(a) to (c).
'upon notice' (to the parties or the tribunal): s.80.

[32A] ## Status

This is a new provision. It is *mandatory*, (s.4(1) and Sched.1).

[32B] **Summary**

This section provides for the determination of a preliminary point of jurisdiction by the court, being one of the ways in which a challenge to the tribunal's jurisdiction may proceed. It is a narrowly drawn and limited procedure. The apparent intention is to encourage parties to adopt the alternative course and permit tribunals to rule on their own jurisdiction pursuant to ss.30 and 31, with the safeguard of a possible challenge to the resulting award under s.67. It may only be in rare cases that an immediate application to the court under the present section would be justified in preference to seeking an award from the tribunal. However, we draw attention to the comments on *Birse Construction Ltd* v. *St David Ltd* in paragraph 30F above. This case suggests that some courts, at least, do not see the policy of the Act in this area in quite the same way as we do and as the DAC did.

Given that the procedure set out here will represent the only available course if the tribunal does not have power to rule on its own jurisdiction, the section has mandatory status. Furthermore, it has been held that where the seat of an arbitration is in England, a party wishing to have any question of the jurisdiction of arbitrators decided must make an application under this section. A declaration at common law is not available: *ABB Lummus Global Ltd* v. *Keppel Fels Ltd* [1999] 2 Lloyd's Rep. 24.

Points

[32C] *Requirements and restrictions – Subss.(1) to (3), (5)*
The hurdles which an applicant must overcome to progress an application under this section are numerous. First, it must raise the challenge as to jurisdiction within the timescales set out in s.31(1) or (2), subject to the tribunal's discretion to admit a challenge made late where the delay is justified. The requirement to raise the challenge very promptly is reinforced by the express reference, in subs.(1), to s.73 (which provides for the loss of the right to challenge in the event of knowing delay). It also features as a specific obligation within the general duty of the parties, (s.40(2)(b)).

Subs.(2) then imposes conditions on the making of the application. The application requires either the agreement of all the parties or, alternatively, the permission of the tribunal. In the latter case, there must also be positive evidence that the application was promptly made (in short that the party applying has complied with its s.40(2)(b) duty); that it will save costs; and that there is good reason why the issue should be decided by the court (the grounds for this latter contention being set out in the application, subs.(3)).

No appeal will lie without leave against a decision as to whether these conditions have been complied with, (subs.5).

[32D] *Stay of proceedings pending application to court*
Whilst an application is pending, the tribunal has a discretion either to stay the arbitral proceedings, or to continue those proceedings and make an award; however, the parties may, by agreement, require the tribunal to stay the arbitral proceedings, (see s.31(5), and our commentary at paragraph 31F).

If the parties do not so require, and the proceedings continue, the tribunal may proceed to an award notwithstanding the pending application, (subs.(4)).

[32E] *Appeals – Subs.(6)*
The right of appeal from a decision of the court as to jurisdiction is severely limited. Not only is leave required, but the court must also identify a point of law of general importance or some other special reason justifying such leave.

Note that if the tribunal is a judge-arbitrator the initial decision on jurisdiction will be made by the Court of Appeal, and any appeal will lie to the House of Lords, (Sched.2, para.2).

[32F] *The application*
The application under s.32(1) is an arbitration application falling within the terms of the Practice Direction on Arbitrations under CPR Part 49. We deal generally with arbitration applications in Part 3, Materials, Section H. Specifically in respect of an application under s.32:

(i)   the application must be on notice to the other parties;
(ii)   the requirement for giving notice to the other parties is to be met by making those parties respondents to the application and serving on them the arbitration claim form and evidence in support;
(iii)   the arbitration claim form, unless the application is made with the agreement of all the other parties to the proceedings, must state the grounds on which it is said that the matter should be decided by the court;
(iv)   the evidence in support of the application must give details either of the agreement in writing of the parties to the making of the application, or of the permission of the tribunal, and must exhibit a copy of any document evidencing that agreement or permission;
(v)   additionally the affidavits or witness statements in support of or in opposition to the application must, where the application is made without the agreement of the other parties but with the permission of the tribunal, set out any evidence relied on by the parties in support of their contentions as to whether the determination of the question is likely to produce substantial savings in costs, the application was made without delay and there is good reason why the matter should be decided by the court;

167

(vi)   where the application is made without the agreement of the other parties but with the permission of the tribunal, after receipt of the evidence the court will initially consider whether or not it will entertain the application, and it will normally do so without a hearing. If necessary, it may direct a hearing;

(vii) when the court hears the substantive application, unless it orders to the contrary, it does so in private.

[32G]   *Model Law*

There is no comparable provision in the Model Law, which by Art.16 provides for a determination of such questions by the tribunal, with a right to apply to the court within 30 days of the decision of the arbitral tribunal for a final (unappealable) decision.

[32H]   *Rules*

As a mandatory section, any conflicting rules are ineffective, and in any event, the section deals with the power of the court. We would accordingly expect arbitration rules to be silent on the matter. Of those we have considered, CIMAR, the ICE Procedure and the CIArb Rules do not cover this ground. Of the international rules, Art.23.4 of the LCIA Rules reflects the Act, whilst the ICC Rules and UNCITRAL Arbitration Rules do not cover the ground directly so that s.32 will operate to supplement them.

The Arbitral Proceedings

Section 33 – General Duty of the Tribunal

33.–(1) The tribunal shall–

(a) act fairly and impartially as between the parties, giving each party a reasonable opportunity of putting his case and dealing with that of his opponent, and

(b) adopt procedures suitable to the circumstances of the particular case, avoiding unnecessary delay or expense, so as to provide a fair means for the resolution of the matters falling to be determined.

(2) The tribunal shall comply with that general duty in conducting the arbitral proceedings, in its decisions on matters of procedure and evidence and in the exercise of all other powers conferred on it.

Definitions

'party': ss.82(2), 106(4).

[33A] ### Status

This section derives from the Model Law, Art.18. It is *mandatory*, (s.4(1) and Sched.1).

[33B] ### Summary

This section sets out the general duty of the tribunal. It stands as a clear statement in broad terms of the way in which the tribunal is to approach its task in order to achieve the objective set out in s.1(a), namely, '... the fair resolution of disputes by an impartial tribunal without unnecessary delay or expense'. It is a vital and central provision, establishing the parameters within which the tribunal should act in order to be seen to be doing justice as between the parties.

Those parameters concern the elements of the tribunal's conduct towards each of the parties and of its evolution and operation of an arbitral format that is well suited to the particular case.

The section is mandatory, so that an arbitration agreement which required or permitted the tribunal to depart from its general duty as here set out would, to that extent, be ineffective. The corresponding mandatory duty imposed on the parties to conduct themselves properly and expeditiously, and to co-operate with the tribunal is to be found in s.40.

The sections which follow (ss.34, 35, 37 to 39, and 41) set out the tribunal's powers; and they must each be read subject to this overall duty.

Points

[33C] *Wording*
It has been suggested that the generality and vagueness of the section may give rise to problems. The DAC took the view (paragraph 151 of its Report) that the variations in styles and types of arbitration made it impossible to set out a complete code of what might, and what might not, be done. Thus it followed that the only practical limits were those set out in the section. As the DAC said, 'It is to be hoped that the courts will take a dim view of those who try to attack awards with suggested breaches of this [section] which have no real substance.'

[33D] *Fairness as between the parties – Subs.(1)(a)*
The keynote of the tribunal's conduct towards the parties is that of fairness and impartiality as between them. An even-handed approach has its most practical expression in giving each party a reasonable opportunity of putting their case and of responding to that of their opponent. A further aspect of this fairness and impartiality is the requirement that tribunals must give parties an opportunity of dealing with any factors which may be likely to affect the tribunal's decision and which have not been the subject of evidence or argument by the parties. Thus, it has been held that where an arbitrator felt influenced in the exercise of his discretion on costs by factors which had not been drawn to the attention of the parties, who had thus not had an opportunity of dealing with them, he did not act fairly and impartially unless he gave such an opportunity to the parties: *Gbangbola* v. *Smith & Sherriff Ltd* [1998] 3 All ER 730.

It is noteworthy that this paragraph has not adopted the term used in the Model Law, Art.18, 'a full opportunity of presenting his case'. Such a term might have given the impression that a party was entitled to take as long as he required to explore every aspect of his case, at absurd length if necessary. The term 'a reasonable opportunity' conveys an objectively viewed balance of what is fair to the party, but is also compatible with expedition and economy.

An arbitrator whose partiality is reasonably in doubt may be removed (under s.23) by the parties acting jointly or by an institution or other person vested with the power, or (under s.24) by the court.

[33E] *Suitable procedures – Subs.(1)(b)*
Rather than simply fall into a set pattern, the tribunal is positively required to use the flexibility and adaptability of arbitration to evolve

an appropriate arbitral format to meet the individual circumstances of each case. In particular, it does this through its decisions on procedural and evidential matters (s.34) and its approach towards dividing the reference up where appropriate into separate 'tranches', each the subject of an award (s.47). The keynote in this process is the avoidance of unnecessary delay and expense. It is now absolutely clear that it is unnecessary for the arbitrator to follow 'court' procedures slavishly or at all.

If these principles are followed, arbitration should avoid becoming prohibitively lengthy and costly. It should therefore provide a means of dispute resolution that is fair not only in terms of the quality of the substantive award, but also in terms of the overall result to the parties.

[33F]   *Power to proceed 'ex parte'*
The s.33 duty is not incompatible with the tribunal acting incisively, firmly and – where appropriate – in a way that shuts out one of the parties. Note the important power given to the tribunal by s.41(4) to proceed in the absence of a party in the event of that party failing to attend or be represented at an oral hearing, or to proceed in the absence of written evidence or written submissions in the event of a party having been required to submit them and failing to do so.

[33G]   *Scope of duty – Subs.(2)*
The tribunal's general duty governs its operation of the arbitral format, decisions as to procedure and evidence and the exercise of all its powers.

[33H]   *Relationship with party autonomy*
A question may arise as to how the tribunal's duty stands in relation to the principle of party autonomy (s.1(b)), and the practical effect of that principle, namely that the parties are able, by agreement, to make decisions in relation to matters of procedure and evidence that will take precedence over the views of the tribunal, see s.34(1). Arbitration tribunals were not – intentionally – given the power to override the will of the parties as that would have meant a fundamental departure from existing law and from the internationally accepted principle of party autonomy. In practice, problems are hardly likely to exist since, in Lord Saville's memorable phrase: 'Parties in dispute are usually unable to agree upon the time of day, let alone that unfair, expensive or unduly lengthy proceedings are to be preferred.' The underlying principle is that arbitration tribunals will have to take control of proceedings from the outset.

In practice, of course, one party at least will usually want to go along with any course of action a tribunal might suggest, so conflicts between a tribunal on the one hand and both parties on the other are unlikely to occur.

Under subs.34(1) the tribunal is obliged to act in accordance with any agreement as to procedure that the parties make. That remains the case even if they agree to adopt procedures which, in the tribunal's view, do not permit it to perform its s.33 duty. In practical terms, if the parties agreed to choose a format that seemed to the tribunal needlessly lengthy or expensive, how could the tribunal proceed, consistently with its duty?

The tribunal might first seek to persuade the parties round to its own point of view. Plainly, an appeal to good sense may, in many cases, avoid a conflict.

If the conflict persisted, then one option would be for the tribunal to give in to the parties' wishes and to implement their chosen procedure, even if it did not accord with the tribunal's perception of its s.33 duty. In such circumstances, it does not appear that the parties could validly complain about the tribunal's failure to follow, or its breach of, the s.33 duty, since that would have been brought about by their own agreement, and no relief would be available in respect of it.

Another option would be for the tribunal to resign. Unless the terms of its appointment permitted resignation, such a step would put the tribunal in breach of contract and might incur liability to the parties. Moreover, s.29(3) expressly excludes any liability incurred by an arbitrator by reason of his resigning from the scope of the immunity that would otherwise protect him.

However, s.25 provides for the possibility of an application to the court on resignation, to seek relief from liability and an order with respect to entitlement to fees and expenses. A tribunal that found itself forced to resign because the agreement of the parties was seen as incompatible with its s.33 duty could apply for such relief. If the court was satisfied that the resignation was reasonable, relief might be granted.

Note that if there were a deadlock in that the parties agreed to reject all the courses the tribunal considered suitable, but were not able positively to agree on a course they wished to pursue, then the tribunal would have no choice but to resign.

In short, if the parties persisted in adopting procedures which did not permit the tribunal to perform its s.33 duty, then the tribunal could either give in to the parties' wishes and not subsequently be liable to the parties for so doing; or it could resign with (if the resignation were reasonable) the protection of a s.25 application.

[33I]   *Model Law*

Art.18 of the Model Law provides:

> 'The parties shall be treated with equality and each party shall be given a full opportunity of presenting his case.'

[33J]   *Rules*

Although the CIArb Rules say nothing about this section, arbitration rules drafted in this country since the Act have generally adopted s.33

directly (e.g. CIMAR) or set out similar wording (ICE Arbitration Procedure Rule 1.1, LCIA Rules Art.14.1). Art.15 of the UNCITRAL Arbitration Rules reflects the 'full' opportunity to present the case to be found in the Model Law and upon which we have commented above: 'Subject to these Rules, the arbitral tribunal may conduct the arbitration in such manner as it considers appropriate, provided that the parties are treated with equality and that at any stage of the proceedings each party is given a full opportunity of presenting his case.' Insofar as there may be a significant difference, s.33 would appear to override this provision. Interestingly, the ICC Rules (Art.15.2) adopt the 'reasonable' opportunity approach.

Section 34 – Procedural and Evidential Matters

34.–(1) It shall be for the tribunal to decide all procedural and evidential matters, subject to the right of the parties to agree any matter.

(2) Procedural and evidential matters include–

(a) when and where any part of the proceedings is to be held;

(b) the language or languages to be used in the proceedings and whether translations of any relevant documents are to be supplied;

(c) whether any and if so what form of written statements of claim and defence are to be used, when these should be supplied and the extent to which such statements can be later amended;

(d) whether any and if so which documents or classes of documents should be disclosed between and produced by the parties and at what stage;

(e) whether any and if so what questions should be put to and answered by the respective parties and when and in what form this should be done;

(f) whether to apply strict rules of evidence (or any other rules) as to the admissibility, relevance or weight of any material (oral, written or other) sought to be tendered on any matters of fact or opinion, and the time, manner and form in which such material should be exchanged and presented;

(g) whether and to what extent the tribunal should itself take the initiative in ascertaining the facts and the law;

(h) whether and to what extent there should be oral or written evidence or submissions.

(3) The tribunal may fix the time within which any directions given by it are to be complied with, and may if it thinks fit extend the time so fixed (whether or not it has expired).

Definitions

'agree': s.5(1).
'party': ss.82(2), 106(4).
'written': s.5(6).

[34A] Status

This section derives from the Model Law, Arts.19, 20, 22, 23 and 24; and from the Arbitration Act 1950, s.12(1) to (3). It is *non-mandatory*, (s.4(2)).

[34B] Summary

The section provides for procedural and evidential matters. The power to make decisions as to such matters is given to the tribunal, subject to the parties agreeing to take a different course.

We are not aware of any court decisions on this section, but we do know that many arbitrators are now taking the far more radical approach which it permits them to adopt, as indicated in the notes that follow.

Points

[34C]   *Effect – Subs.(1)*

The tribunal must exercise its power in accordance with its mandatory duty under s.33. It follows that the requirement to 'adopt procedures suitable to the circumstances of the particular case, avoiding unnecessary delay and expense' should influence its choices in every respect, and prompt it to evolve an arbitral format that is appropriate to the matter in issue.

Where the parties agree to take a different course, that agreement will take precedence over the views of the tribunal. This is a practical expression of the principle of party autonomy, (s.1(b)). Where the parties do not make an agreement in relation to a particular matter, then the tribunal will have the power to decide.

If the parties make an agreement that conflicts with the tribunal's view of its mandatory duty pursuant to s.33, then the tribunal may either give in to the parties' wishes and not subsequently be liable to the parties for so doing; or it can resign with (if the resignation is reasonable) the protection of a s.25 application for relief from liability. This kind of conflict is discussed at paragraph 33H above.

It should be noted that s.34(1) applies to *all* procedural and evidential matters. Subs.(2) which follows lists only certain of such matters.

The non-mandatory nature of the section would appear to mean that the parties might override the tribunal's power either by their own agreement as to individual matters, or by a general agreement to adopt institutional rules. Thus the parties might agree that a third party, such as an institution, might decide how some of the procedural or evidential matters should be dealt with, (s.4(3)).

It is important to remember that any agreement between the parties, if it is effectively to override the tribunal's power to decide procedural and evidential matters, must be in writing (s.5(1).

Points

[34D]   *Procedural and evidential matters – Subs.(2)*

A non-exhaustive list of the procedural and evidential matters falling within the tribunal's power is set out. The individual matters identified and the terminology in which they are described (avoiding technical or court-orientated expressions so far as possible) serve as illustrations and reminders of the general duty to act flexibly, and in response to the requirements of the particular case.

[34E]   *Time and place – Subs.(2)(a)*
The power to decide when and where any part of the proceedings is to be held is, at least in its expression, new. Previously, when an arbitration agreement provided for arbitration in, say, London, it could be argued that all proceedings must take place there (unless the parties agreed otherwise), even though it might be preferable that some steps, at least, should be dealt with elsewhere.

Under this provision, if an arbitrator is satisfied that it is suitable in the particular case to sit elsewhere, and that to do so would be to act fairly and impartially and would not involve unnecessary delay or expense, he may do so. Obvious relevant factors include the presence of witnesses and matters requiring examination away from the designated place of arbitration. The arbitrator's convenience alone would not be a sufficient ground.

[34F]   *Language – Subs.(2)(b)*
Again, this provision is new, at least in its expression. In deciding what language(s) to use care must be taken, as ever, to ensure that the s.33 duties are respected, particularly that of ensuring fairness to all parties.

The power to order the supply of translations of documents is a useful one. In the past, there were often disputes about whether translations had to be provided by a party producing a document, or whether they had to be obtained by the other party; and in the latter case, whether they had to make their translations available to the first party.

[34G]   *Written statements of claim and defence – Subs.(2)(c)*
One of the underlying aims of the Act is to explode the myth that arbitration is or should be a carbon copy of court proceedings. In this subsection appears a good example of the manifestation of that philosophy, for under it arbitrators may decide how parties' cases are to be pleaded – *if at all*.

Unless the parties agree otherwise, therefore, arbitrators are able to direct that appropriate cases (always bearing in mind the s.33 duties) may be heard without any pleadings at all (as used to happen on occasions in the early days of the Commercial Court). If there are to be pleadings, they need not be formal nor ape a court statement of case and defence. Many arbitrators prefer service of a fuller document, sometimes with evidence and submissions attached, and they are now able to give directions accordingly, bearing in mind also the following subsection.

[34H]   *Documents – Subs.(2)(d)*
Subject (as in all these cases) to the parties' agreement and to s.33, arbitrators have complete power over the production of documents. Thus documents may, as suggested above, be produced with the parties' statements of case. Thereafter, for example, the tribunal might

entertain applications by parties to order their opponents to produce further documents, but would not be bound to accede, provided to refuse would not breach s.33. There is no question of lists of documents being required automatically.

Since we wrote the first edition, the courts have moved away from the full discovery that characterised English court procedure. Where appropriate we now regularly order parties in the first place to give standard disclosure in accordance with sub-paragraph (a) and (b) of CPR Rule 31.6. This provides

> '31.6 Standard disclosure requires a party to disclose only –
>> the documents on which he relies; and
>> the documents which –
>>> adversely affect his own case;
>>> adversely affect another party's case; or
>>> support another party's case.'

Where such an order is made, the arbitrator will need to give thought to how much of the rest (if any) of Part 31 he wishes to import. Thus Rule 31.7 sets out the extent of search which should be made by the party concerned, and Rule 31.10 provides a procedure for standard disclosure, including a requirement for a 'disclosure statement'.

[34I] *Questions – Subs.(2)(e)*

This subsection gives a broad power to decide how, if at all, the parties should be questioned. It includes oral questioning or written questioning. Written questions might take the form of interrogatories (now abolished in court procedure) answered either by witness statement or affidavit, a request for further particulars of a statement of case, defence or counterclaim, or a request for further information akin to CPR Part 18 (but arbitrators using the latter need to be aware of the somewhat complex procedure set out in the Practice Direction which accompanies it).

We would also draw readers' attention to Rule 35.6 of CPR which has introduced a procedure under which a party may put written questions to an expert, and we commend the use of similar procedures in arbitration. It should obviously be read in conjunction particularly with the following three sub-sections, which are all to some extent concerned with evidence and ascertaining the facts.

[34J] *Evidence – Subs.(2)(f)*

Since the first edition of this book and the coming into effect of the Civil Evidence Act 1995, hearsay evidence has more generally become admissible in court proceedings in England so that the possible differences between court proceedings and arbitration are eroded. By subs.2(f) the arbitrator has a total discretion over whether or not he will apply any rules of evidence and which ones; in practice we think most arbitrators will admit all relevant evidence, thereafter taking submissions as to the relevance and weight to be applied to it. It should be noted that insofar as arbitrators were ever exercised by the rules of court applying to the service of hearsay evidence (as part of the rules of evidence), they can be regarded as a thing of the past; CPR Part 32 now provides that adequate notice of hearsay evidence is given by its incorporation in a witness statement and we suggest that arbitrators should be no more demanding.

[34K]   *Inquisitorial procedure – Subs.(2)(g)*
This provision enables tribunals to act in an inquisitorial manner. Once again, it removes the possibility of debate as to whether arbitrators could so act.

It is not in the tradition of common lawyers to espouse inquisitorial procedures, and generally English arbitration has hitherto been run on purely adversarial lines. However, that is not always an appropriate method of proceeding, particularly where (as so often in arbitration) the tribunal has relevant expertise, and where (as will now be the case) tribunals are under an obligation to adopt suitable procedures which, *inter alia*, avoid unnecessary delay and expense.

In many instances it will undoubtedly prove more satisfactory all round for a tribunal to proceed inquisitorially, even if only in respect of some aspects of a case, and either as to fact or law (though, if lawyers are involved for the parties, the latter subject will often best be left to them to deal with adversarially).

As with all these provisions, a tribunal must be mindful of its s.33 duties, particularly, for example, by giving all the parties an opportunity of commenting on any evidence it obtains or any law it thinks applies.

[34L]   *Oral hearing – Subs.(2)(h)*
Until now, most practitioners have thought – probably correctly – that any party requiring an oral hearing in an English arbitration was entitled to one as a matter of natural justice. Many recalcitrant parties used this belief to delay proceedings, and others used it as a way of putting pressure upon opponents by threatening them with a substantial costs burden.

Under this provision, it is for tribunals to decide whether, and if so to what extent, there should be oral (or indeed written) evidence or submissions. That must be the case not only in respect of substantive issues, but also procedural ones, so that tiresome and expensive oral applications for directions are now avoidable. Of course, there are many instances where matters are most easily resolved in oral hearings, but equally that has often not been the case in the past.

In cases involving modest amounts, having regard to the s.33 duties, it is often appropriate to deal with all matters on documents alone, even if one party wants a hearing. And in other cases, indeed, it may similarly be best so to proceed. The touchstone must always be s.33. Is it fair to the parties and appropriate to the case, will it give each party a reasonable opportunity to put its case, and will it avoid unnecessary delay and expense to proceed without a hearing? If so, the arbitrator may do that. If not, a hearing should take place.

However, the situation should not be seen as an 'all or nothing' one. There is no reason why only such parts of a case as require an oral hearing should be subject to one, the rest being dealt with in writing.

[34M]   **Time for compliance with directions – Subs.(3)**
The power here given to tribunals to fix the time (and extend it) for compliance with its directions hardly needed to be stated; but it should be noted that each party is under an obligation to do all things necessary for the proper and expeditious conduct of the arbitration, a duty which (at s.40(2)(a)) specifically includes compliance with determinations on procedural and evidential matters, and with orders and peremptory orders in the event of non-compliance.

Note that where the parties have agreed a matter, so that the tribunal does not give a direction on it or fix the time for completion, on a strict analysis of the wording of this subsection it would appear that the power of extension will lie with the court (s.79) unless the parties agree to give the tribunal power to extend time. We anticipate that this will happen in most cases, by implication in the parties' agreement, if not expressly.

[34N]   **Model Law**
The Model Law covers this ground at Arts.19 to 24.

We note that unlike s.34(2)(c) the Model Law specifically contemplates a procedure using written statements of claim and defence. Amendments to these documents may be made, subject only to refusal on grounds of delay. Under the Act refusal in other circumstances is possible.

Under the Model Law a hearing must be held if required. This is not so under the Act, but, if it refuses a hearing, the tribunal must be careful nonetheless to comply with its s.33 duties.

[34O]   **Rules**
CIMAR deals specifically with the ground covered by s.34 at Rule 5, and sets out procedures at Rules 6 to 9. The ICE Arbitration Procedure deals with the ground covered by s.34 at Rule 7.3 (and more), and sets out procedures at Rules 6 and 8. In addition there are short procedures (Rule 15) and a special procedure for experts (Rule 17). The CIArb Rules effectively repeat subs.(1) and (2) in Art.6.1 and set out more detailed provisions as to procedures in Art.6.2–6.5 and, particularly, Art.8. Further, Art.7.2 allows the arbitrator to limit the number of expert witnesses to be called by a party or to direct that no expert evidence be called on any issue or issues, or that it may only be called with the arbitrator's permission.

As might be expected, the international rules do not follow the pattern of s.34. Nonetheless over a series of articles and rules, each covers broadly similar ground generally and, subject to the agreement of the parties, allows the tribunal a wide discretion as to the conduct of the proceedings.

Section 35 – Consolidation of Proceedings and Concurrent Hearings

35.-(1) The parties are free to agree–

(a) that the arbitral proceedings shall be consolidated with other arbitral proceedings, or
(b) that concurrent hearings shall be held,

on such terms as may be agreed.

(2) Unless the parties agree to confer such power on the tribunal, the tribunal has no power to order consolidation of proceedings or concurrent hearings.

Definitions

'agree': s.5(1).
'party': ss.82(2), 106(4).

[35A] Status

This is a new provision. It is *non-mandatory*, (s.4(2)).

[35B] Summary

The section provides for consolidation and concurrent hearings. In accordance with the principle of party autonomy (s.1(b)), the tribunal has no power to order the consolidation of proceedings (that is to say, the combining of different, but related claims into a single proceeding) or to order concurrent hearings (where the claims remain separate, but are heard together for the sake of convenience and cost saving) other than with the agreement of the parties. Although not spelt out in this section, the court also has no power so to order, in any circumstances.

Points

[35C] *Effect*
The structure of the section (separating 'The parties are free to agree . . .' in subs.(1) from 'to confer such power on the tribunal' in subs.(2)) reflects the fact that the parties may *themselves* agree to consolidate arbitrations or have concurrent hearings either at a stage prior to the appointment of the tribunal or after the appointment of one or more tribunals; or alternatively they may agree to confer the relevant powers on the tribunals.

[35D]   *Rationale*

The rationale behind this approach is that the parties should not have to find their agreed procedure for the private resolution of their own disputes being used to deal with other parties and their disputes, or to find themselves part of someone else's arbitration, unless they specifically so agree. Consolidation in the absence of agreement could operate as a disincentive to arbitrate. On an international level it might equally result in the inability to enforce an award, where a tribunal, for instance, had been imposed on an unwilling party. The DAC Report (para.180) also noted that difficulties over discovery might arise.

With the agreement of the parties, however, there is no problem about the tribunal operating the power to consolidate or order concurrent hearings. Some institutions and associations concerned with situations in which such a power may be desirable (for example, where there are likely to be many connected contracts and subcontracts) include suitable clauses in their standard terms, and arbitral institutions may include such provision in their rules.

[35E]   *Model Law*

There is no comparable provision in the Model Law, but in Hong Kong (domestic arbitrations only) and British Columbia, the Model Law has been amended so as to allow consolidation in certain circumstances.

[35F]   *Rules*

In a complex series of provisions (within Rules 2 and 3) CIMAR has taken up the challenge offered by s.35(1). Thus the provisions of Rule 2 are designed to lead to the appointment of the same arbitrator in related arbitral proceedings unless there are good grounds to avoid such an appointment. Rule 3.7 provides:

> 'Where the same arbitrator is appointed in two or more related arbitral proceedings on the same project each of which involves some common issue, whether or not involving the same parties, the arbitrator may, if he considers it appropriate, order the concurrent hearing of any two or more such proceedings or of any claim or issue arising in such proceedings.'

The ICE Arbitration Procedure is less ambitious in that it does not seek to procure the appointment of the same arbitrator. Nonetheless, by Rule 9:

> '9.1 Where disputes or differences have arisen under two or more contracts each concerned wholly or mainly with the same subject matter and the resulting arbitrations have been referred to the same Arbitrator he may with the agreement of all the parties concerned or upon the application of one of the parties being a party to all the contracts involved order that the whole or any part of the matters at issue shall be heard together upon such terms or conditions as the Arbitrator thinks fit.'

The CIArb Rules are perhaps the most comprehensive in this area. They provide, in Art.7 (after giving, in Art.7.1, the arbitrator the powers contained in this section):

'7.3 Where the same arbitrator is appointed under these Rules in two or more arbitrations which appear to raise common issues of fact or law, whether or not involving the same parties, the arbitrator may direct that such two or more arbitrations or any specific claims or issues arising therein be consolidated or heard concurrently.

7.4 Where an arbitrator has ordered consolidation of proceedings or concurrent hearings he may give such further directions as are necessary or appropriate for the purposes of such consolidated proceedings or concurrent hearings and may exercise any powers given to him by these Rules or by the Act either separately or jointly in relation thereto.

7.5 Where proceedings are consolidated the arbitrator will, unless the parties otherwise agree, deliver a consolidated award or awards in those proceedings which will be binding on all parties thereto.

7.6 Where the arbitrator orders concurrent hearings the arbitrator will, unless the parties otherwise agree, deliver separate awards in each arbitration.

7.7 Where an arbitrator has ordered consolidation or concurrent hearings he may at any time revoke any orders so made and give such further orders or directions as may be appropriate for the separate hearing and determination of each arbitration.'

Of the international rules, Art.22.1(h) of the LCIA Rules gives tribunals power:

'To allow, only upon the application of a party, one or more third persons to be joined in the arbitration as a party, provided any such third person and the applicant party have consented thereto in writing, and thereafter to make a single final award or separate awards, in respect of all parties so implicated in the arbitration.'

Neither the ICC Rules nor the UNCITRAL Arbitration Rules appear to permit consolidation or concurrent hearings.

Section 36 – Legal or Other Representation

36.   Unless otherwise agreed by the parties, a party to arbitral proceedings may be represented in the proceedings by a lawyer or other person chosen by him.

Definitions

'agreed': s.5(1).
'party': ss.82(2), 106(4).

[36A] Status

This is a new provision. It is *non-mandatory*, (s.4(2)).

[36B] Summary

The section provides for the representation of a party by a lawyer or other person of his choice, subject only to contrary agreement.

Points

[36C] *'Other person'*
The phrase 'other person' is very wide and clearly embraces consultants other than lawyers, indeed anyone.

[36D] *Effect*
The wording of the section ('... a lawyer or other person chosen by him') is intended to preserve the right to representation, but at the same time to prevent the section from being used as a pretext for delay by being construed as entitling a party to insist on his first choice of representative (who may not be available for a considerable time). If the first choice is not available to a party, then (so the DAC suggested at para.18(4) of the Report) a necessary substitute will still be a 'lawyer or other person chosen by him'.

Whilst in the first edition we said that we were not as convinced as the DAC that the wording would not be subject to the very debate it attempted to avoid, we are not aware that any problems have in fact arisen of the kind we feared.

[36E] *Rules*
Rule 14.1 of CIMAR amplifies s.36: 'A party may be represented in the proceedings by any one or more persons of his choice and by different

persons at different times'. Art.11.1 of the CIArb Rules provides: 'Any party may be represented by any one or more person or persons of their choice subject to such proof of authority as the arbitrator may require.' The ICE Arbitration Procedure is silent on the matter. Art.18 of the LCIA Rules permits representation by legal practitioners or any other representatives; Art.21 of the ICC Rules permits the parties to appear in person or through duly authorised representatives (and the parties may be assisted by advisers); and the UNCITRAL Rules (Art.4) provide that the parties may be represented or assisted by persons of their choice.

Section 37 – Power to Appoint Experts, Legal Advisers or Assessors

37.–(1) Unless otherwise agreed by the parties–

(a)   the tribunal may–

(i)   appoint experts or legal advisers to report to it and the parties, or
(ii)   appoint assessors to assist it on technical matters,
and may allow any such expert, legal adviser or assessor to attend the proceedings; and

(b)   the parties shall be given a reasonable opportunity to comment on any information, opinion or advice offered by any such person.

(2) The fees and expenses of an expert, legal adviser or assessor appointed by the tribunal for which the arbitrators are liable are expenses of the arbitrators for the purposes of this Part.

Definitions

'agreed': s.5(1).
'arbitrator': s.82(1).
'party': ss.82(2), 106(4).

[37A]   Status

This section derives from the Model Law, Art.26. Subs.(1) is *non-mandatory*; subs.(2) is *mandatory*.

[37B]   Summary

The section deals with the power of the tribunal to appoint experts, legal advisers and assessors, and their fees and expenses. Such power can be excluded by contrary agreement, such agreement being in writing in accordance with s.5.

Points

[37C]   *Power to appoint – Subs.(1)*
Subject to the contrary agreement of the parties, subs.(1) permits the tribunal to appoint its own experts or legal advisers, where appropriate. In so doing, arbitrators must have in mind their s.33 duty to adopt procedures suitable to the circumstances of the case, and to avoid unnecessary delay and expense. A case for appointing an expert might arise, for example, where the tribunal, pursuant to its s.33 duty, adopted an inquisitorial procedure and engaged its own expert to

report on certain issues. The tribunal may also seek the support of technical assessors.

Consistent with another aspect of its s.33 duty – to act fairly and give each party a reasonable opportunity of putting his case – the tribunal is bound to give the parties a reasonable opportunity to comment on the contributions such persons may make. Where there is a legal adviser, such comment is likely to take the form of submissions by the parties' representatives, orally or in writing. Where there is a technical expert or assessor, it is more probable, in a complex case at least, that the parties would be given an opportunity to present their own expert evidence on controversial matters. However under s.34(1)(h) the tribunal could direct that such evidence be in writing, unless the parties agreed otherwise.

This subsection is non-mandatory, (s.4(2)), and accordingly, the parties may agree that the tribunal should not have the power it sets out, or that they should not be given any opportunity to comment.

[37D]   *Comparison with legal proceedings*
Under the CPR, in legal proceedings the court may now direct that where two or more parties wish to submit expert evidence on a particular issue, the evidence on that issue is to be given by one expert only. Whilst the Act does not spell this out specifically, we see no difficulty in an arbitrator adopting a similar approach in a suitable case, and always assuming no contrary agreement between the parties.

[37E]   *Instructing the expert*
Arbitrators appointing their own expert or legal adviser will need to adopt a procedure for instructing him. In our view it is preferable that the arbitrator seeks the input of the parties, either by directing the parties to agree draft instructions (insofar as they are able), or by drafting such instructions himself and submitting them to the parties for comment. It may also be possible to adopt the approach taken by Rule 35.8 of the CPR, namely that each party gives instructions to the expert. However, unless the arbitrator adds his own instructions, his opportunity to direct the expert's attention towards matters he considers relevant will be limited.

[37F]   *Choice of expert*
Often the arbitrator will have his own view. In such a case we suggest that the arbitrator confirms with the parties that they have no objection to the suggested person and, in the absence of objection, proceeds. In other circumstances, we suggest that the arbitrator asks the parties to confer and, if possible, reach a joint conclusion. If there is no agreement, the arbitrator may invite names from the parties and select the most suitable, taking account, of course, of the necessity for independence and impartiality.

[37G]   *Expenses of the arbitrators – Subs.(2)*
If, however, the tribunal does have the power to appoint its own experts and advisers in this way, then the fees and expenses of the persons so appointed become expenses of the arbitrators, and the parties may not agree otherwise, this subsection being mandatory, (s.4(1) and Sched.1).

Such expenses would be covered by s.28, which concerns the joint and several liability of the parties to the arbitrators for their fees and expenses. Arbitrators would be well advised to ensure the reasonableness of such sums, probably ideally by obtaining the agreement of the parties before incurring the expenditure.

We have explained the operation of s.28 above. Contractual arrangements aside, the parties are only liable to pay the arbitrators' 'reasonable' expenses. The parties would not, therefore, be liable for unreasonably high sums paid to experts, legal advisers or assessors; nor for any sums incurred pursuant to a decision to employ such an expert, legal adviser or assessor which was inappropriate and unreasonable in the light of the s.33(1)(b) duty to 'adopt procedures suitable to the circumstances of the particular case, avoiding unnecessary . . . expense'.

[37H]   *Model Law*
Art.26 of the Model Law (Expert appointed by arbitral tribunal) provides:

> '(1) Unless otherwise agreed by the parties, the arbitral tribunal
> (a) may appoint one or more experts to report to it on specific issues to be determined by the arbitral tribunal;
> (b) may require a party to give the expert any relevant information or to produce, or to provide access to, any relevant documents, goods or other property for his inspection.
> (2) Unless otherwise agreed by the parties, if a party so requests or if the arbitral tribunal considers it necessary, the expert shall, after delivery of his written or oral report, participate in a hearing where the parties have the opportunity to put questions to him and to present expert witnesses in order to testify on the points at issue.'

We note that there is no directly comparable power in the Act to require a party to give a tribunal-appointed expert relevant information, documents, goods or other property. However, the parties do have a general duty of co-operation (s.40), and the arbitrators have power under s.34, in subss.(d), (e) and (g), to order disclosure and production of documents, and they may put questions and elicit facts themselves. Property owned by, or in the possession of, a party to the proceedings may also be directed to be inspected, photographed, experimented upon, and so on, by an expert pursuant to an order of the tribunal under s.38(4) below.

The Model Law specifically gives the parties the right to put questions to the expert and to present their own expert witnesses on the points in issue. The Act is more cautious, requiring only a 'reasonable opportunity to comment'. In paragraph 37C above we have suggested that in practice it is likely that where there is a tribunal-appointed expert, the arbitrators would, at least in a complex case, permit the parties to call their own experts. We would like to think, however, that the Act allows greater flexibility than the Model Law so that the tribunal can determine in each case how best to proceed in accordance with its s.33 duties.

The Model Law does not deal specifically with legal advisers. Perhaps that is because in practice, most international arbitrators are lawyers.

[371] *Rules*

Rule 4.2 of CIMAR specifically gives the arbitrator the power under s.37. Rule 11 of the ICE Arbitration Procedure is in different terms and in our view would substitute for s.37 as an agreement 'otherwise'. It provides:

> '11.1 The Arbitrator may:
> (a)   appoint a legal technical or other assessor to assist him in the conduct of the arbitration. The arbitrator shall direct when such assessor is to attend hearings of the arbitration;
> (b)   seek legal technical or other advice on any matter arising out of or in connection with the proceedings.'

The CIArb Rules are silent on this topic. Art.21 of the LCIA Rules and Art.20.4 of the ICC Rules have broadly similar effect to s.37, but (in our view) would substitute for it. The same is probably true of the detailed provisions at Art.27 of the UNCITRAL Arbitration Rules. It is to be noted that Art.27.4 specifically provides that at the hearing either party may present expert witnesses in order to testify on the points at issue.

Section 38 – General Powers Exercisable by the Tribunal

38.–(1) The parties are free to agree on the powers exercisable by the arbitral tribunal for the purposes of and in relation to the proceedings.

(2) Unless otherwise agreed by the parties the tribunal has the following powers.

(3) The tribunal may order a claimant to provide security for the costs of the arbitration.

This power shall not be exercised on the ground that the claimant is–

(a) an individual ordinarily resident outside the United Kingdom, or
(b) a corporation or association incorporated or formed under the law of a country outside the United Kingdom, or whose central management and control is exercised outside the United Kingdom.

(4) The tribunal may give directions in relation to any property which is the subject of the proceedings or as to which any question arises in the proceedings, and which is owned by or is in the possession of a party to the proceedings–

(a) for the inspection, photographing, preservation, custody or detention of the property by the tribunal, an expert or a party, or
(b) ordering that samples be taken from, or any observation be made of or experiment conducted upon, the property.

(5) The tribunal may direct that a party or witness shall be examined on oath or affirmation, and may for that purpose administer any necessary oath or take any necessary affirmation.

(6) The tribunal may give directions to a party for the preservation for the purposes of the proceedings of any evidence in his custody or control.

Definitions

'agree', 'agreed': s.5(1).
'claimant': s.82(1).
'costs of the arbitration': s.59.
'party': ss.82(2), 106(4).

[38A] ## Status

This section derives from the Arbitration Act 1950, s.12(1) to (3), and replaces what was s.12(6)(a) of that Act. See also the Model Law, Art.17. It is *non-mandatory*, (s.4(2)).

[38B] ## Summary

The section sets out certain general powers exercisable by the tribunal. It complements s.34 which deals with procedural and evidential matters which the tribunal may decide.

Points

[38C] *Effect*

The section essentially provides that the parties are at liberty to agree what general powers should be exercisable by the tribunal; and that certain powers are to be exercisable in the absence of contrary agreement. These are powers to order security for costs, (subs.(3)); to give directions in relation to property which is the subject of the proceedings and which is owned or in the possession of a party, (subs.(4)); to direct the examination of parties or witnesses upon, and to administer, oaths and affirmations, (subs.(5)); and to give directions for the preservation of evidence in the custody or control of a party, (subs.(6)).

To some extent, (but, significantly, not in respect of security for costs) these powers run in parallel with the corresponding powers of the court (as to which, see s.44), the scheme of the Act being, so far as possible, to enable the tribunal to act rather than require the parties to submit to the inconvenience and expense of an application to the court.

The court's powers are generally more extensive and, in particular, extend to certain areas where it would not be appropriate or possible for the tribunal to act. However, they are also subject to a number of restrictions designed to prevent application to the court rather than the tribunal (where the latter is competent to act) and to leave the control of the arbitral process in the hands of the tribunal so far as possible. In particular, s.44(5) provides that the court should only act where the tribunal (or third party vested with power) has no power or is unable for the time being to act effectively.

[38D] *Agreement by the parties – Subss.(1), (2)*

The parties may agree on the powers exercisable by the arbitral tribunal. Unless they expressly or impliedly agree that the powers they have decided upon supersede any or all of the powers granted in subss.(3) to (6), so that the tribunal will not have those powers, or unless they agree that one or more of the powers in subss.(3) to (6) simply will not apply, then subss.(3) to (6) will apply.

[38E] *Security for costs – Subs.(3)*

Unlike those powers which may be exercised both by the tribunal and by the court, the tribunal alone has power to order security for costs, the court having no such power. This is a very significant change from the situation prior to the Act where *only* the court could order security for costs, unless the parties agreed otherwise. Note that now, if the parties agree that the tribunal should not have the power to order security, it will not be available from any source. The question has been raised whether, by agreeing to rules which do not specifically give power to order such security, parties exclude the possibility of it being ordered having regard to s.4(3). We think not: see paragraph 4G above.

The wording 'The tribunal may order a claimant to provide

security...' is, in the light of the definition of 'claimant' in s.82(1), sufficiently wide to permit an order to be made against a counter-claiming respondent, as well as a true claimant.

'Costs of the arbitration' for the purposes of security are defined in s.59. They are not restricted to the legal costs of the parties. Thus, under this provision, security may also be required for the arbitrators' fees and expenses, the fees and expenses of any arbitral institution con-cerned and the costs (other than legal) of the parties. The 'other costs of the parties' might, for example, include the costs incurred by a party in complying with an order which required the preservation of property under s.38. We doubt, though, that the costs of court proceedings to obtain security would be included. The costs of the proceedings to determine the amount of recoverable costs are included (see s.63).

It appears that the tribunal may exercise this power of its own motion, as well as upon the application of a party. However to do so spontaneously, at least in relation to a party's costs, might well breach the s.33 obligation of fairness and impartiality in that the tribunal could be said to be looking after the interests of one of the parties at the expense of the other, and of its own volition. But we do not see why the same should be true as regards security for the tribunal's own fees and expenses. Presumably, the tribunal might also exercise the power in support of an arbitral institution concerned in respect of its own fees.

[38F]   *Discretion to order security for costs*

The only provision as to how the power is to be exercised is a negative stipulation. It may *not* be exercised on the ground that a party who is an individual is ordinarily resident outside the United Kingdom, or that a party that is a company or association was incorporated or formed, or in either case is managed or controlled, from outside the United Kingdom. In short, security may not be ordered on the basis that the claimant is foreign. (In the court context, ordinary residence outside the jurisdiction is precisely one of the grounds on which the power to order security *is* exercisable).

The reason for this provision lies in the very different arbitration context and, in particular, in the anxiety not to deter foreign parties from choosing to arbitrate in England. They might (not unreasonably) fear that a provision for security on the ground of residence abroad would operate unfairly and discriminate against them. Problems of possible conflict with European Community law are also avoided. The Act does not, of course, preclude the possibility of security being ordered against a foreign claimant on other grounds. It does not, therefore, meet the fundamental objection found in some jurisdictions to the very concept of security for costs. However by ensuring that the court may not order security it does, at least, meet the objection to the decision in *S.A. Coppée Lavalin NV* v. *Ken-Ren Chemicals and Fertilisers Ltd.* [1995] 1 AC 38, where, notwithstanding that ICC arbitrators had no power to order security for costs, the court did so.

It follows, however, that no positive indication is given by the Act as to how the tribunal's discretion should be exercised (apart, of course, from the obligation to comply with the mandatory duty prescribed in s.33). In particular, any reference to the rules and existing case-law relating to security for costs in the court context has been avoided. Whereas early drafts of the Act made the tribunal's power co-extensive with that of the court, stating, 'The tribunal shall exercise its power on the same principles as the court', the deliberate omission of any such statement makes it clear that the power given to arbitrators is not intended to be so restricted.

The inference would seem to be that arbitrators may exercise their discretion very flexibly, and in a manner that may well diverge from that which the court would adopt, so long as what they do is in keeping with their s.33 duty, in particular to act fairly and impartially. Thus, for example, we do not believe that an applicant for security is obliged to show 'by credible testimony' that the other party will be unable to pay the costs if the applicant is successful (see s.726 of the Companies Act 1985). The approach to be adopted is left to the tribunal. In practice, it seems likely that the exercise of the discretion will centre upon an assessment of the accounts of, or other financial information concerning, the claimant, at least where it is the financial soundness of the claimant that is primarily in issue.

It may also centre on the location of the assets of a party. If an English claimant only had assets abroad, a tribunal might reasonably order security for costs. If a foreign claimant only had assets outside his own country (e.g. an EC claimant whose assets were outside the EC), the same might be true. We would venture to suggest that a tribunal might also find it appropriate to make an order where a foreign claimant's assets were within his own country, but outside the jurisdiction. This would not infringe the prohibition in subs.(3)(a) since it would not be an order based on *residence*, but on the location of assets. (Of course, this would effectively circumvent the prohibition, but it is difficult to see the difference in principle between the first two situations postulated here, and the third.)

In *Azov Shipping Co.* v. *Baltic Shipping Co. (No. 2)* [1999] 1 All ER (Comm.) 716 the court commented, on an application for security for the costs of a challenge to jurisdiction under s.67, that in the light of s.1(a) 'the cases will be rare in which a court or indeed an arbitrator would think it right to order security for costs if an applicant for relief has sufficient assets to meet any order for costs and if those assets are available for satisfaction of any such order for costs'.

It will normally be necessary to receive evidence in some form or other (to be decided under s.34) in order to make the necessary assessment. Note, for instance, CIMAR Rule 5.7 under which the arbitrator is required to receive affidavit evidence or at least that 'some other formal record of the evidence be made'.

[38G]   *Sealed offers and security*

An application for security may well involve the claimant, who will be resisting the application, wishing to inform the tribunal of the existence of an offer of settlement of which it would not normally learn until after its award on the merits. In short the claimant is saying to the tribunal: 'The respondent obviously thinks I have a good case because it has made me an offer. Please don't deprive me of my just desserts because of my impecuniosity . . .'

In the court context, judges will permit such sealed offers to be shown to them in appropriate cases and to be used as part of the evidence. This does not present a problem for them, since the security application can be dealt with by a judge different from the one who will deal with the substantive hearing. Under the law prior to this Act, in situations where both the arbitrators and the court had power to order security (such as under the various JCT conditions) and one party wished to refer to a sealed offer, it was held appropriate for the application to be made to the court, but under this Act that alternative is not possible.

Either the arbitral tribunal will have to hear of the sealed offer, and to put such knowledge out of its mind when it comes to try the merits of the claim, or another solution, such as putting security applications to a third party (by agreement) will be needed. Whilst this is not altogether satisfactory, it is one of the consequences that flows from parties choosing their own tribunal, and it is debatable whether Parliament should have sought to help them out of this difficulty.

[38H]   *Dismissal of claim where security not given*

It should be noted that a specific sanction exists for a party failing to provide security after a peremptory order has been made. S.41(6) permits the tribunal, in those circumstances, to make an award dismissing the claim. Thereafter the party in default will be unable to revive the claim (perhaps at a much later date) by providing the required security, as they would be able to do if the only sanction were a stay on the proceedings.

We note here that neither s.38 nor s.41 empower arbitrators to impose stays on proceedings where security for costs is not provided. Since we do not consider that arbitrators had that power under the previous legislation, in the absence of any special agreement, we do not believe that they have it under this Act. Some rules may, however, give arbitrators such power: see, for example, the LMAA Terms (1997), and the LCIA Rules (see below).

[38I]   *Directions in relation to property – Subs.(4)*

The tribunal is given substantial powers in respect of property the subject of the proceedings or as to which a question arises, where the property is owned by or is in the possession of a party. The powers may be exercised for the benefit of the tribunal, an expert (including an expert appointed by the tribunal under s.37) or a party. The tribunal may act on its own initiative: for example, at the prompting of its own appointed expert.

It is important to note that the tribunal has no such power in relation to property over a non-party, but the court *does* have such power under s.44(2)(c).

[38J]   *Model Law*

Art.17 (Power of arbitral tribunal to order interim measures) provides:

> Unless otherwise agreed by the parties, the arbitral tribunal may, at the request of a party, order any party to take such interim measure of protection as the arbitral tribunal may consider necessary in respect of the subject-matter of the dispute. The arbitral tribunal may require any party to provide appropriate security in connection with such measure'.

We note that the Model Law power must be exercised at the bidding of a party. We think the Act goes further, at least in certain respects. The Model Law power is limited to 'interim measure[s] of protection'. Again we think the Act goes rather further. The arbitral tribunal may order security to be paid by the party requesting the interim measure of protection. As we have said in paragraph 38E above, we think the definition of costs in s.59 is sufficiently wide to embrace a similar order under the Act in respect of costs. However it is quite arguable that under the Model Law a tribunal could order a party to provide security for losses which might be suffered by another party whose property was the subject matter of a protection order. We do not think that the Act would allow arbitrators to order such security.

[38K]   *Rules*

Under CIMAR, by Rule 4.3 the arbitrator has the powers set out in subss.(4) to (6). Rules 4.4 and 4.5 supplement these powers. Rule 4.6 deals with security, unlike the Act setting out the basis upon which the power is to be exercised; the security may be ordered if the arbitrator is 'satisfied that the claimant is unlikely to be able to pay those costs if the claim is unsuccessful. In exercising this power, the arbitrator shall consider all the circumstances including the strength of the claim and any defence, and the stage at which the application is made'. The power is nonetheless subject to paragraphs (a) and (b) of subs.(3).

By Rule 7.5 of the ICE Arbitration Procedure the arbitrator has a wide power to 'make an order for security for costs in favour of one or more of the parties'. Monies must be paid into a stakeholder bank account. By Rule 7.6 the tribunal is given powers broadly similar to subss.(4) and (6), but this is supplemented by (c) under which the arbitrator may 'give directions for the detention storage sale or disposal of the whole or any part of the subject matter of the dispute at the expense of one or both of the parties'. It is our view that Rules 7.5 and

7.6 operate in place of s.38, save for the ground covered by s.38(5) which is not separately dealt with.

The matter is not covered in the CIArb Rules. Accordingly the arbitrator has the powers set out in s.38.

Art.25.2 of the LCIA Rules provides that the tribunal may order security for costs 'by way of deposit or bank guarantee or in any other manner and upon such terms as the Arbitral Tribunal considers appropriate'. This includes a cross-indemnity. Non-compliance with the order entitles the tribunal to stay or dismiss the claim or counter-claim. Security for costs being something embodied in common law procedure, it is not surprising that neither the ICC nor the UNCITRAL Arbitration Rules provide for it.

Section 39 – Power to Make Provisional Awards

39.-(1) The parties are free to agree that the tribunal shall have power to order on a provisional basis any relief which it would have power to grant in a final award.

(2) This includes, for instance, making–

(a) a provisional order for the payment of money or the disposition of property as between the parties, or
(b) an order to make an interim payment on account of the costs of the arbitration.

(3) Any such order shall be subject to the tribunal's final adjudication; and the tribunal's final award, on the merits or as to costs, shall take account of any such order.

(4) Unless the parties agree to confer such power on the tribunal, the tribunal has no such power.

This does not affect its powers under section 47 (awards on different issues, &c.).

Definitions

'agree': s.5(1).
'costs of the arbitration': s.59.
'party': ss.82(2), 106(4).

[39A] ## Status

This is a new provision. It is *non-mandatory*, (s.4(2)).

[39B] ## Summary

The section provides for arbitrators to be given, by agreement only, the power to make provisional orders. This is a power to make orders for any relief which could be granted in the final award in the reference on a provisional or temporary basis. Such orders are distinct from interim awards, see paragraph 39E below. The orders are subject to adjustment on the final adjudication and are brought into account in the final award, (subs.(3)). Subject to the fact that they must be limited to relief which could be granted in the final award, their types are not specified save that (unusually) subs.(2) gives examples.

Points

[39C] *Discretion to make provisional orders – Subs.(1)*
If this power is conferred on the tribunal, no positive indication is given by the Act as to how the tribunal's discretion as to whether or not to

grant an order should be exercised (apart, of course, from its obligation to comply with the mandatory duty prescribed in s.33). In particular, any reference to the rules and case-law relating to 'interim orders' (as they are there known) in the court context has been avoided. The inference would seem to be that arbitrators may exercise their discretion very flexibly, and in a manner that may well diverge from that which the court would adopt, so long as what they do is in keeping with their s.33 duty.

We suggest, however, that tribunals should generally be cautious about using any such power the parties may give them, for fear of causing injustice in the long term. For instance, a provisional order may be made for the payment of money, and may be complied with. In the final adjudication it may be found that in fact a smaller sum was due, but the respondent cannot obtain reimbursement because the claimant has since become impecunious. And if liability is not clear at the time a provisional order is sought, it is difficult to imagine circumstances in which it would be acting 'fairly and impartially' to make one.

[39D] *Examples – Subs.(2)*

Subs.(2) provides some examples of ways in which the power might be used. It is clearly not exclusive.

One context in which we think provisional orders could prove useful is that of rent review arbitrations. If, for example, it were plain that there was going to be an increase in the future rent, although the extent of that increase was in issue, a provisional order as to an undisputed level of increase could well be appropriate. Rent review clauses in leases may begin to include reference to the tribunal having, by agreement, the power to make provisional orders.

[39E] **Provisional orders and awards on different issues – Subs.(4)**

Subs.(4) specifically distinguishes between provisional orders (which are subject to later adjustment, if necessary) and awards on different issues made pursuant to s.47 (which may also be made before the end of the arbitration, but are nevertheless final as to the matters which they determine).

It follows that although the margin note to the Act refers to 'provisional *awards*', what the section in fact concerns (as its text indicates) should properly be termed 'provisional *orders*'. The terminology is important since the section specifically contemplates orders which are subject to later adjustment, (subs.(3)), whereas awards, unless otherwise agreed, are 'final and binding' in their effect, (s.58(1)).

The word 'provisional' has no doubt been carefully chosen to avoid the use of 'interim', with its connotation of an 'interim award', which is nevertheless final in respect of the matters which it determines.

[39F] **Formalities and reasons**

Since the power is to make an 'order', it is not, in our view, necessary for the tribunal to comply with the formalities of s.52 concerning awards, including that of giving reasons.

[39G] **Model Law**

There is no comparable provision in the Model Law.

[39H] **Rules**

The tribunal does not have power to make a provisional order unless it is given it. CIMAR gives the tribunal such power at Rule 10. The arbitrator may exercise it of his own motion or on the application of a party. There must be formal evidence. The arbitrator may give such reasons for his order as he thinks appropriate. Payment may be to a stakeholder. Rules 19.3 to 19.5 of the ICE Arbitration Procedure similarly give the arbitrator this power.

Art.7.1 of the CIArb Rules gives the arbitrator the powers contained in s.39 and Art.7.8 gives him power to grant a provisional order (a) for the payment of money or the disposition of property as between the parties; (b) for interim payment on account of the costs of the arbitration and (c) for the grant of any relief claimed in the arbitration; in short, all the matters anticipated in s.39. He may act of his own motion or on application, provided he gives notice of his intention to the parties who must have a chance to make representations (Art.7.9); payment may be made or property delivered to a stakeholder (Art.7.10) and the arbitrator (or any subsequent arbitrator having jurisdiction) may confirm, vary or revoke any such order (Art.7.11).

Of the international rules, Art.25.1 (c) of the LCIA Rules provides that the tribunal may:

> 'order on a provisional basis, subject to final determination in an award, any relief which the Arbitral Tribunal would have power to grant in an award, including a provisional order for the payment of money or the disposition of property as between any parties.'

Art.23 of the ICC Rules gives tribunals the power to order 'any interim or consequential measure it deems appropriate'. It is not clear whether 'interim ... measure' covers provisional orders as anticipated by this section. On the whole we think it does not. 'Interim' is not the same as 'provisional', and given the dangers inherent in ordering provisional relief we think clear words would be required before a court would hold that a tribunal had power so to do.

Section 40 – General Duty of Parties

40.–(1) The parties shall do all things necessary for the proper and expeditious conduct of the arbitral proceedings.

(2) This includes–

(a) complying without delay with any determination of the tribunal as to procedural or evidential matters, or with any order or directions of the tribunal, and

(b) where appropriate, taking without delay any necessary steps to obtain a decision of the court on a preliminary question of jurisdiction or law (see sections 32 and 45).

Definitions

'the court': s.105.
'party': ss.82(2), 106(4).
'question of law': s.82(1).

[40A] Status

This is a new provision. It is *mandatory*, (s.4(1) and Sched.1).

[40B] Summary

This section sets out the general duty of the parties in respect of the arbitration. It complements, so far as the parties are concerned, the general duty of the tribunal expressed in s.33.

The section sets out, in general terms, the obligations of the parties to progress the proceedings with expedition, and to co-operate with the tribunal (although plainly, such co-operation will only be required within the parameters of the powers which the parties have agreed that the tribunal should have or which are given by the Act). Because of its mandatory status, the parties are not able to avoid the effect of the section by agreement.

The provision is given teeth by other sections of the Act such as s.41 (which provides for the powers of the tribunal in the event of a party's default); s.42 (which gives the court power to enforce peremptory orders of the tribunal); and s.73 (which provides for the loss of the right to raise objections after knowing delay).

Since the first edition of this book the CPR has provided, by Rule 1.3, that in litigation the parties are required to help the court to further the overriding objective, i.e. dealing with cases justly and managing them actively. Case law has yet to develop this requirement, but on its face it is not quite as burdensome a duty as is placed on those who arbitrate (although the sanctions for default may be severe). This perhaps reflects the consensual nature of arbitration.

Points

[40C] *General duty – Subs.(1)*
This subsection sets out the general duty. Exactly what is 'necessary' for the proper and expeditious conduct of the proceedings is open-ended, but some of the relevant matters can be derived from subs.(2) and from s.41. They include, notably, pursuance of the claim without inordinate or inexcusable delay; attendance at hearings; making written submissions when required; compliance with orders and directions generally; compliance with determinations in respect of procedure and evidence; and taking, without delay, steps to obtain the decision of the court on preliminary questions of jurisdiction or law.

[40D] *Examples – Subs.(2)*
This subsection sets out two examples of what must be done by the parties to comply with their general duty.

The first is that the parties must comply without delay with any determination of the tribunal as to procedural or evidential matters, or with any order or directions of the tribunal. Where a party fails to comply with such order or directions, the tribunal may make a per-emptory order to the same effect, prescribing a time for compliance (s.41(5)). Non-compliance with a peremptory order may result in any of the range of consequences at the hands of the tribunal (s.41(6) and (7)) or the court (s.42).

The second is that the parties must take without delay any necessary steps to obtain a decision of the court on a preliminary question of jurisdiction or law. The Act specifically cross-refers to ss.32 and 45. Under the former the court may determine a question of substantive jurisdiction at the request of a party, but only either with the agreement of the other party or parties, or with permission of the tribunal *and* the court's satisfaction as to a number of conditions, which include the absence of delay. S.32(1)(b) thus makes specific the general requirement set out in this subsection. S.45(2) does the same in the context of the determination of a preliminary point of law by the court.

As we have noted above, the general duty of assisting the proper and expeditious course of the proceedings is carried through into s.73. Objections in the nature of a challenge to the arbitral process which might be raised at any point may not be raised later if known about at an earlier time.

[40E] *Model Law*
The Model Law contains no comparable provision.

[40F]   *Rules*

Since this is a mandatory provision, it is unnecessary that it be reflected in rules, except perhaps to draw attention to it. CIMAR does this at Rule 1.2 and the ICE Arbitration Procedure does it at Rule 1.1, whilst the CIArb Rules are silent. The international rules are also silent on the matter, so that the duty falls to be implied by law.

Section 41 – Powers of Tribunal in Case of Party's Default

41.–(1) The parties are free to agree on the powers of the tribunal in case of a party's failure to do something necessary for the proper and expeditious conduct of the arbitration.

(2) Unless otherwise agreed by the parties, the following provisions apply.

(3) If the tribunal is satisfied that there has been inordinate and inexcusable delay on the part of the claimant in pursuing his claim and that the delay–

(a) gives rise, or is likely to give rise, to a substantial risk that it is not possible to have a fair resolution of the issues in that claim, or
(b) has caused or is likely to cause, serious prejudice to the respondent,

the tribunal may make an award dismissing the claim.

(4) If without showing sufficient cause a party–

(a) fails to attend or be represented at an oral hearing of which due notice was given, or
(b) where matters are to be dealt with in writing, fails after due notice to submit written evidence or make written submissions,

the tribunal may continue the proceedings in the absence of that party or, as the case may be, without any written evidence or submissions on his behalf, and may make an award on the basis of the evidence before it.

(5) If without showing sufficient cause a party fails to comply with any order or directions of the tribunal, the tribunal may make a peremptory order to the same effect, prescribing such time for compliance with it as the tribunal considers appropriate.

(6) If a claimant fails to comply with a peremptory order of the tribunal to provide security for costs, the tribunal may make an award dismissing his claim.

(7) If a party fails to comply with any other kind of peremptory order, then, without prejudice to section 42 (enforcement by court of tribunal's peremptory orders), the tribunal may do any of the following–

(a) direct that the party in default shall not be entitled to rely upon any allegation or material which was the subject matter of the order;
(b) draw such adverse inferences from the act of non-compliance as the circumstances justify;
(c) proceed to an award on the basis of such materials as have been properly provided to it;
(d) make such order as it thinks fit as to the payment of costs of the arbitration incurred in consequence of the non-compliance.

Definitions

'agree', 'agreed': s.5(1).
'claimant': s.82(1).
'costs of the arbitration': s.59.
'the court': s.105.
'notice (or other document)': s.76(6).

'party': ss.82(2), 106(4).
'peremptory order': s.82(1),
  and see s.41(5).
'written', 'in writing': s.5(6).

[41A] Status

This section derives in part from the Arbitration Act 1950, s.13A; and in part from the Model Law, Art.25. It is *non-mandatory*, (s.4(2)).

[41B] Summary

The section deals with the powers which the tribunal has in case a party fails to do what is required of it to progress the arbitration.

As with most other non-mandatory sections, the parties are essentially at liberty to agree on these powers. Those that are set out are the ones that will apply if the parties do not agree otherwise. They cover the situations of delay by the claimant, failure by either party to attend the proceedings or to make submissions, and failure to comply with the tribunal's orders.

The section reinforces the tribunal's general powers set out in s.38; and reflects s.40 which requires the parties to do 'all things *necessary for the proper and expeditious conduct* of the arbitral proceedings' – a phrase that is repeated in subs.(1). It is also related to s.42, under which the court may supplement the tribunal's power to make parties comply with its orders.

Points

[41C] *Delay by the claimant – Subs.(3)*

The subsection preserves the tribunal's previous power under s.13A of the 1950 Act to strike out a claim if it is not properly prosecuted. The approach here formulated is in line with current case law as applied by the courts on dismissal for want of prosecution.

Thus the power to make an award dismissing the claim arises if there has been first, inordinate and inexcusable delay on the part of the claimant or his representatives. 'Inordinate' in this context means 'significantly longer than what would normally be regarded as acceptable'. 'Inexcusable' in this context means not excused by such objective factors as illness, accident or the conduct of the respondent.

In addition, such delay must also cause 'prejudice' (which in this context means 'detriment') to the respondent. The prejudice may affect the proceedings by creating a risk, or potential risk, that a 'fair resolution of the issues' (reflecting the 'fair resolution of disputes' principle in s.1(a)) is impossible. That will be so where, for example, due to the lapse of time, witnesses have died or disappeared, or where their memory of important events may have faded and become untrustworthy.

Alternatively, the prejudice may concern the respondent's own circumstances. For example, if the respondent is a professional person, excessively prolonged arbitration proceedings might prove both professionally damaging and personally worrying, and this could amount to prejudice. A small business might be similarly affected. Another example would be that of a third party, from whom the respondent might seek an indemnity, becoming insolvent.

[41D]   *Failure of party to attend – Subs.(4)*
The tribunal has the power simply to carry on with the arbitration and proceed to an award without the participation of a party if that party (or their representatives) fails to attend an oral hearing, or, primarily in the context of a 'documents only' arbitration, fails to put in evidence or submissions, and no good reason is shown for the failure or omission.

The party must have had due notice of what was required. Plainly, the tribunal must ensure that such notice has been received before exercising this power, if it is to avoid the risk of a challenge to the award on the ground of serious irregularity affecting the proceedings, (s.68).

This power to carry on 'ex parte' is drastic, and before exercising it the tribunal must have very much in mind its duty, under s.33, to give each party 'a reasonable opportunity of putting his case and dealing with that of his opponent...'. It is a fundamental principle that each party has a right to be 'heard' (although not necessarily orally, see s.34(2)(h)). It is for the tribunal carefully to consider whether, due to a failure to co-operate, insistence on that right has become unreasonable.

In our view, in the majority of cases it will be wise for the tribunal to give a further, if brief, notice after the first default before proceeding 'ex parte', if only to be seen to be utterly fair. This may be particularly important in the context of international arbitrations, to avoid possible enforcement problems.

[41E]   *Peremptory orders – Subss.(5) to (7)*
These subsections set out a scheme for enforcing the tribunal's orders by making them peremptory. That is done by making a further order to the same effect, putting a time limit for compliance on it and making it clear that if there is a failure to comply within that time limit, certain consequences will follow. Such a procedure is familiar from the practice of the courts to make 'unless' orders, so-called because they set out the consequences to the party concerned 'unless' it complies within a specified time.

There will no doubt be some debate as to the form peremptory orders should take. We strongly suggest that tribunals make it abundantly clear that the order they are making is peremptory. This could be done by specific inclusion of the word 'peremptory' or 'peremptorily' in the order or by specific reference to s.41(5), with or without use of the 'unless' formula. We think, too, it will almost invariably be appropriate to spell out the intended sanction in the peremptory order rather than it being decided and imposed later. In the light of the wide discretion as to such sanction this will be fairest to the party concerned, and may encourage the recalcitrant to act.

For example, the original order may provide that :

> 'The parties shall exchange experts' reports on or before [a certain date]'.

If a party fails to offer their expert's report for exchange in time without good reason, the ensuing peremptory order may then state:

> 'By way of peremptory order, unless the [claimant/respondent] offers his expert's report for exchange on or before [a certain date] he shall not be entitled to rely upon any expert evidence;' or,
> 'It is peremptorily ordered that the [claimant/respondent] shall offer his expert's report for exchange on or before [a certain date], failing which he shall not be entitled to rely upon any expert evidence.'

Subs.(7) sets out the range of sanctions that may be applied by the tribunal in relation to most peremptory orders. Note that the power simply to make an award against the party in default dismissing his claim is not included as this was considered too drastic a step. Only in relation to the failure to provide security for costs (see s.38(3)) does this power exist (subs.(6)), so that the proceedings may be brought to an end and not simply stayed, with the risk that they might be revived by the provision of security at a much later date.

As indicated in paragraph 38H above, we do not think that, in the absence of a special agreement, arbitrators have any power to stay proceedings for want of provision of security for costs.

We consider that now that there is clear statutory authority for these powers, arbitrators should use them without hesitation.

[41F]   *Model Law*

Art.25 (Default of a party) provides:

> 'Unless otherwise agreed by the parties, if, without showing sufficient cause,
> (a) the claimant fails to communicate his statement of claim in accordance with article 23(1), the arbitral tribunal shall terminate the proceedings;
> (b) the respondent fails to communicate his statement of defence in accordance with article 23(1), the arbitral tribunal shall continue the proceedings without treating such failure in itself as an admission of the claimant's allegations;
> (c) any party fails to appear at a hearing or to produce documentary evidence, the arbitral tribunal may continue the proceedings and make the award on the evidence before it.'

We note that s.41(3) is wider than Art.25(a) in that it goes to delay in general and not simply in relation to the statement of claim, if there is to be one. Moreover, the Model Law does not provide for what is to happen in the event of procedural default, other than in respect of statements of claim and defence, non-appearance or the non-production of evidence.

[41G]   *Rules*
Rules 11.1 to 11.4 of CIMAR give the arbitrator all the powers in s.41(3) to s.41(7). Under Rule 8 of the ICE Arbitration Procedure each party must serve its statement of case, disclose documents as may be ordered, and provide clarification of its case as ordered. Rule 8.4 then provides for a peremptory order to be made for failure to comply with an order under this rule with consequences similar to s.41(7) for non-compliance. S.41(3) is substantially reproduced with one difference: the inordinate and inexcusable delay may be by either party in pursuing its claim. Whilst this seeks to make it clear that a counterclaim may be struck down for a similar reason, that 'clarification' is unnecessary since s.82 makes it plain that 'claimant' includes 'counterclaimant'. Rule 13.9 has similar effect to s.41(4)(a). The rules do not appear to provide for failure to comply with an order to provide security, so s.41(6) would appear to apply by default. The CIArb Rules merely confirm, through Arts.7.1 and 8.1, that the arbitrator has the s.41 powers.

Of the international rules, the LCIA Rules do not deal with delay in pursuing a claim directly, but do deal with failure to submit pleadings and documents: 'The arbitral tribunal may nevertheless proceed with the arbitration and make an award'. There is no peremptory order procedure expressed. By Art.25.2, where there is a failure to comply with an order for security, there may be a stay or the defaulting party's claims may be dismissed.

In the ICC Rules, Art.21(2) provides:

> 'If any of the parties, although duly summoned, fails to appear without valid excuse, the Arbitral Tribunal shall have the power to proceed with the hearing.'

The other ground of the section is not covered by the Rules and, of course, there is no power to order security for costs.

The UNCITRAL Arbitration Rules specifically deal with default at Art.28. Failure by the claimant to communicate his claim timeously leads to termination, whilst failure by the respondent to communicate his defence leads to an order that the proceedings continue without it. The balance of the article has broadly the same effect as s.41(3). There is no peremptory order regime.

We do not think that the effect of incorporating into an arbitration agreement rules which are silent on one or more topics covered by s.41 is to deprive a tribunal of the powers this section purports to give: see our commentary in paragraph 4G above.

Powers of Court in Relation to Arbitral Proceedings

Section 42 – Enforcement of Peremptory Orders of Tribunal

42.–(1) Unless otherwise agreed by the parties, the court may make an order requiring a party to comply with a peremptory order made by the tribunal.

(2) An application for an order under this section may be made–

(a) by the tribunal (upon notice to the parties),
(b) by a party to the arbitral proceedings with the permission of the tribunal (and upon notice to the other parties), or
(c) where the parties have agreed that the powers of the court under this section shall be available.

(3) The court shall not act unless it is satisfied that the applicant has exhausted any available arbitral process in respect of failure to comply with the tribunal's order.

(4) No order shall be made under this section unless the court is satisfied that the person to whom the tribunal's order was directed has failed to comply with it within the time prescribed in the order or, if no time was prescribed, within a reasonable time.

(5) The leave of the court is required for any appeal from a decision of the court under this section.

Definitions

'agreed': s.5(1). .
'available arbitral process': s.82(1).
'the court': s.105.
'notice (or other document)': s.76(6).
'party': ss.82(2), 106(4).

'peremptory order': s.82(1), and see s.41(5).
'upon notice' (to the parties or the tribunal): s.80.

[42A] **Status**

This is a new section. It may be seen as replacing s.5 of the Arbitration Act, 1979. It is *non-mandatory*, (s.4(2)).

[42B] **Summary**

The previous section (s.41) gives the tribunal the power to make peremptory orders, and gives it sanctions which can be applied if the peremptory orders are not complied with. This section permits the court to supplement the sanctions available to the tribunal by applying those sanctions that are available to the court for breach of a court order. For example, the court would be able to fine a party, or send him to prison for contempt.

At para. 212 of their Report, the DAC said : 'In our view there may well be circumstances where in the interests of justice, the fact that the court has sanctions which in the nature of things cannot be given to arbitrators (e.g. committal to prison for contempt) will assist the proper functioning of the arbitral process'. For our part, we have difficulty in envisaging circumstances where it will be necessary for tribunal or party to look beyond the powers available to the tribunal in s.41.

A possible example might be where a party refused to comply with a peremptory order for discovery and was prepared to suffer such sanctions as the tribunal could impose; however the continuing non-availability of the documents affected another party's right to recover. Only the threat of imprisonment might actually produce the documents.

Nevertheless, we doubt whether in practice, much use will be made of this section.

Points

[42C] *Availability – Subss.(1) to (4)*
The court may make orders under this section unless it is agreed that it cannot. Nonetheless, certain restrictions are placed on the court's power.

Under subs.(2), the power can only be invoked at the behest of the tribunal, or by a party with the tribunal's permission, unless the parties have positively agreed that the power will be available. We find it difficult to see why arbitrators should themselves apply and thus expose themselves to costs risks. If no party wants the court's help, why should the arbitrators?

Next, the court is only entitled to act if courses available without reference to the court have been exhausted, (subs.(3)).

Finally, the court may not exercise its discretion unless satisfied that the party affected is indeed in default, (subs.(4)).

The provision, 'or, if no time was prescribed, within a reasonable time', initially seems odd since a peremptory order is hardly ever likely to be made without a time limit, and indeed the essence of s.41(5) is that a peremptory order prescribes 'such time for compliance with it as the tribunal considers appropriate'. The provision may be intended to cover a 'forthwith' order. This contains no specific time limit, but because absolutely instantaneous compliance is impossible, *some* time must be allowed (even if only hours or days); and the effect of the provision is that such time must be reasonable.

[42D]   ***Housing Grants, Construction and Regeneration Act 1996***
By virtue of SI 1998 No.649, Sched., Part I, Art.24, this section applies to
the Scheme for Construction Contracts set up under the above Act,
with the following modifications:

> '(a) in subsection (2) for the word "tribunal" wherever it appears
> there shall be substituted the word "adjudicator",
> (b) in subparagraph (b) of subsection (2) for the words "arbitral
> proceedings" there shall be substituted the word "adjudication",
> (c) subparagraph (c) of subsection (2) shall be deleted, and
> (d) subsection (3) shall be deleted.'

[42E]   ***Judge-arbitrators***
Note that if the tribunal consists of or includes a judge-arbitrator, he
may exercise the court's power under this section concurrently with
the High Court, see Sched.2, para.4. However, under s.93 judges may
only be appointed as sole arbitrators or umpires, and the wording of
s.21 suggests that, prior to disagreement, the arbitrators are the tribunal
and therefore the umpire is not part of it. On this approach an arbitral
tribunal could never 'include' a judge-arbitrator within Sched.2, para.4.
We believe, though, that this paragraph would be read more broadly,
so that its plain intention might be given effect.

[42F]   ***The application***
The application under s.42(2) is an arbitration application falling
within the terms of the Practice Direction on Arbitrations under CPR
Part 49. We deal generally with arbitration applications in Part 3,
Materials, Section H. Specifically in respect of an application under
s.42:

(i)   the application may be made by the tribunal; or
(ii)   the application may be made by a party with the permission of
        the tribunal, or without that permission if the parties have
        positively agreed that the powers of the court are available;
(iii)   an application by the tribunal must be on notice to the parties;
(iv)   an application by one party must be on notice to the other
        parties, at least as regards an application under subs.(b);
(v)   the requirement for giving notice to the other party or parties is
        to be met by making those parties respondents to the applica-
        tion and serving on them the arbitration claim form and evi-
        dence in support;
(vi)   the arbitration claim form must show the applicant to have
        exhausted any available arbitral process in respect of the failure
        to comply with the tribunal's order;
(vii)   the evidence in support of the application must give details of
        any permission of the tribunal to the making of the application,
        and exhibit a copy of any document evidencing that agreement
        or permission. Similarly, where the parties have positively

agreed that the power of the court under the section will be available, the party relying on the agreement will need to provide evidence of it to the court. The evidence must also show that the party concerned has failed to comply with an order of the tribunal within the time prescribed or a reasonable time, as appropriate;
(viii) unless otherwise ordered, the hearing will be in private.

[42G] *Appeal – Subs.(5)*
An appeal from the court's decision may only be made with leave.

[42H] *Model Law*
There is no comparable provision.

[42I] *Rules*
CIMAR embraces s.42 at Rule 11.5, but the application may only be made by or with the permission of the arbitrator. The ICE Arbitration Procedure and the CIArb Rules are silent, so the effect is the same as CIMAR. Of the international rules, there is generally no procedure for peremptory orders, so s.42 is inapplicable.

Section 43 – Securing the Attendance of Witnesses

43.–(1) A party to arbitral proceedings may use the same court procedures as are available in relation to legal proceedings to secure the attendance before the tribunal of a witness in order to give oral testimony or to produce documents or other material evidence.

(2) This may only be done with the permission of the tribunal or the agreement of the other parties.

(3) The court procedures may only be used if–

(a) the witness is in the United Kingdom, and
(b) the arbitral proceedings are being conducted in England and Wales or, as the case may be, Northern Ireland.

(4) A person shall not be compelled by virtue of this section to produce any document or other material evidence which he could not be compelled to produce in legal proceedings.

Definitions

'agreement': s.5(1).
'the court': s.105.
'legal proceedings': s.82(1).
'party': ss.82(2), 106(4).

[43A] Status

This section derives from the Model Law, Art.27 and from the Arbitration Act 1950, s.12(4), (5). It is *mandatory*, (s.4(1) and Sched.1).

[43B] Summary

This section makes available to a party to arbitral proceedings conducted in England and Wales or Northern Ireland the court procedures that would be available to parties to proceedings in court to compel the attendance of witnesses who are located in the United Kingdom. Their attendance may be compelled for the purpose of giving oral evidence, or for the purpose of producing documents, or both. The procedure is to issue a witness summons, see CPR Part 34.

Points

[43C] *Availability – Subs.(2)*
The court procedures may only be used with the permission of the tribunal or the agreement of the parties. This ensures that if, for

example, the tribunal has decided that there should be no oral evidence, or that certain documentary evidence should not be admitted, these procedures may not be used to make that decision ineffective, unless the parties agree otherwise.

Arbitrators should not unreasonably refuse permission to issue a witness summons, bearing in mind the s.33 duty to afford each party a reasonable opportunity to put its case.

[43D] *Procedures*

The party who wishes to obtain a witness summons must apply to the Admiralty and Commercial Registry in London or, if the attendance of the witness is required within the district of a District Registry, he may apply at that Registry. He must file a witness statement or affidavit which shows that the application is made with the permission of the tribunal or the agreement of the other parties. For further information, see CPR Part 34 and the Practice Direction which supplements it entitled: 'Depositions and Court Attendance by Witnesses'.

[43E] *Privilege – Subs.(4)*

As in legal proceedings, a witness cannot be compelled to produce evidence which is protected by privilege, see CPR Rule 34.2(5). For example, advice given to a party by his lawyers; communications between a party (or his lawyers) and witnesses concerning the arbitration; and 'without prejudice' correspondence between the parties, need not be disclosed.

[43F] *Judge-arbitrators*

Note that if the tribunal consists of or includes a judge-arbitrator, he may exercise the court's power under this section concurrently with the High Court, see Sched.2, para.4. See also our comment at paragraph 42E.

[43G] *Model Law*

Art.27 (Court assistance in taking evidence) is couched in rather wider terms than s.43, but embraces the compulsion of attendance by witnesses:

> 'The arbitral tribunal or a party with the approval of the arbitral tribunal may request from a competent court of this State assistance in taking evidence. The court may execute the request within its competence and according to its rules on taking evidence.'

This Article also contemplates assistance in taking evidence from witnesses overseas by 'Letters of Request' addressed to a foreign court.

[43H] *Rules*

Since the section is mandatory it will be implied in all rules. We would therefore expect English domestic rules since the Act to be silent on the point and CIMAR, the ICE Arbitration Procedure and the CIArb Rules are indeed so. The various international rules we have considered are also silent on this matter, so that no conflict with s.43 arises.

Section 44 – Court Powers Exercisable in Support of Arbitral Proceedings

44.–(1) Unless otherwise agreed by the parties, the court has for the purposes of and in relation to arbitral proceedings the same power of making orders about the matters listed below as it has for the purposes of and in relation to legal proceedings.

(2) Those matters are–

(a) the taking of the evidence of witnesses;
(b) the preservation of evidence;
(c) making orders relating to property which is the subject of the proceedings or as to which any question arises in the proceedings–

  (i) for the inspection, photographing, preservation, custody or detention of the property, or
  (ii) ordering that samples be taken from, or any observation be made of or experiment conducted upon, the property;

  and for that purpose authorising any person to enter any premises in the possession or control of a party to the arbitration;

(d) the sale of any goods the subject of the proceedings;
(e) the granting of an interim injunction or the appointment of a receiver.

(3) If the case is one of urgency, the court may, on the application of a party or proposed party to the arbitral proceedings, make such orders as it thinks necessary for the purpose of preserving evidence or assets.

(4) If the case is not one of urgency, the court shall act only on the application of a party to the arbitral proceedings (upon notice to the other parties and to the tribunal) made with the permission of the tribunal or the agreement in writing of the other parties.

(5) In any case the court shall act only if or to the extent that the arbitral tribunal, and any arbitral or other institution or person vested by the parties with power in that regard, has no power or is unable for the time being to act effectively.

(6) If the court so orders, an order made by it under this section shall cease to have effect in whole or in part on the order of the tribunal or of any such arbitral or other institution or person having power to act in relation to the subject-matter of the order.

(7) The leave of the court is required for any appeal from a decision of the court under this section.

Definitions

'agreed': s.5(1).
'agreement in writing': s.5(2) to (5).
'the court': s.105.
'legal proceedings': s.82(1).
'notice (or other document)': s.76(6).

'party': ss.82(2), 106(4).
'premises': s.82(1).
'upon notice' (to the parties or the tribunal): s.80.

[44A]   **Status**

This section derives in part from the Model Law, Art.9, in part from the Arbitration Act 1950, s.12(6)(d) to (h), and in part it is new. It is *non-mandatory*, (s.4(2)).

[44B]   **Summary**

This section gives powers to the court in support of an arbitration such as it has for the purposes of court proceedings, (subs.(1)). To some extent, these powers run parallel to corresponding powers of the tribunal (as to which, see s.38). Generally, the court's powers are wider, and in particular, extend to certain areas where it would not be appropriate or possible for the tribunal to act.

   One significant exception is the power to order security for costs, which is conferred exclusively on the tribunal, (s.38(3)).

Points

[44C]   *Agreement by the parties – Subs.(1)*
The court has the powers set out in subs.(2) unless the parties specifically agree that the court shall not have any or certain of these powers, or they make an agreement which is in some other way incompatible or inconsistent with the court having these powers. The parties may, of course, exclude certain of these powers, whilst retaining others.

   Obviously, they cannot by agreement confer *different* powers on the court, as they can in relation to the tribunal under s.38.

[44D]   *Evidence – Subss.(2)(a), (2)(b)*
Under subs.(2)(a) the court may make orders in respect of the taking of the evidence of witnesses. Whilst, under s.43, a party to an arbitration may use the witness summons (CPR Part 34) process to obtain testimony or documents from a witness provided the witness is in the United Kingdom and the proceedings are being conducted in England, Wales or Northern Ireland, that will not meet the requirements of all situations. In particular, it may be necessary to secure the evidence of witnesses overseas. The court has power in this respect under CPR Part 34.

   Under subs.(2)(b) the court has the power to order the preservation of evidence. This is essentially a reference to measures involving the urgent search of a party's premises and the seizure of materials found there (formerly known as an 'Anton Piller' order, after the case of that name, and now embodied in the Civil Procedure Act 1997 s.7 and Part 25 of the CPR). The power will often be used without notice to the other party, but the applicant will have to justify making the application 'without notice' to the court.

[44E]   *Property – Subs.(2)(c)*
Under this subsection the court has power to make orders in respect of property the subject of proceedings or as to which any question arises in the proceedings. These powers extend to property in the hands of third parties, and the court's powers in this respect are to be found in the CPR Rule 25.1 and s.34(3) of the Supreme Court Act 1981. Note that the court can authorise entry by 'any person' into 'premises' in the possession or control of a party to the arbitration. By virtue of s.82(1), 'premises' includes 'vehicles, vessels, aircraft and hovercraft'.

In a case where arbitrators had ordered the sellers of a yacht which was the subject of an arbitration to store it, insure it and not deal with it commercially until after the publication of their award, the court supported this order by making a similar order itself against an undertaking in damages from the applicant: *Copsa Enterprises Ltd* v. *Tecnomarine SpA* (unreported, Cresswell J, 25 September 1998).

Contrast the tribunal's corresponding powers, under s.38(4), which are restricted to directions in relation to property in the hands of the parties, which only the tribunal, an expert or a party may inspect, photograph, and so on.

[44F]   *Sale of goods – Subs.(2)(d)*
Under this subsection, the court may order the sale of perishable goods. The power is to be found in the CPR Rule 25.1.

[44G]   *Interim injunction/receiver – Subs.(2)(e)*
Under this subsection the court may grant an interim injunction or appoint a receiver.

The circumstances in which a court may grant an injunction are, of course, too numerous to set out here. Of most significance is the 'freezing' injunction, derived from the 'Mareva' injunction, designed to secure and protect assets so as to prevent their dissipation within and outside the jurisdiction, resulting in the frustration of any eventual award, see CPR Rule 25.1.

The court's jurisdiction in respect of ancillary, as distinct from substantive proceedings, for example the granting of freezing orders (formerly known as 'Mareva' injunctions) is not excluded by an arbitration clause which provides for an arbitration tribunal to 'have exclusive jurisdiction': *Re Q's Estate* [1999] 1 Lloyd's Rep. 931; nor, unsurprisingly, by rules which prohibited legal action 'other than to obtain security for any claim': *A. Meredith Jones & Co. Ltd* v. *JSC Innovatsia* (unreported, Toulson J, 17 February 1999). On the other hand, in *Tsakos Shipping & Trading SA* v. *Orizon Tanker Co. Ltd* [1998] CLC 1003 the court was plainly troubled as to whether there was jurisdiction under s.44 to make an order for inspection of property (a ship) where no cause of action might have arisen, but did not have to decide the point. It did, however, refuse to exercise its discretion to require such an inspection because to do so would have been likely to amount in effect to re-writing the parties' contract, and that already contained carefully negotiated terms as to inspection.

As for the appointment of receivers, this power was available under s.12(6)(h) of the 1950 Act. It is rarely used, and is almost solely confined to partnership disputes, the essence of the receiver's function being to deal impartially with property which the court considers should not be in the control of either party, pending the outcome of the dispute. The court's powers in respect of receivers are to be found in Order 30 of the Supreme Court Rules, which is retained by the CPR in amended form (see CPR Schedule 1).

[44H]   *Availability – Subss.(3) to (5)*
The court's powers are subject to a number of restrictions designed to prevent application to the court rather than the tribunal where the latter is competent to act, and to leave the control of the arbitral process in the hands of the tribunal so far as possible.

In cases of urgency the court may act to preserve evidence or assets, even where the arbitration has not been started, (subs.(3)). Indeed, this power would be particularly relevant before the tribunal has been appointed. Where, however, the situation is not urgent, subs.(4) – which presupposes that the arbitration has started – provides that the court is only entitled to act with the permission of the tribunal or the agreement of the parties.

In any event, the court is only entitled to act to the extent that the tribunal and any arbitral institution is unable to do so effectively, or at all, (subs.(5)). So if the tribunal is constituted, has a power corresponding to that of the court and can use it effectively, the parties are bound to apply to the tribunal.

We consider that the powers granted by this section are designed to be used, as the section's title implies, in support of arbitral proceedings and not, for example, in relation to securing a party's claim. Thus, for example, we do not think that a tribunal's permission is required under s.44(4) before a party may register a caveat against the release of a ship which is under arrest, such caveat being filed in an attempt to secure that party's claim.

[44I]   *Tribunal assuming control – Subs.(6)*
After the court's intervention, control may be handed over to the tribunal (or any relevant institution or person) by the court giving the tribunal (institution or person) the power to put an end to its order. Such a procedure might be appropriate, for example, where a tribunal had been temporarily unable to act (because it had not yet been constituted, or because a member was missing) so that the court was required to intervene. When the tribunal was later constituted, or again complete, and had the relevant power to act, it could assume control in place of the court.

[44J]   *The application*

The application under s.44 is an arbitration application falling within the terms of the Practice Direction under CPR Part 49. We deal generally with arbitration applications in Part 3, Materials, Section H. Specifically in respect of an application under s.44:

(i)   unless the case is one of urgency, the application must be on notice to the tribunal and the other parties and with the permission of the tribunal or the agreement in writing of the other parties;

(ii)   the requirement for giving notice to the other parties is to be met by making those parties respondents to the application and serving on them the arbitration claim form and evidence in support;

(iii)   the requirement for giving notice to the tribunal is met by sending copies of the arbitration claim form to the arbitrators for their information at their last known addresses together with copies of the evidence in support;

(iv)   the arbitration claim form must show that the statutory requirements of ss.44(4) and (5) have been complied with;

(v)   where the application is made with the written agreement of the other parties or the permission of the tribunal, the affidavit or witness statement in support must give details of the agreement or permission, and exhibit a copy of any document evidencing it;

(vi)   in cases of urgency, the application may be made without notice and before issue of a claim form. In such a case the affidavit in support must include the reasons why the application is made without notice, and why it was not practicable to obtain the permission of the tribunal or agreement of the other parties (if that permission or agreement has not been obtained);

(vii)   in all cases the affidavit must further state why the witness believes the tribunal (or other body normally vested with relevant power) has no power or is unable for the time being to act effectively;

(viii) unless otherwise ordered, the hearing will be in private.

[44K]   *Judge-arbitrators*

Note that if the tribunal consists of or includes a judge-arbitrator, he may exercise the court's power under this section concurrently with the High Court, see Sched.2, para.4. See also our comment at paragraph 42E.

[44L]   *Appeal – Subs.(7)*

An appeal from the court's decision may only be made with leave.

[44M]   *Model Law*

Art.9 (Arbitration agreement and interim measures by court) provides:

'It is not incompatible with an arbitration agreement for a party to request, before or during arbitral proceedings, from a court an interim measure of protection and for a court to grant such measure.'

We note that for English lawyers this provision is vague; whereas the Act defines what the interim measures may be. As with the Act, the Model Law measures are available, to some extent, before the arbitration is under way, although the Act appears to confine such orders to those necessary to preserve evidence or assets.

[44N] *Rules*

CIMAR, the ICE Arbitration Procedure and the CIArb Rules are unsurprisingly silent on the matter so that s.44 applies. Neither the LCIA Rules (Art.25.3) nor the ICC Rules (Art.23.2) place any bar on applications to court for interim or conservatory measures (except in relation to security for legal costs – see Art.25.3 of the LCIA Rules) but consider them exceptional after formation of the tribunal. Art.26.3 of the UNCITRAL Rules provides (in line with the Model Law) that: 'A request for interim measures addressed by any party to a judicial authority shall not be deemed incompatible with the agreement to arbitrate, or as a waiver of that agreement.' Since there is no agreement otherwise, s.44 applies.

Section 45 – Determination of Preliminary Point of Law

45.-(1) Unless otherwise agreed by the parties, the court may on the application of a party to arbitral proceedings (upon notice to the other parties) determine any question of law arising in the course of the proceedings which the court is satisfied substantially affects the rights of one or more of the parties.

An agreement to dispense with reasons for the tribunal's award shall be considered an agreement to exclude the court's jurisdiction under this section.

(2) An application under this section shall not be considered unless-

(a) it is made with the agreement of all the other parties to the proceedings, or
(b) it is made with the permission of the tribunal and the court is satisfied-

   (i) that the determination of the question is likely to produce substantial savings in costs, and
   (ii) that the application was made without delay.

(3) The application shall identify the question of law to be determined and, unless made with the agreement of all the other parties to the proceedings, shall state the grounds on which it is said that the question should be decided by the court.

(4) Unless otherwise agreed by the parties, the arbitral tribunal may continue the arbitral proceedings and make an award while an application to the court under this section is pending.

(5) Unless the court gives leave, no appeal lies from a decision of the court whether the conditions specified in subsection (2) are met.

(6) The decision of the court on the question of law shall be treated as a judgment of the court for the purposes of an appeal.

But no appeal lies without the leave of the court which shall not be given unless the court considers that the question is one of general importance, or is one which for some other special reason should be considered by the Court of Appeal.

Definitions

'agreement', 'agreed': s.5(1).
'the court': s.105.
'notice (or other document)': s.76(6).
'party': ss.82(2), 106(4).

'question of law': s.82(1).
'upon notice' (to the parties or the tribunal): s.80.

[45A]   Status

This section derives from the Arbitration Act 1979, s.2 with some important changes. It is *non-mandatory*, (s.4(2)). Note the absence of restriction in respect of domestic arbitration (see our comment at paragraph 45J below), so that the section applies to both domestic and international arbitrations.

[45B] **Summary**

This section preserves the existing power of the court to intervene during the course of an arbitration and before an award has been made, for the purpose of making a preliminary determination of a point of law.

The usefulness of the section would arise, for example, where a particular point of law is central to the arbitration, and an authoritative decision one way or the other will effectively determine the whole or a large part of the dispute between the parties. It may also arise where a major event affects a large number of arbitrations, and early and definitive consideration by the court would assist a large number of parties to different proceedings, subject of course to the proviso that there must be a substantial effect on the rights of one or more parties to the arbitration in question.

Points

[45C] *Question of law – Subs.(1)*
Note that a 'question of law' means, for a court in England and Wales, a question of the law of England and Wales; and for a court in Northern Ireland, a question of the law of Northern Ireland, (s.82(1)).

[45D] *Exclusion agreement – Subs.(1)*
The nature of the section coupled with the general principle that court intervention should be kept to the minimum (s.1(c)), means that the court's power is subject to a number of restrictions, and may be excluded altogether if the parties so agree. They are taken to have so agreed if they agree to dispense with reasons for the tribunal's award (which step would also have the effect of precluding an appeal against the award on a point of law, see s.69(1)).

[45E] *Availability – Subss.(1) to (3)*
An application, which in all cases must identify the question of law to be determined (subs.(3)), must be made either with the agreement of all parties or with the permission of the tribunal (subs.(2)). In the latter case, the court must additionally be satisfied that the determination sought is likely to produce substantial savings in costs and the application has been made without delay (subs.(2)). The application must also state the grounds on which it is said the question of law should be decided by the court (subs.(3)). No appeal lies without leave from the court's decision as to whether the subs.(2) conditions have been met.

Note that with regard to delay, s.40(2)(b) refers to this section as a specific example of the general duty of the parties to proceed with expedition. Delay is likely to be measured from the time when the question of law might first be identified, such as the close of any pleading stage. If substantial progress has since been made in the arbitration, and particularly if there have been steps towards the determination of the question of law in the course of the arbitration itself, it is, in our view, likely that the court will refuse the request to hear the application.

In all cases – and this is an important change from s.2 of the 1979 Act – the court must be satisfied that the question of law substantially affects the rights of one or more of the parties, (subs.(1)). There is thus introduced into this procedure a criterion that formerly applied (s.1 of the 1979 Act) and continues to apply (s.69(3)(a)) to an appeal against an award on a question of law.

It is noteworthy, in our view, that there is no provision allowing such applications where the question of law thrown up is a matter of 'public interest' or 'importance', but has only a limited impact on the parties to the arbitration in question. Such a question of law would not qualify for preliminary determination by the court.

[45F] *Continuing arbitration – Subs.(4)*
The tribunal may continue the arbitral proceedings and make an award while an application under this section is pending, although the parties may agree that it should not do so. The tribunal might seek to continue the proceedings if it did not concur with an application made by the parties (in which case they would presumably agree that it should not do so), or if there were aspects of the case which could conveniently be dealt with regardless of the answer to the question of law.

[45G] *The application*
The application under s.45 is an arbitration application falling within the terms of the Practice Direction on arbitrations under CPR Part 49. We deal generally with arbitration applications in Part 3, Materials, Section H. Specifically in respect of an application under s.45:

(i)   the application must be on notice to the other parties;
(ii)   the requirement for giving notice to the other parties is to be met by making those parties respondents to the application and serving on them the arbitration claim form and evidence in support;
(iii) the arbitration claim form must identify the question of law to be determined and, unless made with the agreement of all other parties to the proceedings, must state the grounds on which it is said that the question should be decided by the court;
(iv)   where the application is made with the written agreement of all the other parties to the arbitral proceedings or with the permission of the tribunal, the evidence in support must give details of the agreement or permission and exhibit a copy of any document evidencing the agreement or permission;
(v)   where the application is made without the agreement of the

other parties, but with the permission of the tribunal the affidavits or witness statements filed by either party must include
such evidence as is relied upon by either in support of their
respective contentions as to whether the determination of the
question is likely to produce substantial savings in costs and
whether the application was made without delay;

(vi)   where the application is made without the agreement of the
other parties, but with the agreement of the tribunal, after receipt
of the evidence, the court will consider initially whether or not it
will entertain the application, and it will normally do so without
a hearing. If necessary, it may direct a hearing;

(vii)   when the court hears the substantive application, unless it
orders to the contrary, it does so publicly.

[45H]   *Appeals – Subss.(5), (6)*

The opportunities for appeal from decisions of the court in relation to
this section are highly restricted. Not only is leave required on all
issues, but where the appeal is from the determination of the question
of law, the court must additionally affirm that there is a question of
general importance or some other special reason justifying the consideration of the Court of Appeal.

Note that if the tribunal is a judge-arbitrator the initial determination
of the question of law will be made by the Court of Appeal, and any
appeal will lie to the House of Lords, (Sched.2, para.2).

[45I]   *Model Law*

There is no comparable provision in the Model Law.

[45J]   *Rules*

Had s.87 been enacted it would have been impossible to exclude the
effect of this section in domestic arbitrations unless the agreement to
exclude the jurisdiction of the court was made after the commencement
of the arbitral proceedings in which the question arose. Accordingly it
is no surprise that CIMAR, the ICE Arbitration Procedure and the
CIArb Rules do not contain exclusion agreements. However, the
CIArb's Controlled Costs Arbitration Rules do contain such a clause.

Of the international rules, Art.26.9 of the LCIA Rules contains an
irrevocable waiver of the right to appeal but does not apparently
exclude the right to ask the court to rule on a preliminary issue. Neither
the ICC Rules nor the UNCITRAL Rules contain any exclusion, so that
s.45 applies.

The Award

Section 46 – Rules Applicable to Substance of Dispute

46.–(1) The arbitral tribunal shall decide the dispute–

(a) in accordance with the law chosen by the parties as applicable to the substance of the dispute, or

(b) if the parties so agree, in accordance with such other considerations as are agreed by them or determined by the tribunal.

(2) For this purpose the choice of the laws of a country shall be understood to refer to the substantive laws of that country and not its conflict of laws rules.

(3) If or to the extent that there is no such choice or agreement, the tribunal shall apply the law determined by the conflict of laws rules which it considers applicable.

Definitions

'agreement', 'agree', 'agreed': s.5(1).
'dispute': s.82(1).
'party': ss.82(2), 106(4).

[46A] ### Status

This section is derived from the Model Law, Art.28. It is *non-mandatory*, (s.4(2)).

[46B] ### Summary

This section concerns what law or rules the tribunal should apply in order to decide the actual issues in the dispute. This broadly depends upon whether the parties have reached agreement on the question; and if so, whether they have chosen a system of law or some other set of rules.

Points

[46C] *Choice of law – Subss.(1)(a), (2)*
One possibility is that the parties will have chosen a particular system of law to govern the substantive dispute. If so, then the tribunal must apply that system of law, (subs.(1)(a)).

In this context, the system of law means the substantive laws of a country (for example, the English law relating to contracts) and not its conflict of laws rules (for example, the English rules relating to which system of law governs individual aspects of contracts when there is more than one possibility), (subs.2). If the reference were to conflict of laws rules, then those rules might themselves point to a different system of law from that which the parties had chosen.

Note that in *Wealands* v. *CLC Contractors Ltd* [1999] Lloyd's Rep. 739 the Court of Appeal held that there was no reason why s.46(1) did not expressly preserve and confer the same general jurisdiction, powers and duty as arbitrators were recognised by the courts to possess by implication under the previous legislation, including the power to award contribution, in *Chandris* v. *Isbrandtsen Moller* [1951] 1 KB 240 and *President of India* v. *La Pintada Cia Nav. SA* [1985] AC 104.

[46D] **'Other considerations' – Subs.(1)(b)**

The phrase 'in accordance with such other considerations as are agreed by them' contemplates the possibility of the parties agreeing that the dispute should be decided not in accordance with any system of law, but in accordance with, for example, equity and good conscience. The tribunal would be permitted to apply its general sense of what is fair and just, rather than strict legal rules.

Such an approach is variously described as arbitration in accordance with an equity clause, or *ex aequo et bono*, or with the tribunal acting as 'amiable compositeur'. The parties may leave it to the tribunal to determine the form of such an equity-based approach.

Since such an approach applies no system of law, it follows that no question of law would arise for decision, preliminary determination by the court, (s.45), or appeal, (s.69).

Note that in the context of statutory arbitrations, this subsection does not apply, (s.96(4)).

[46E] *A note of caution regarding subs.(1)(b)*

Legal advisers and potential parties should think carefully about the consequences of an agreement other than one which requires the tribunal to decide in accordance with law. Not only will there be no appeal, but there is in our view substantial scope for divergence as between parties and legal advisers as to precisely what may be intended. For instance, the expression 'amiable compositeur' may be well known, but it will mean slightly different things to different people depending on their legal background. Since it is defined in some jurisdictions, it may be thought by some to mean something very specific. We would advise the greatest of caution in the use of this subsection.

If it is to be used, we further suggest that it would be wise to avoid (as did the DAC) the Latin and French formulas. Agreements under this section should set out clearly the considerations to be taken into account, including (and perhaps particularly) the extent to which *any* law is to be applied, and the precise extent to which the *tribunal* may determine what the considerations are to be.

It should also be noted that the Commencement Order made under s.109(2), which brought the Act into effect on 31 January 1997, provided that in the application of subs.(1)(b) to an arbitration agreement made before 31 January 1997, the agreement is to have effect in accordance with the rules of law (including any conflict of laws rules) as they stood immediately before that date.

[46F] **Default situation – Subs.(3)**

Insofar as the parties have made no positive choice either of a system of law or of an equity-based approach, then the tribunal must apply the system of law which it considers applicable. In this context, the system of law means that determined by the relevant conflict of laws rules. The tribunal must characterise the dispute or issues which it has to determine, apply the relevant conflict of laws rules and thereby arrive at the substantive law that is to be applied.

Plainly, this may be a complex and difficult task, and there is much in favour of the parties making a choice of substantive law or agreeing on an alternative approach.

[46G] **Model Law**

Art.28 (Rules applicable to substance of dispute) provides:

> '(1) The arbitral tribunal shall decide the dispute in accordance with such rules of law as are chosen by the parties as applicable to the substance of the dispute. Any designation of the law or legal system of a given State shall be construed, unless otherwise expressed, as directly referring to the substantive law of that State and not to its conflict of laws rules.
> (2) Failing any designation by the parties, the arbitral tribunal shall apply the law determined by the conflict of laws rules which it considers applicable.
> (3) The arbitral tribunal shall decide ex aequo et bono or as amiable compositeur only if the parties have expressly authorised it to do so.
> (4) In all cases, the arbitral tribunal shall decide in accordance with the terms of the contract and shall take into account the usages of the trade applicable to the transaction.'

We note that apart from para.(4), the effect of Art.28 is similar to that of s.46. The Model Law uses the Latin and French formulas *ex aequo et bono* and amiable compositeur. The DAC specifically decided not to use these expressions, and s.46(b) clearly does not confine the parties to any pre-existing formulas or definitions.

The Model Law requires the tribunal *in all cases* to take into account 'the usages of the trade applicable to the transactions'. There is some danger that these might override the law otherwise applicable which the parties intended should apply. Furthermore, of course, the Act can and often will apply to non-commercial matters where there is no relevant 'trade'. The Act, quite correctly, in our view, omits this provision.

[46H] *Rules*

CIMAR, the ICE Arbitration Procedure and the CIArb Rules are, not surprisingly, silent on the ground covered by s.46. Arts.22.3 and 22.4 of the LCIA Rules and Art.17 of the ICC Rules cover the ground to broadly similar effect. Art.33 of the UNCITRAL Arbitration Rules is broadly similar to Art.28 of the Model Law above, and again has much the same effect as s.46.

Section 47 – Awards on Different Issues etc.

47.–(1) Unless otherwise agreed by the parties, the tribunal may make more than one award at different times on different aspects of the matters to be determined.

(2) The tribunal may, in particular, make an award relating–

(a)   to an issue affecting the whole claim, or
(b)   to a part only of the claims or cross-claims submitted to it for decision.

(3) If the tribunal does so, it shall specify in its award the issue, or the claim or part of a claim, which is the subject matter of the award.

Definitions

'agreed': s.5(1).
'party': ss.82(2), 106(4).

[47A]   Status

This section is derived from the Arbitration Act 1950, s.14. It is *non-mandatory*, (s.4(2)).

[47B]   Summary

This section preserves the tribunal's right, which existed prior to the Act, to make awards on different issues, often referred to as 'interim' awards.

It is important to appreciate that the awards to which this section refers, although made prior to the conclusion of the arbitral proceedings, are nonetheless final as to the matters which they determine, (s.58(1)). They should be distinguished from 'provisional orders', (s.39). The latter are made on a temporary basis only, and are subject to adjustment on the final adjudication.

It is open to the parties to agree that the tribunal should not have such a power, (subs.(1)).

Points

[47C]   *Effect*
The facility to make awards on different issues is an important tool with which the arbitrator can implement his duty to 'adopt procedures suitable to the circumstances of the particular case ...', (s.33(1)(b)). It is also very flexible. We list some examples.

The procedure may be applied in the early determination of an issue

229

fundamental to the whole claim, (subs.(2)(a)), which may enable the parties to go away and settle their remaining differences. It may be applied to deal progressively with parts of the claim that are capable of being decided at an early stage, (subs.(2)(b)), thus reducing and clarifying the issues remaining at large at the hearing stage. It may be applied to one stage of a series of arguments, the answer possibly eliminating the need to progress to the remaining points. And so on.

In a case of the *Kostas Melas* type ([1981] 1 Lloyd's Rep. 18) where money withheld is prima facie due to the claimant and the respondent cannot show it is acting bona fide in seeking to set off, it may also be appropriate for the tribunal to make an award on the claim ahead of the award on the counterclaim.

[47D] *Drafting award – Subs.(3)*
Note that if the tribunal makes interim awards it is required clearly to identify the subject matter of each interim award, within that award.

[47E] *Model Law*
There is no comparable provision in the Model Law, although it is clearly contemplated by Art.32(1) that there may be more than one award: 'The arbitral proceedings are terminated by the final award...'.

[47F] *Rules*
CIMAR provides that the arbitrator has the powers set out in s.47 (Rule 12.1). In addition the arbitrator may, by Rule 12.2, decide what are the issues or questions to be determined and decide whether or not to give an award on part of the claims submitted. He may also make an order for provisional relief (s.39). Rule 19.1 of the ICE Arbitration Procedure confirms that the arbitrator may 'at any time' make an award, otherwise reproducing s.47(1). The CIArb Rules do not specifically cover the topic, but provide in Art.7.1 that the arbitrator shall have all the powers given by the Act, which must include those under this section.

Of the international rules, LCIA Rules Art.26.7 permits the tribunal to make separate awards on different issues at different times. Art.2 of the ICC Rules defines 'Award' to include an interim, partial or final award although there is no specific provision dealing therewith. Art.32.1 of the UNCITRAL Arbitration Rules states that: 'In addition to making a final award, the arbitral tribunal shall be entitled to make interim, interlocutory, or partial awards.' The terminology in the ICC and UNCITRAL Arbitration Rules, in our experience, can cause confusion. We regard the power under s.47 as equivalent to one to make a partial award and see no restriction on the application of the power as expressed in the Act in any of the international rules.

Section 48 – Remedies

48.–(1) The parties are free to agree on the powers exercisable by the arbitral tribunal as regards remedies.

(2) Unless otherwise agreed by the parties, the tribunal has the following powers.

(3) The tribunal may make a declaration as to any matter to be determined in the proceedings.

(4) The tribunal may order the payment of a sum of money, in any currency.

(5) The tribunal has the same powers as the court–

(a) to order a party to do or refrain from doing anything;
(b) to order specific performance of a contract (other than a contract relating to land);
(c) to order the rectification, setting aside or cancellation of a deed or other document.

Definitions

'agree', 'agreed': s.5(1).
'the court': s.105.
'party': ss.82(2), 106(4).

[48A] Status

This section derives in part from the Arbitration Act 1950, s.15. It is *non-mandatory*, (s.4(2)).

[48B] Summary

This section deals with the powers of the tribunal so far as the remedies which it may order are concerned. It essentially provides that the parties are at liberty to agree what powers the tribunal should exercise. These powers are not restricted to those that are available to the court in court proceedings. It is therefore possible for the parties to agree that the tribunal should have different, and even greater, powers than the court.

They may, for example, agree that the tribunal should be able to use a remedy on different grounds from those on which it would be available to the court; or that the tribunal should be able to adopt remedies that are known only in other jurisdictions – for example, punitive damages; or that the tribunal should be able to adopt remedies suitable to the type of contract – such as the power to 'open up, review and revise certificates' found in many building contracts. This is a significant instance of the parties to an arbitration being '... free to agree how their disputes are resolved...' (s.1(b)), although, to some extent at least, it reflects the previous position.

The section also sets out the powers which the tribunal is to have if the parties do not agree otherwise. They correspond to certain of the more familiar powers of the court. An agreement 'otherwise' may be in respect of some or all of the powers.

Points

[48C]   *Declaration – Subs.(3)*
A declaration is a determination by the tribunal of contested rights as between the parties. The parties may, for example, dispute the effect of part of a contract, or of a notice given under a contract, or they may contest the meaning of an important clause. A declaration will rule on the essential issue thereby determining the parties' entitlements.

A declaratory award might well be made under s.47(2) in respect of an issue affecting the whole claim (subs.(a)), or in respect of part only of the claims or counterclaims submitted for decision (subs.(b)).

[48D]   *Payment of money – Subs.(4)*
An order for the payment of money will, of course, be the most common remedy used by the tribunal. As with the court, the tribunal has power to order payment in any currency. (It has had such power since *The Kozara* (1973) 2 Lloyd's Rep. 1.)

[48E]   *Injunctions – Subs.(5)(a)*
The tribunal has the same powers as the court in certain other respects. It may order a party positively to do, or to refrain from doing something. This important new power corresponds to the court's power to order a 'mandatory', or positive, injunction, or a 'prohibitory', or negative, injunction. Such remedies are discretionary and, particularly in their positive or 'mandatory' form, require consideration of a number of issues.

Thus before a positive order requiring a respondent party to do something may be made, it must usually be found that the applicant is very likely to suffer serious damage; that monetary compensation would not be an adequate remedy if such damage did happen; and that the cost to the respondent party of carrying out the positive order is not disproportionate to the likely damage to the applicant (unless the respondent party has behaved unreasonably, in which case the cost to him may be irrelevant). If the tribunal does decide to make the order, it must be careful to set out clearly what the respondent party must do: *Redland Bricks* v. *Morris* [1970] AC 652.

It is to be questioned whether a tribunal may act in all respects in the same way as a court can when it issues an injunction, e.g. without notice to the other party and in its absence (*ex parte*); and whether it would have all the court's ancillary powers, e.g. to make temporary orders equivalent to interim injunctions, and to require undertakings in damages. Given the wording of the subsection and the fact that it appears in the part of the Act dealing with the award, we think these questions would be answered in the negative.

[48F]   *Specific performance – Subs.(5)(b)*
Specific performance is a discretionary remedy by which a party in breach of contract is ordered to complete its performance. Its useful-ness arises when the subject matter of the contract is unique or not readily available elsewhere. Examples might be a rare book or a commodity that is in short supply. Monetary compensation would not be an adequate remedy to the claimant, and specific performance may therefore be ordered.

Orders for specific performance are not usually made in the context of contracts for building work, for example, because they would require a high degree of specification of what precisely the respondent contractor had to do and a high level of supervision over a long period of time to ensure that he complied with the order. In addition, monetary com-pensation would usually be an adequate remedy since the claimant could use the money to pay a different contractor to carry out the work.

The position prior to this Act, that specific performance was not available in respect of a contract relating to land, is preserved, (see the Arbitration Act 1950, s.15).

[48G]   *Other orders – Subs.(5)(c)*
The power of rectification of a contract is confirmed. Hitherto this depended upon the wording of the agreement (see *Ashville* v. *Elmer* [1988] 3 WLR 867). Now the situation will be clear.

Rectification is a remedy by which a written contract that does not set out the true agreement between the parties in some important respect may be amended to reflect that agreement. It therefore operates to correct a mistake on the part of the relevant parties that is common to them all.

The remedies of setting aside or cancellation of contracts may be appropriate where other kinds of mistake are alleged. Such remedies may also apply where agreements are challenged by allegations of misrepresentation, duress, illegality and so on.

[48H]   *Model Law*
The Model Law contains no comparable provision, leaving, by Art.28, the dispute to be decided in accordance with the rules of law chosen by the parties as applicable to the substance of the dispute.

[48I]   *Rules*
By Rule 12.6 CIMAR gives the arbitrator the power to order the remedies as set out in s.48. Because of the difficulties that may be

encountered where a party is ordered positively to do something such as carry out work, it additionally provides for supervision:

> '12.7: Where an award orders that a party should do some act, for instance carry out specified work, the arbitrator has the power to supervise the performance or, if he thinks it appropriate, to appoint (and reappoint as may be necessary) a suitable person so to supervise and to fix the terms of his engagement and retains all powers necessary to ensure compliance with the award.'

It will thus be difficult to maintain an argument against specific performance on the basis that it is not possible to supervise the order.

Rule 19.2 of the ICE Arbitration Procedure includes the same powers as are set out in s.48 in different format. The qualification 'same powers as the court' in respect of the subs.(5) matters is missing, but we doubt whether this makes any practical or legal difference.

The CIArb Rules do not specifically deal with the topic, but provide in Art.7.1 that the arbitrator shall have all the powers given by the Act, which must include those under this section.

Of the international rules, the LCIA Rules grant the tribunal specific power to rectify a contract to correct a common mistake but only to the extent permitted by the law of the contract or arbitration agreement (Art.22.1). There is also a specific power to express the award in any currency (Art.26.6). The ICC Rules and the UNCITRAL Arbitration Rules are silent on the remedies that may be granted, so that the s.48 remedies apply.

Section 49 – Interest

49.–(1) The parties are free to agree on the powers of the tribunal as regards the award of interest.

(2) Unless otherwise agreed by the parties the following provisions apply.

(3) The tribunal may award simple or compound interest from such dates, at such rates and with such rests as it considers meets the justice of the case–

(a) on the whole or part of any amount awarded by the tribunal, in respect of any period up to the date of the award;

(b) on the whole or part of any amount claimed in the arbitration and outstanding at the commencement of the arbitral proceedings but paid before the award was made, in respect of any period up to the date of payment.

(4) The tribunal may award simple or compound interest from the date of the award (or any later date) until payment, at such rates and with such rests as it considers meets the justice of the case, on the outstanding amount of any award (including any award of interest under subsection (3) and any award as to costs).

(5) References in this section to an amount awarded by the tribunal include an amount payable in consequence of a declaratory award by the tribunal.

(6) The above provisions do not affect any other power of the tribunal to award interest.

Definitions

'agree', 'agreed': s.5(1).
'party': ss.82(2), 106(4).

[49A] Status

This section derives from the Arbitration Act 1950, ss.19A, 20. It is *non-mandatory*, (s.4(2)).

[49B] Summary

This section concerns the tribunal's powers as to interest both up to the date of the award and beyond that date. The section essentially provides that the parties are at liberty to agree what powers the tribunal should have regarding the award of interest. It sets out the powers which the tribunal will have if the parties do not agree otherwise.

The section preserves the tribunal's powers prior to the Act, but with some important changes, notably that of giving the tribunal power to award compound interest.

Points

[49C] *Background and effect*

The position as to whether interest may be awarded at all varies widely around the world. Thus in Muslim jurisdictions it is generally prohibited, whilst in some civil law countries it is compulsory.

In England, in the absence of a statutory provision, there is no power to award interest as (general) damages, (see *London, Chatham and Dover Railway Co.* v. *South Eastern Railway Co.* [1893] AC 429). Arbitrators were first given such a statutory power in s.19A of the 1950 Act, as inserted into that Act by the Administration of Justice Act 1982.

By this Act the tribunal has now acquired the power to award compound interest where it was previously confined to awarding simple interest. In our view this is remarkable, since, as the House of Lords has recently confirmed in *Westdeutsche Landesbank Girozentrale* v. *Islington London Borough Council*, [1996] AC 669, the courts do not have this power except under their equitable jurisdiction, in very limited circumstances. The possibility of awarding compound interest may have the effect of making arbitration a very attractive option for potential claimants who have the alternative. On the other hand, potential respondents may only be persuaded to arbitrate, where litigation is the primary option, if there is agreement to delete this power.

Another significant change is that interest will no longer automatically accrue on sums awarded, as it did under s.20 of the 1950 Act. So arbitrators will have to remember to consider awarding such interest: in our view they should generally do so.

If the substantive contract between the parties includes terms as to interest, the tribunal must give effect to those terms and award interest pursuant to contract, (subs.(6)). In such cases, relevant contractual terms are not affected by this section.

[49D] *Interest up to the award – Subs.(3)*

The tribunal may award interest on sums claimed but paid before the award, up to the date of payment. It may also award interest on all or any part of the amount awarded, up to the date of the award. In both cases, the questions of simple or compound interest, rates and rests are within the tribunal's discretion, as also is the period of the award of interest, within the parameters of the subsection.

[49E] *Interest beyond the award – Subs.(4)*

The outstanding amount of the award, together with interest up to the date of the award and costs, may all be the subject of an award of interest from the date of the award (or any later date) until payment. The questions of simple or compound interest, rates and rests are again within the tribunal's discretion, as also is the period of the award of interest, within the parameters of the subsection.

As we have already mentioned, because there is no longer any automatic accrual of such interest, it will be essential that arbitrators positively consider and deal with it. The court cannot do so when entering judgment in terms of an award: *Walker* v. *Rome* (unreported, Aikens J, 10 September 1999).

There appears to be nothing to prevent an arbitrator who does not deal with post-award interest in his substantive award from making a separate award in respect of it later. However, for practical reasons if for no others, we do not think that this should be the usual practice. Paying parties should be entitled to know the extent of their liability from the moment an award is notified to them and should not have to speculate as to whether a tribunal will later award post-award interest and if so at what rate, etc.

The subsection prompts the arbitrator to clarify, both for his benefit and that of the parties, the date from when interest will run on the award. Under the previous legislation, a question often arose of the effect of an award which said that payment of a principal sum should be made within, say, 28 days. If payment was made within that period, did interest accrue during that period? Under this provision, the tribunal's power to direct that interest runs, say, from the end of the 28 day period is made clear. We advise arbitrators to indicate very clearly in their award from when interest runs.

It is useful to note that an award of interest on costs may only be made in respect of the period *after* the award. This, although somewhat unfavourable to a successful party, reflects the position in litigation as to period but also improves on it since the award may be of compound interest.

[49F]   *Drafting awards*

We suggest that awards should therefore normally contain orders covering interest to the date of the award on the principal sum or sums; interest on costs; and interest from the date of the award (or from some later date) on the outstanding amount of the award (which will probably include principal and interest).

The operative part of a typical award might read as follows:

'And I order and direct that
(1) The respondent shall pay to the claimant within 14 days the sum of £25,000 together with interest thereon calculated at 10% per annum compounded at quarterly rests from [date] to the date of this award;' [In some cases it may be appropriate to calculate and insert the total interest thus awarded, but the parties can often be left to do the calculation.]
'(2) The respondent shall pay the claimant's costs of the arbitration which, unless agreed, shall be determined by me under s.63(3) of the Arbitration Act 1996 ('the Act') on the basis set out in s.63(5) of the Act, together with interest thereon at 10% per annum

compounded at quarterly rests calculated from the date of publication of this award until payment;

(3) The respondent shall pay my fees and expenses in the sum of
£4,000 plus VAT together with interest thereon at 10% per annum
compounded at quarterly rests calculated from the date of publication of this award until payment, provided that if, in the first
instance, the claimant shall have paid any amount in respect of my
fees and expenses, he shall be entitled to an immediate refund
from the respondent of the sum so paid together with interest
thereon as aforesaid but from the date of the claimant's payment;

(4) In the event that the respondent does not pay the sum of
£25,000 together with interest thereon in accordance with paragraph (1) above within 14 days, compound interest shall thereafter
be payable upon the total sum due under paragraph (1) or such
part of it as remains unpaid at the rate of 10% per annum compounded at quarterly rests until payment.'

See also paragraph 61E.

[49G] *Interest on declaratory award – Subs.(5)*
Where the tribunal has made a declaration, (s.48(3)), rather than an
order for the payment of money, (s.48(4)), it may nevertheless make an
award of interest in respect of the amount that is payable in consequence of the declaration.

[49H] *Other powers – Subs.(6)*
Other powers of the tribunal to award interest are not affected. The two
main other powers are likely to be a contractual power; and the power
to award interest as special damages, for example, 'Minter' interest as
part of an award of loss and expense under a building contract. We
should add that for practical purposes, we largely see this latter form of
interest as superseded by the new statutory provision.

[49I] *Model Law*
There is no comparable provision in the Model Law.

[49J] *Rules*
By Rule 12.8 CIMAR gives the arbitrator all the powers of this section.
As we have indicated in the main text, the arbitrator must be mindful
to deal with both subs.(3) and subs.(4) matters in his awards. The ICE
Arbitration Procedure has a much simpler provision at Rule 19.6 but
we think its effect no different from the Act. The CIArb Rules are silent
on interest, so the section applies.

Of the international rules, Art.26.6 of the LCIA Rules reflects s.49.
Neither the ICC Rules nor the UNCITRAL Arbitration Rules deal with
the matter so that s.49 will apply.

Section 50 – Extension of Time
for Making Award

50.–(1) Where the time for making an award is limited by or in pursuance of the arbitration agreement, then, unless otherwise agreed by the parties, the court may in accordance with the following provisions by order extend that time.

(2) An application for an order under this section may be made–

(a)   by the tribunal (upon notice to the parties), or
(b)   by any party to the proceedings (upon notice to the tribunal and the other parties),

but only after exhausting any available arbitral process for obtaining an extension of time.

(3) The court shall only make an order if satisfied that a substantial injustice would otherwise be done.

(4) The court may extend the time for such period and on such terms as it thinks fit, and may do so whether or not the time previously fixed (by or under the agreement or by a previous order) has expired.

(5) The leave of the court is required for any appeal from a decision of the court under this section.

Definitions

'agreement', 'agreed': s.5(1).   'notice (or other document)': s.76(6).
'arbitration agreement': ss.6, 5(1).   'party': ss.82(2), 106(4).
'available arbitral process': s.82(1). 'upon notice' (to the parties or the
'the court': s.105.   tribunal): s.80.

[50A]   Status

This section derives from the Arbitration Act 1950, s.13(2). It is *non-mandatory*, (s.4(2)).

[50B]   Summary

This section preserves the court's power prior to this Act to extend the tribunal's time for making an award where there is a time limit imposed by or pursuant to the arbitration agreement.

The parties may, by agreement, exclude the court's power in this regard, (subs.(1)).

Points

[50C] *Effect*

The court's power is discretionary, and an extension may be granted even though the time limit (under the arbitration agreement or an earlier order) has passed, (subs.(4)).

However, there are certain new qualifications to the power. The tribunal or any party may apply for an order but only after exhausting any other procedure available without recourse to the court, (subs.(2)). Moreover the court must only grant an extension if satisfied that a substantial injustice would result if it did not do so, (subs.(3)).

Thus, under certain institutional rules (the ICC Rules are a good example) awards must be made within certain time limits. Provision is made for extensions to be granted by the institution. It will be rare, therefore, for such a matter to come before the court since it will normally have been considered by an institution first. Subs.(3) may be read in this light: it will, we suggest, be a very rare case indeed in which the court in effect reverses the decision of the institution.

[50D] *Relationship with s.79*

It is difficult to see how this section differs in its effect from s.79, which gives the court a general power to extend time limits relating to arbitral proceedings. We are not clear as to why this separate power to extend time for making an award was thought necessary.

[50E] *The application*

The application under s.50 is an arbitration application falling within the terms of the Practice Direction under CPR Part 49. We deal generally with arbitration applications in Part 3, Materials, Section H. Specifically in respect of an application under s.50:

(i) the application may be made either by the tribunal (upon notice to the parties), or by any party (upon notice to the tribunal and the other parties);

(ii) the requirement for giving notice to the parties in either case is to be met by making those parties respondents to the application and serving on them the arbitration claim form and evidence in support;

(iii) the requirement for giving notice to the tribunal is met by sending copies of the arbitration claim form to the arbitrators for their information at their last known addresses together with copies of the evidence in support;

(iv) the arbitration claim form must show that any available arbitral process for obtaining an extension of time has been exhausted;

(v) unless otherwise ordered, the hearing will be in private.

[50F] *Appeal – Subs.(5)*

Any appeal from a decision of the court may only be made with leave.

[50G]   *Judge-arbitrators*

Note that if the tribunal is a judge-arbitrator, he may exercise the power conferred by this section himself, in which case any appeal from his decision will lie to the Court of Appeal with the leave of that court, (Sched.2, para.5).

[50H]   *Model Law*

There is no comparable provision.

[50I]   *Rules*

Slow delivery of awards is inevitably an annoyance to the parties. CIMAR places no time limit so that s.50 is of no relevance, but by Rule 12.3 requires the arbitrator to inform the parties of a target date for delivery. Thereafter he must inform the parties of any reason that prevents him making that date. The ICE Arbitration Procedure and the CIArb Rules are silent on the point.

Of the international rules, the ICC Rules (Art.24) require the final award to be rendered within six months of the Terms of Reference. The ICC Court may extend time 'pursuant to a reasoned request from the Arbitral Tribunal or on its own initiative if it decides it is necessary to do so'. A further application under s.50 would therefore be a possibility. The LCIA and UNCITRAL Arbitration Rules are silent on the point.

Section 51 – Settlement

51.–(1) If during arbitral proceedings the parties settle the dispute, the following provisions apply unless otherwise agreed by the parties.

(2) The tribunal shall terminate the substantive proceedings and, if so requested by the parties and not objected to by the tribunal, shall record the settlement in the form of an agreed award.

(3) An agreed award shall state that it is an award of the tribunal and shall have the same status and effect as any other award on the merits of the case.

(4) The following provisions of this Part relating to awards (sections 52 to 58) apply to an agreed award.

(5) Unless the parties have also settled the matter of the payment of the costs of the arbitration, the provisions of this Part relating to costs (sections 59 to 65) continue to apply.

Definitions

'agreed': s.5(1).
'costs of the arbitration': s.59.
'dispute': s.82(1).
'party': ss.82(2), 106(4).

[51A] ## Status

This section is derived from the Model Law, Art.30. It is *non-mandatory*, (s.4(2)).

[51B] ## Summary

The section provides for the making of an agreed award if the parties settle their dispute. It formalises the existing practice of making a 'consent award' in such circumstances.

The section permits the parties, on settlement, to request the tribunal to make an agreed award recording the terms of the settlement, (subs.(2)). It expressly provides that the agreed award shall state that it is an award of the tribunal, and that it shall have the same status and effect as any other award made by the tribunal on the merits of the case, (subs.(3)). The agreed award will therefore be capable of enforcement as a judgment or order of the court in the same manner as any other award, (see s.66).

The parties are capable of excluding or modifying the provisions of the section by agreement.

Points

[51C]   *'During arbitral proceedings' – Subs.(1)*
The provisions apply (subject to contrary agreement) 'during arbitral proceedings'. By virtue of s.14 arbitral proceedings may well commence before an arbitrator or arbitrators are appointed, and sometimes substantially before then. However the wording of s.51 presupposes a tribunal. If a matter is settled before the tribunal has been appointed, in our view the section cannot apply.

[51D]   *Tribunal's right to object – Subs.(2)*
It is noteworthy that the tribunal is given the right to object to making an agreed award, (subs.(2)). The purpose of this right is to allow the tribunal to protect its own position and to refuse to make an award if the settlement contains some objectionable feature that might compromise the tribunal. This could, for example, be an arrangement designed to defraud the tax authorities. Or it might be the inclusion of matters that cannot properly form the subject of arbitration, such as an agreed declaration concerning the marital status of one of the parties.

[51E]   *'An award of the tribunal' – Subs.(3)*
The provisions of this subsection are important to obtain enforcement under s.66(1), which refers specifically to an award 'made by the tribunal'. In our view this requirement will be satisfied by reciting the appointment of the arbitrator(s) and the terms of the award, and by the signature(s) of the arbitrator(s).
    The provisions of this subsection are also likely to be important to assist enforcement overseas. In this context we draw particular attention to Arts. III, IV and V of the New York Convention, parts of which are set out in paragraph 51I below.

[51F]   *Other provisions applying to agreed awards – Subs.(4)*
A number of other sections of this Part of the Act relating to awards (ss.52-58) are expressly made applicable to agreed awards. It should particularly be noted that by virtue of s.52 (which leaves the parties free to agree on the form of the award) an agreed award need not call itself an agreed award. It may therefore outwardly appear the same as an award on the merits, apart from the fact that it will probably not contain reasons, (s.52(4)). Whilst there are often perfectly legitimate reasons for parties requiring an agreed award not to want their consent to appear on the face of the award, this provision does give rise to the possibility of abuse, e.g. in aid of money-laundering operations.

[51G]   *Costs – Subs.(5)*
Where costs have not formed part of the settlement, the provisions of this Part of the Act relating to costs (ss.59-65) continue to apply.

[51H] *Model law*

Art.30 (Settlement) provides:

> '(1) If, during arbitral proceedings, the parties settle the dispute, the arbitral tribunal shall terminate the proceedings and, if requested by the parties and not objected to by the arbitral tribunal, record the settlement in the form of an arbitral award on agreed terms.
>
> (2) An award on agreed terms shall be made in accordance with the provisions of article 31 [Form and contents of award] and shall state that it is an award. Such an award has the same status and effect as any other award on the merits of the case.'

We note that the Model Law and the Act are in similar terms, save that Art.30 would appear to be mandatory.

[51I] *New York Convention*

To obtain enforcement in Convention countries, there must be an award. That award may not fall into certain categories where enforcement may be refused, (Art.V). See s.103 for more detail, but here it is useful to note:

> Art. III
>
> 'Each contracting state shall recognise arbitral awards as binding and enforce them in accordance with the rules of procedure of the territory where the award is relied upon...'
>
> Art. IV
>
> 'To obtain the recognition and enforcement mentioned in the preceding article, the party applying for recognition and enforcement shall, at the time of the application, supply:
>
> (a) The duly authenticated original award or a duly certified copy thereof...'

[51J] *Rules*

Rule 14.4 of CIMAR and Art.11.4 of the CIArb Rules require the parties promptly to inform the arbitrator of any settlement. S.51 then applies. The ICE Arbitration Procedure is silent on the point so, again, s.51 applies. Of the international rules we have considered, each contains a provision to similar effect as s.51(2). In addition, the LCIA Rules have a specific provision at Art.26 providing for the situation where the parties do not seek a consent award. To the extent that the international rules do not cover the matters dealt with in subss.(3) to (5) those subsections will be applicable.

Section 52 – Form of Award

52.–(1) The parties are free to agree on the form of an award.

(2) If or to the extent that there is no such agreement, the following provisions apply.

(3) The award shall be in writing signed by all the arbitrators or all those assenting to the award.

(4) The award shall contain the reasons for the award unless it is an agreed award or the parties have agreed to dispense with reasons.

(5) The award shall state the seat of the arbitration and the date when the award is made.

Definitions

'agreement', 'agree', 'agreed': s.5(1). 'seat of the arbitration': s.3.
'arbitrator': s.82(1). 'in writing': s.5(6).
'party': ss.82(2), 106(4).

Status

[52A]

This section derives from the Model Law, Art.31. It is *non-mandatory*, (s.4(2)).

Summary

[52B]

The section essentially provides that the parties are at liberty to agree on the form an award should take. Insofar as they do not do so, it makes a number of provisions which, for the first time in English law, set out formal requirements for an award, in the process fundamentally changing one aspect of the law prior to this Act by requiring the parties to opt out of (rather than opt into) a *reasoned* award.

Points

[52C] *Writing – Subs.(3)*
Unless the parties agree otherwise, the award must be in writing. This is a new requirement of English law, an oral award, however unusual, still being enforceable under the law prior to the Act.

The award must also be signed either by all the arbitrators, or by all those assenting to the award. Where, for example, the award is made by a majority of the tribunal (as to which, see s.22), it is possible for only that majority (i.e. those assenting to the award) to sign. Alternatively, all the arbitrators may wish to sign, in which case it will be impossible

245

for the parties to discover which of them formed the majority and which dissented, unless (as often happens) the award otherwise identifies the dissentient.

[52D] *Reasons – Subs.(4)*
There is a general requirement for the award to contain reasons. This is a fundamental reversal of the law prior to this Act, in which reasons did not have to be given unless they were specifically requested. The requirement may reflect a growing tendency for the courts to require reasons when reviewing the actions of public bodies which make important and binding decisions affecting parties' rights and obligations. It is increasingly recognised that those making such significant decisions are expected to explain the reasons for them.

There are two exceptions to the requirement for reasons. Reasons need not be given in respect of an agreed award, (as to which see s.51). In addition, reasons need not be given if the parties have agreed that they are not required. In some fields of arbitration (for example, maritime) where it has been the norm *not* to incorporate reasons in the award unless specifically requested, some arbitration rules may need to be amended if practices prior to the Act are to be perpetuated. The LMAA Terms, for example, have been amended so that para.23(b) now provides that the parties have agreed to dispense with reasons in all cases where no notice shall have been given to the tribunal under para.23(a) requiring reasons. A note explains that the effect of this is to exclude the court's jurisdiction under s.69.

It should be noted that an agreement not to have reasons will exclude the jurisdiction of the court to determine a preliminary point of law, (s.45(1)), or to entertain an appeal on a question of law arising out of the award, (s.69(1)).

[52E] *Seat and date – Subs.(5)*
The award is required to state the seat of the arbitration, as to which see s.3. This is a new requirement to domestic arbitrators. It is, however, only significant in relation to international arbitrations or the enforcement of the award abroad, and then only in respect of countries where a statement of the seat of the arbitration is required by national law.

In relation to domestic arbitrations, an omission to state the seat of the arbitration will probably be of little consequence, since there will be no serious irregularity, that is to say an irregularity causing or potentially causing substantial injustice within the meaning of s.68, that would make the award open to challenge under that section.

The award is required to state the date when it was made. As to this, see s.54.

[52F] *Failure to comply with formalities*
Whilst a failure to comply with the formalities prescribed by this section may be of consequence in any attempt to enforce the award

outside England and Wales or Northern Ireland, there is unlikely to be any such consequence within those areas unless the failure constitutes a serious irregularity, that is one causing or potentially causing substantial injustice to an applicant who seeks to challenge the award under s.68.

In our view, too, most failures to comply would not prevent enforcement being obtainable with leave of the court under s.66, since it would make little sense for the court to apply different criteria under that section from those applicable under s.68.

Note also the unreported decision in *Ranko Group* v. *Antarctic Maritime SA* (1998) LMLN 492 where an arbitrator's letter finding and declaring that there was a contract which included an arbitration agreement, that he had been correctly appointed under that agreement and that all the disputes under the contract had been referred to arbitration, was held to constitute an award as to his substantive jurisdiction even though it did not state the seat of the arbitration or his reasons.

[52G]   *Model Law*

Art.31(Form and contents of award) provides:

> '(1) The award shall be made in writing and shall be signed by the arbitrator or arbitrators. In arbitral proceedings with more than one arbitrator, the signatures of the majority of all members of the arbitral tribunal shall suffice, provided that the reason for any omitted signature is stated.
> (2) The award shall state the reasons upon which it is based, unless the parties have agreed that no reasons are to be given or the award is an award on agreed terms under article 30.
> (3) The award shall state its date and the place of arbitration as determined in accordance with article 20(1). The award shall be deemed to have been made at that place.'

We note that unlike the Act, the Model Law specifically requires the reason for an omitted signature to be stated. It also refers to the 'place' of arbitration, which may be chosen by the parties under Art.20, or determined by the tribunal, having regard to the circumstances of the case, including the convenience of the parties. S.3 of the Act defines 'seat' in fairly similar terms.

[52H]   *New York Convention*

It is useful to note that Art.IV.1 requires a 'duly authenticated original award or a duly certified copy' to obtain enforcement. Art.IV.2 requires a translation if the award is not in an official language of the country in which it is to be relied upon. Under Art.V.1 enforcement will be refused if the award has not yet become binding.

[52I]   *Rules*

Rule 12.4 of CIMAR requires the award to be in writing, signed and dated. At Rule 12.5 it attempts to deal with the difficult question of the extent of the reasons:

'An award should contain sufficient reasons to show why the arbitrator has reached the decisions contained in it unless the parties otherwise agree or the award is agreed.'

Art.9.1 of the CIArb Rules is to similar effect, and Rule 20.1 of the ICE Arbitration Procedure broadly reproduces s.52.

Of the international rules, Art.26.1 of the LCIA Rules has similar effect to s.52. Art.25 of the ICC Rules requires the award to state the reasons on which it is based. It is deemed to be made at the place of the arbitration and on the date stated therein. Note that by Art.27 the award must be submitted in draft form to the ICC Court, who may lay down modifications as to the form of the award. Art.32 of the UNCITRAL Arbitration Rules requires the award to be in writing and reasoned. It should be signed, and must state the reason for the absence of any signature.

Section 53 – Place Where Award Treated as Made

53. Unless otherwise agreed by the parties, where the seat of the arbitration is in England and Wales or Northern Ireland, any award in the proceedings shall be treated as made there, regardless of where it was signed, despatched or delivered to any of the parties.

Definitions

'agreed': s.5(1).
'party': ss.82(2), 106(4).
'seat of the arbitration': s.3.

[53A] Status

This section derives from the Model Law, Art.31(3). It is *non-mandatory*, (s.4(2)).

[53B] Summary

The section concerns an award made in an arbitration whose seat is in England and Wales or Northern Ireland. (As to the seat of the arbitration, see s.3.) Such an award is deemed to be made in England and Wales or Northern Ireland.

Thus the fact that the award may, arbitrarily, have been signed in, or sent to the parties from, a different jurisdiction (simply because that is where the arbitrator happened to be at the time) must be disregarded. It will not characterise the award as having been made in that different jurisdiction, with the unintended consequences that might flow from such a result.

The section has the effect of reversing the judgment of the House of Lords in the case of *Hiscox* v. *Outhwaite* [1992] 1 AC 562, in which an award was held to have been made in Paris, France, purely because it was signed there, even though the arbitration (which had been conducted in London) had no connection with France.

The parties may, by agreement, exclude or modify this provision.

Points

[53C] *Effect on practice*

Domestic arbitrators do not normally indicate in their awards where they are made, whereas international arbitrators often do, so as to assist enforcement. In particular, under the New York Convention, recognition and enforcement of an award may be refused under Art.V in circumstances where it may be necessary for the court dealing with the question to know where the award was made. We set out relevant passages from the Convention below.

It will now be important for both domestic and international practitioners to adhere to the same practice and, for both, a change must be made. Neither should now state the *place* where the award was made; but both should state the *seat of the arbitration* (under s.52(5)). For domestic and international purposes, the place where the award was made is thereby automatically determined, by virtue of s.53.

We would caution that potential difficulties will be created by an award which states both the seat *and* the place where the award is made (if they differ), although it may be hoped that foreign courts construing s.53 will understand its effect as being that an award signed, for example, as 'made in Milan', must be taken as having been 'made in England' where there is also a statement in the award that the seat of the arbitration was London.

We also note that there could be difficulties where arbitration rules require the 'place' of arbitration to be stated. We think 'place' and 'seat' are clearly synonymous under ICC Rules and probably so under the UNCITRAL Arbitration Rules.

[53D]   **New York Convention awards**
Note the corresponding provision in s.100(2)(b) relating to foreign awards whose recognition or enforcement is sought in England and Wales or Northern Ireland under the New York Convention.

[53E]   **Model Law**
Art.31(3) provides:

> 'The award shall state its date and the place of arbitration as determined in accordance with article 20(1). The award shall be deemed to have been made at that place.'

We note that the Act requires a statement of the date when the award was made by s.52(5). Under s.3 'seat' has a more complex definition than 'place' under Art.20. Nonetheless the effect is similar. Both the Act and the Model Law designate the place where the award is deemed to be made by reference to the place or seat of the arbitration.

[53F]   **New York Convention**
As we have stated above, it is of particular importance in connection with enforcement under the Convention to be able to identify where the award is made. Art.V provides:

> '1. Recognition and enforcement of the award may be refused, at the request of the party against whom it is invoked, only if that party furnishes to the competent authority where the recognition and enforcement is sought, proof that:
> (a) The parties to the agreement ... were, under the law applicable to them, under some incapacity, or the agreement is not valid under the law to which

the parties have subjected it or, failing any indication thereon, under the law of the country where the award was made; or . . . . .

(e) The award has not yet become binding on the parties, or has been set aside or suspended by a competent authority of the country in which, or under the law of which, that award was made.'

[53G]   *Rules*

As might be expected, CIMAR, the ICE Arbitration Procedure and the CIArb Rules do not contain any agreement in conflict with s.53. Of the international rules, the LCIA Rules provide (at Art.16.2) that any award shall be treated as made at the seat of the arbitration. The ICC Rules (Art.25.3) similarly provide that the Award shall be deemed made at the place of the arbitration. Both these provisions have similar effect to s.53. The UNCITRAL Arbitration Rules (by Art.32.4) require a statement of the place where the award was made, and (by Art.16.4) that the award shall be made at the place of arbitration. It is our view that in the Model Law and the UNCITRAL Arbitration Rules 'place' and 'seat' should be treated as having similar meaning, so that s.53 will operate to supplement the provisions of the Rules.

Section 54 – Date of Award

54.–(1) Unless otherwise agreed by the parties, the tribunal may decide what is to be taken to be the date on which the award was made.

(2) In the absence of any such decision, the date of the award shall be taken to be the date on which it is signed by the arbitrator or, where more than one arbitrator signs the award, by the last of them.

Definitions

'agreed': s.5(1).
'arbitrator': s.82(1).
'party': ss.82(2), 106(4).

[54A] ## Status

This section is new. It is *non-mandatory*, (s.4(2)).

[54B] ## Summary

The section provides for the date when the award was made. In the first instance, and unless the parties agree otherwise, this will be a matter for the tribunal to decide, (subs.(1)). If no such decision is made, then the date of signing by the sole or last member of the tribunal will be taken to be the date of the award, (subs.(2)).

Points

[54C] *Decision as to the date – Subs.(1)*
Arbitrators must state in their award the date when the award is made, (s.52(5)). Such a statement would presumably comprise a 'decision' for the purposes of subs.(1).

The 'decision' may not be straightforward in all cases. Where there is more than one arbitrator, the award is likely to circulate, and may even go overseas. Signatures may be days or weeks apart. It will usually be most practical for the arbitrators either to ask the last signatory to date the award (as has in fact been common practice hitherto), or to return it to the chairman, who will do so. In view of our comments in paragraph 54E below, and despite possible temptation to the contrary, arbitrators would be well advised to consider a later, rather than earlier, date in order to give the parties a fair time in which to consider challenges or appeals, and in order to avoid applications under s.80(5) for extensions of time for appealing.

In the absence of a positive decision as to the date, subs.(2) applies. Given this default provision, it is unlikely that failure to state the date of an award could ever be a serious irregularity within the meaning of s.68(2)(h).

[54D] **Signing in front of each other – Subs.(2)**

It appears no longer to be a requirement that arbitrators execute an award in the presence of each other. The rule in *Wade* v. *Dowling* (1854) 4 E&B 44 that, per *Mustill and Boyd* (2nd Edition) page 383, 'where the award is made by more than one arbitrator, the arbitrators should execute it in the presence of each other, otherwise the award is invalid', is therefore abrogated. This is a small but welcome change in the law.

In practice, of course, most arbitrators have not in fact met to sign awards, and it is probably as a result of this that the habit of having arbitrators' signatures witnessed has grown up.

[54E] **Significance of the date**

In conjunction with s.70(3), the section has the important effect of linking any challenge or appeal to the date of the award, rather than to its publication, as under the regime prior to the Act. Under s.70(3) any application under s.67 (challenge to substantive jurisdiction), s.68 (challenge for serious irregularity) or s.69 (appeal on a point of law) must be brought 'within 28 days of the date of the award', subject to the court's power of extension (s.80(5)). This is a change to the previous position (although in practice, of course, the date of publication was usually that of the award).

The section also interacts with s.49 (interest). Thus the tribunal may award interest on sums awarded by it 'in respect of any period up to the date of the award', (s.49(3)(a)), and may award interest on the outstanding amount of the award, interest and costs 'from the date of the award (or any later date)', but not from an earlier one, (s.49(4)).

[54F] **Model Law**

Art.31(3) provides:

'The award shall state its date...'

[54G] **Rules**

CIMAR Rule 12.2 and Art.9.1 of the CIArb Rules require the award to be dated. The ICE Arbitration Rules are silent on the matter so that s.54 applies. Of the international rules, Art.26.1 of the LCIA Rules and Art.32(4) of the UNCITRAL Arbitration Rules require the award to state the date it was made. Art.25 of the ICC Rules provides that the award is deemed made on the date stated therein.

Section 55 – Notification of Award

55.–(1) The parties are free to agree on the requirements as to notification of the award to the parties.

(2) If there is no such agreement, the award shall be notified to the parties by service on them of copies of the award, which shall be done without delay after the award is made.

(3) Nothing in this section affects section 56 (power to withhold award in case of non-payment).

Definitions

'agreement', 'agree': s.5(1).
'party': ss.82(2), 106(4).
'service' (of notice or other document): s.76(6).

[55A] Status

This is a new section. It is *non-mandatory*, (s.4(2)).

[55B] Summary

The section essentially provides that the parties are at liberty to agree what requirements there should be for the tribunal to notify them of the award. If there is no such agreement, the tribunal must notify the parties by serving on them copies of the award without delay after it is made, subject to its right to withhold an award until payment, (s.56).

For the date of the award (which is the same as the date on which it was made), see s.54.

Points

[55C] *Significance of notification*
Notification is significant because certain important time limits are linked to the date of the award. The time limits for applying for corrections to the award and for corrections to be made on the tribunal's own initiative (s.57), and for challenges and appeals (s.70(3)), run from the date of the award, although they may be extended by the court, (ss.79 and 80(5)).

Rather than allowing receipt of the award to be the first intimation of its date, arbitrators should always seek in practice to give notice of an award's availability on the day it is dated.

[55D]   **Notification – Subs.(2)**
The tribunal is bound to notify *all* the parties. It should therefore no longer be possible (as the DAC remarked at para.255 of their Report) for a single party to obtain the award and deliberately not inform other parties of its contents until important time limits had passed. In the authors' own experience, however, examples of that were rare, simply because the practice of arbitrators was to send copies of an award to non-collecting parties once one had taken it up.

[55E]   **Power to withhold award not affected – Subs.(3)**
The tribunal's power to withhold an award until payment, set out in s.56, is not affected by any duty under this section. In particular, we do not consider that there would be any merit in an argument that whilst s.56(1) permits the tribunal to withhold an *award* pending payment, it could nevertheless be required to provide *copies* of the award to the parties even though the arbitrators' fees had not been paid. That would render the tribunal's lien nugatory.

  However once the tribunal has secured payment or been ordered by the court to deliver the award (s.56(2)(a)), it should deliver the award, and copies, promptly.

[55F]   **Model Law**
Art.31(4) provides:

> '(4) After the award is made, a copy signed by the arbitrators ... shall be delivered to each party.'

We note that the principle here stated is reflected in the Act. The Act, of course, emphasises the need to notify the award without delay for the reasons set out in paragraphs 55C and 54E above.

[55G]   **Rules**
CIMAR does not cover notification of the award. In contrast, the ICE Arbitration Procedure provides, at Rule 21.2, that 'when the Arbitrator has made his award ... he shall so inform the parties in writing and shall specify how and where it may be taken up upon full payment of his fees and expenses'. We take the view that this substitutes for s.55.

  The CIArb Rules are silent on the topic though Art.9.2 allows an arbitrator to 'notify an award to the parties as a draft or proposal and may in his discretion consider any further submissions or proposal put to him by any party but subject to any time limit which he may pro-pose'. This idea, which reflects practice in Scotland, has its attractions but may lead to an arbitrator going unpaid unless he has secured his fees in advance!

  Of the international rules, Art.26.5 of the LCIA Rules requires the sole arbitrator or chairman to deliver the award to the LCIA court which then transmits certified copies to the parties provided the costs have been paid. Art.28 of the ICC Rules is to similar effect, notification being carried out by the ICC Secretariat. Art.32(6) of the UNCITRAL Arbitration Rules requires that 'copies of the award signed by the arbitrators shall be communicated to the parties by the arbitral tribunal'. Use of this rule would appear to be an agreement within subs.(1), so that subs.(2) is rendered inapplicable, the effect being that the express requirement for speed of communication is thereby removed.

Section 56 – Power to Withhold Award in Case of Non-payment

56.–(1) The tribunal may refuse to deliver an award to the parties except upon full payment of the fees and expenses of the arbitrators.

(2) If the tribunal refuses on that ground to deliver an award, a party to the arbitral proceedings may (upon notice to the other parties and the tribunal) apply to the court, which may order that–

(a) the tribunal shall deliver the award on the payment into court by the applicant of the fees and expenses demanded, or such lesser amount as the court may specify,

(b) the amount of the fees and expenses properly payable shall be determined by such means and upon such terms as the court may direct, and

(c) out of the money paid into court there shall be paid out such fees and expenses as may be found to be properly payable and the balance of the money (if any) shall be paid out to the applicant.

(3) For this purpose the amount of fees and expenses properly payable is the amount the applicant is liable to pay under section 28 or any agreement relating to the payment of the arbitrators.

(4) No application to the court may be made where there is any available arbitral process for appeal or review of the amount of the fees or expenses demanded.

(5) References in this section to arbitrators include an arbitrator who has ceased to act and an umpire who has not replaced the other arbitrators.

(6) The above provisions of this section also apply in relation to any arbitral or other institution or person vested by the parties with powers in relation to the delivery of the tribunal's award.

As they so apply, the references to the fees and expenses of the arbitrators shall be construed as including the fees and expenses of that institution or person.

(7) The leave of the court is required for any appeal from a decision of the court under this section.

(8) Nothing in this section shall be construed as excluding an application under section 28 where payment has been made to the arbitrators in order to obtain the award.

Definitions

'agreement': s.5(1).
'arbitrator': s.82(1).
'available arbitral process': s.82(1).
'the court': s.105.

'notice (or other document)': s.76(6).
'party': ss.82(2), 106(4).
'upon notice' (to the parties or the tribunal): s.80.

[56A]   **Status**

This section derives from the Arbitration Act 1950, s.19, but differs slightly in its effect. It is *mandatory*, (s.4(1) and Sched.1).

[56B]   **Summary**

The section concerns the tribunal's power to withhold an award until payment. It expressly provides for the tribunal to have a right of lien over its award for the purpose of securing payment of its fees, (subs.(1)). The mandatory status of the section has the effect that the parties cannot, by agreement between themselves, deprive the tribunal of this protection. However the section gives the parties a means of obtaining the award whilst still challenging the tribunal's fees, yet without depriving the tribunal of its security.

Points

[56C]   *Delivery of award against payment into court – Subss.(2) to (4)*
Apart from the possibility of paying the sums demanded by the tribunal and then making application to the court for an adjustment under s.28 (see paragraph 56I below), if a party considers that the tribunal's claim to fees is too high, he may apply to the court to secure the release of the award and, at the same time, seek an adjustment to the fees and expenses claimed.

On the application, the court may order the award to be released against the applicant paying into court the full amount claimed, or some lesser amount, to deal with the possibility of a grossly excessive claim by the tribunal making it impossible for a party to obtain the award, (subs.(2)(a)). The latter is a change from the position prior to the Act.

The court will then direct how the amount of the fees properly payable is to be determined (subs.(2)(b)), possibly by a third party, but more likely by the court's own Taxing Masters. The amount will either be such sum as the applicant is liable to pay under any agreement with the arbitrators, or such reasonable fees and expenses as he is determined to be liable to pay under s.28 (which deals with the liability of a party for arbitrators' fees and expenses), (subs.(3)).

The money paid into court is then applied to meet the properly payable fees and expenses of the tribunal, with any surplus being refunded to the applicant, (subs.(2)(c)).

Note that no such application to the court may be made where there is a possibility of appeal or review of the arbitrators' fees and expenses by some other arbitral process, for instance by applying to a relevant institution or appellate arbitral tribunal, (subs.(4)).

[56D]   *'Arbitrators' – Subs.(5)*

In this section, 'arbitrators' includes an umpire who has not replaced the other arbitrators, and arbitrators who have ceased to act, either because they have themselves been replaced by an umpire or because they have 'ceased to hold office' by reason of revocation of authority (s.23); removal by the court (s.24); resignation (s.25); or death (s.26).

[56E]   *Others withholding the award – Subs.(6)*

The section applies to persons or institutions that may be entitled, by virtue of the parties' agreement, to withhold an award in place of the tribunal itself. In such a case, the fees and expenses of that person or institution will also fall within the scope of the section.

[56F]   *The application*

The application under s.56 is an arbitration application falling within the terms of the Practice Direction on Arbitrations under CPR Part 49. We deal generally with arbitration applications in Part 3, Materials, Section H. Specifically in respect of an application under s.56:

(i)   the application must be on notice to the other parties and the tribunal;

(ii)   the requirement for giving notice to both the other parties and the tribunal is to be met by making those parties and the tribunal respondents to the application and serving on them the arbitration claim form and evidence in support;

(iii) the arbitration claim form must show that there is no available arbitral process for appeal or review of the amount of the fees or expenses demanded;

(iv)   unless otherwise ordered, the hearing will be in private.

[56G]   *The arbitrator and the application*

The arbitrator (or arbitrators) will be made respondents to the proceedings, and it is likely that the applicant will seek costs in his claim form against an arbitrator whom he believes to be seeking excessive fees. The arbitrator is therefore likely to acknowledge service and, thereafter, consider the extent to which he wishes to put in evidence and participate at the hearing. He may also wish to take steps to protect his position by making an offer that is expressed to be 'without prejudice except as to costs'.

[56H]   *Appeal – Subs.(7)*

Leave of the court is required for any appeal.

[56I]   *Later adjustment of fees – Subs.(8)*

Immediate payment of the tribunal's full claim in order to obtain delivery of the award does not preclude a later application to the court under s.28 for an adjustment to the arbitrators' fees and expenses in accordance with what is reasonable. The court may then order the repayment of any excessive amount paid, but only if such an order is reasonable in the circumstances, (s.28(3)). Parties thus have a choice between an application to the court under subs.(2), or the payment of the full amount to the arbitrators and subsequent challenge under s.28.

A challenge to the arbitrators' fees under s.28 may be made by *any* party, so that the gap in the procedure prior to this Act identified by Mr Justice Donaldson in *Rolimpex* v. *Hadji E. Dossa* [1971] 1 Lloyd's Rep. 380 – namely that a party other than the one taking up the award had no route to challenge the tribunal's fees where they were taxed in the award – is now filled.

[56J]   *Judge-arbitrators*

Note that where the tribunal is a judge-arbitrator, a similar procedure to that set out in this section is available, with the significant difference that the application to secure release of the award is made to the judge-arbitrator rather than the court, (see Sched.2, paras.6, 7).

[56K]   *Model Law*

The Model Law contains no comparable provision.

[56L]   *Rules*

S.56 is a mandatory section, so it is implied into all arbitration agreements. CIMAR and the CIArb Rules do not cover the ground. As we have noted in our comments at paragraph 55G above, the ICE Arbitration Procedure contemplates (at Rule 21.2) that the tribunal will require full payment of fees and expenses before delivery of the award, but does not trespass into the related court powers. We have also indicated under s.55 above that both Arts.26.5 of the LCIA Rules and Art.28 of the ICC Rules require the fees and expenses of the arbitrators and administrative costs to have been paid before release of the award (provided for by subs.(6)). The UNCITRAL Arbitration Rules do not cover this ground; but it is probable, in any event, that pursuant to Art.41 the tribunal will be secured for its fees, so the question will not arise.

Section 57 – Correction of Award
or Additional Award

57.–(1) The parties are free to agree on the powers of the tribunal to correct an award or make an additional award.

(2) If or to the extent there is no such agreement, the following provisions apply.

(3) The tribunal may on its own initiative or on the application of a party–

(a)   correct an award so as to remove any clerical mistake or error arising from an accidental slip or omission or clarify or remove any ambiguity in the award, or

(b)   make an additional award in respect of any claim (including a claim for interest or costs) which was presented to the tribunal but was not dealt with in the award.

These powers shall not be exercised without first affording the other parties a reasonable opportunity to make representations to the tribunal.

(4) Any application for the exercise of those powers must be made within 28 days of the date of the award or such longer period as the parties may agree.

(5) Any correction of an award shall be made within 28 days of the date the application was received by the tribunal or, where the correction is made by the tribunal on its own initiative, within 28 days of the date of the award or, in either case, such longer period as the parties may agree.

(6) Any additional award shall be made within 56 days of the date of the original award or such longer period as the parties may agree.

(7) Any correction of an award shall form part of the award.

Definitions

'agreement', 'agree': s.5(1).
'party': ss.82(2), 106(4).

[57A]   ## Status

This section is derived from the Arbitration Act 1950, ss.17 and 18(4); and from the Model Law, Art.33. It is *non-mandatory*, (s.4(2)).

[57B]   ## Summary

The section essentially provides that the parties are at liberty to agree as to the powers of the tribunal either to correct an award or to make an additional award. Insofar as there is no such agreement, it gives powers to the tribunal that are more extensive than those available prior to this Act.

Points

Effect – Subs.(3)

Acting either on its own initiative or on the application of a party, the tribunal may correct an award so as to remove any clerical mistake, or any error arising from an accidental slip or omission, (subs.(3)(a)). Errors in computation, typographical errors and the like may be readily corrected under this provision. In addition, the tribunal may clarify or remove any ambiguity in the award. This latter is a new power. Its effect is that the tribunal has the opportunity to explain or amend a particular aspect of an award whose meaning is, or is thought to be, unclear. Any correction of an award forms part of the original award, (subs.(7)).

The tribunal may also remedy a more fundamental oversight. If a claim has been presented to it, but the tribunal has omitted to adjudicate on that claim in its award, it may put matters right by making an additional award, (subs.(3)(b)). Whereas prior to this Act, an additional award could only be made – by virtue of express statutory provision – with respect to costs, the power under this section extends to a claim of any kind, including interest or costs. This probably in fact does no more than reflect the previous common law position, since a tribunal that failed to adjudicate upon a claim was not *functus officio* as regards that claim and could, therefore, make a further award upon it.

The tribunal must give 'the other parties' a reasonable opportunity to make representations before exercising any of these powers. The meaning of 'the other parties' is plain where the application is made by one party. It presumably means 'all parties' where the tribunal acts on its own initiative. This requirement accords with the tribunal's duty under s.33(1)(a).

We should add that, in our view, subs.(3)(b) is only applicable where a claim *should have been dealt with* in an award. Thus if the parties have agreed that the tribunal should make an interim (or partial) award under s.47 – for instance one exclusive of costs or interest – it would be appropriate for the arbitrators to proceed to a further award on the remaining aspects without reference to this section.

A difficulty may arise where there is a mistake or error falling within subs.(3) but the tribunal fails or refuses to effect a correction. Any solution to this problem will probably depend on s.68(2)(a). When commenting on that sub-section at para.68C (see below) we deal with a recent case which may provide some guidance.

In another recent matter it has been held that if there are ambiguities or uncertainties in an award those must first be dealt with by the tribunal in the light of s.57 (and s.70(2)(b)) before any application or appeal relating to them is taken to the court: *Gbangbola* v. *Smith & Sherriff Ltd* [1998] 3 All ER 730.

[57D]   ***Time limits: applications by parties – Subss.(4) to (6)***
Applications by a party for a correction or for an additional award
must normally be made within 28 days of the date of the award. For the
'date of the award' see s.54. This may not always be the date of the (last)
signature, but it will often be a date earlier than receipt of notification
of the award, and will almost always be earlier than receipt of a copy of
it. In practice, the applicant will thus have fewer than 28 days in which
to apply. The time may, however, be extended by agreement between
the parties or by the court under s.79.

Under subs.(5) any correction of the award must be made within 28
days from receipt of the application. By contrast, under subs.(6) there
are 56 days in which to make an additional award, that period running
from the date of the original award. In either case the parties may
extend the time by agreement, or application may be made by a party
or the tribunal to the court under s.79.

We note that in neither case is there substantial time available for the
parties to make representations to the tribunal under subs.(3).

[57E]   ***Time limits: tribunal acting on own initiative – Subss.(5), (6)***
The tribunal may both correct an award or issue an additional award
on its own initiative. Subject to agreement on an extension of time, or a
court order under s.79, corrections must be made within 28 days of the
date of the award, and an additional award must be made within 56
days of that date. For the 'date of the award', see s.54.

[57F]   ***Judge-arbitrators***
Note that where the tribunal is a judge-arbitrator the time limits set out
in subss.(4) to (6) for application or exercise of these powers do not
apply, (Sched.2, para.8).

[57G]   ***Tribunal's fees for corrections, etc.***
Nothing is said in the Act about any entitlement of arbitrators to charge
fees for corrections or additional awards. Our view is that where such
matters have to be dealt with because of a tribunal's oversight, and
indeed in most cases, it will be inappropriate to charge. Art.40(4) of the
UNCITRAL Arbitration Rules is to this effect, (see paragraph 61I
below).

[57H]   ***Model Law***
Art.33 provides:

> '(1) Within thirty days of receipt of the award, unless another period of time
> has been agreed upon by the parties:
> (a) a party, with notice to the other party, may request the arbitral tribunal to
> correct in the award any errors in computation, any clerical or typographical
> errors or any errors of similar nature;
> (b) if so agreed by the parties, a party, with notice to the other party, may
> request the arbitral tribunal to give an interpretation of a specific point or
> part of the award.

If the arbitral tribunal considers the request to be justified, it shall make the correction or give the interpretation within thirty days of receipt of the request. The interpretation shall form part of the award.

(2) The arbitral tribunal may correct any error of the type referred to in paragraph (1)(a) of this article on its own initiative within thirty days of the date of the award.

(3) Unless otherwise agreed by the parties, a party, with notice to the other party, may request, within thirty days of receipt of the award, the arbitral tribunal to make an additional award as to claims presented in the arbitral proceedings but omitted from the award. If the arbitral tribunal considers the request to be justified, it shall make the additional award within sixty days.

(4) The arbitral tribunal may extend, if necessary, the period of time within which it shall make a correction, interpretation or an additional award under paragraph (1) or (3) of this article.

(5) The provisions of article 31 shall apply to a correction or interpretation of the award or to an additional award.'

We note that the Model Law regime is similar to that set out in the Act, although there is no power in the Act for the tribunal to extend time limits. The Model Law contains a specific statement requiring corrections, interpretations or additional awards to comply with the formal requirements as to writing, signature, etc. set out in Art.31. Since an additional award is nonetheless an award, it would seem to us that such formalities apply under the Act by virtue of s.52. We note also that the Act contains no power for arbitrators to give 'interpretations' of awards. They may, though, clarify and remove any ambiguity.

[57I]   *Rules*

CIMAR gives the arbitrator, by Rule 12.9, the powers set out in subs.(3) to (6). Like Art.9.2 of the CIArb Rules (see paragraph 55G), CIMAR, by Rule 12.10, also provides that an award (or part of one) may be notified as a draft or proposal, thus reducing, we would hope, the scope for the use of s.57. The ICE Arbitration Procedure contains a power in similar, but not identical terms, at Rule 21.3. In particular, the additional award may be in respect of any 'matter' not dealt with in the award, which may be wider than any 'claim'. The former may cover an argument not dealt with, whereas the latter might not if the claim itself was encompassed.

Of the international rules, Art.27 of the LCIA Rules gives a broad power of correction and the making of additional awards but does not specifically include the power to correct an award to remove ambiguity. Under Art.29 of the ICC Rules the tribunal may correct slips. Additionally a party may apply to the secretariat for an interpretation of the award by the tribunal. The tribunal must then give the other party an opportunity to respond, after which it must submit its decision to the ICC Court before transmittal to the parties. Within the UNCITRAL Arbitration Rules, Arts.36 and 37 provide respectively and comprehensively for corrections and additional awards. In our view, the international rules each constitute an agreement overriding subss.(3) to (7).

The international rules generally contain time limits which are more generous than s.57, but probably more appropriate to international arbitration. Whilst we have noted that we think that the international rules are generally wide enough to displace s.57, we do not think adoption of these rules would preclude the possibility of extension of time limits under s.79.

Section 58 – Effect of Award

58.–(1) Unless otherwise agreed by the parties, an award made by the tribunal pursuant to an arbitration agreement is final and binding both on the parties and on any persons claiming through or under them.

(2) This does not affect the right of a person to challenge the award by any available arbitral process of appeal or review or in accordance with the provisions of this Part.

Definitions

'agreed': s.5(1).
'arbitration agreement': ss.6, 5(1).
'available arbitral process': s.82(1).
'party': ss.82(2), 106(4).

[58A] ## Status

This section derives from the Arbitration Act 1950, s.16. It is *non-mandatory*, (s.4(2)).

[58B] ## Summary

The section re-states the law prior to this Act as to the final and binding effect of an award made pursuant to an arbitration agreement as between the parties and those claiming through or under them.

One implication of this effect is that neither party may subsequently re-open an issue decided in an award by bringing further arbitral or court proceedings against the other party. Another is that an award may be enforced in the same way as a judgment of the court, (s.66).

For the sake of clarity, subs.(2) makes it clear that this final and binding quality does not affect challenges to the award, made either without recourse to the court or pursuant to the other provisions of this Part of the Act (for example, ss.67 to 69).

As this is a non-mandatory provision, the parties may agree that an award should have a different effect. Non-binding awards may be useful to permit the parties, for example, to arbitrate initially before a sole arbitrator, on the basis that the losing party may require the issues to be dealt with again by a multiple tribunal if he pays a proportion of the original (non-binding) award.

Points

[58C]   *The corollary*

The DAC Report referred, in para.263, to a suggestion that the corollary of this section should be spelt out, namely that whatever the parties might or might not agree, the award is of no substantive or evidential effect against a third party or anyone who is not claiming through or under a party. It was decided not to adopt this suggestion since it was very difficult to construct such a provision, given, for instance, the difficulties with non-parties such as insurers, whose rights may in fact be affected by awards.

[58D]   *Model Law*

This section is also important for enforcement internationally. It is unlikely that any foreign court would enforce an award that is not demonstrably binding. That an award has not yet become binding, or has ceased to be binding (because set aside or suspended) is a ground for refusal to enforce under the New York Convention, and under Art.36 of the Model Law. Additionally, Art.35 of the latter provides:

> '(1) An arbitral award, irrespective of the country in which it was made, shall be recognised as binding . . .'

[58E]   *Rules*

Rule 12.4 of CIMAR provides that (subject to provisional relief) s.58 applies. The CIArb Rules and the ICE Arbitration Procedure are silent, so that again s.58 applies. Of the international rules, Art.26.9 of the LCIA Rules provides that all awards are final and binding on the parties agreeing to arbitration under the Rules, Art.28.6 of the ICC Rules provides that every award shall be binding, and Art.32.2 of the UNCITRAL Arbitration Rules provides that the award is final and binding on the parties.

Costs of the Arbitration

Section 59 – Costs of the Arbitration

59.–(1) References in this Part to the costs of the arbitration are to–

(a) the arbitrators' fees and expenses,
(b) the fees and expenses of any arbitral institution concerned, and
(c) the legal or other costs of the parties.

(2) Any such reference includes the costs of or incidental to any proceedings to determine the amount of the recoverable costs of the arbitration (see section 63).

Definitions

> 'arbitrator': s.82(1).
> 'costs of the arbitration': s.59.
> 'party': ss.82(2), 106(4).
> 'recoverable costs': ss.63, 64.

[59A] Status

This is a new section. It is *non-mandatory*, (s.4(2)), but, in our view, since it contains definitions, it could not be excluded by the agreement of the parties.

[59B] Summary

The section defines what is meant by the 'costs of the arbitration'. It is particularly relevant to ss.60 to 65 which deal with how the costs of the arbitration are to be allocated and borne as between the parties themselves.

For the question of costs as between the parties and the arbitrators, see s.28.

Points

[59C] *'Costs of the arbitration'*
For the purposes of ss.59 to 65 at least, the formal distinction previously made between 'costs of the reference' and 'costs of the award' does not remain, arbitrators now being required to deal with the single new concept of 'costs of the arbitration'. The latter is, however, defined by reference to the three categories of costs stated in subs.(1).

The definition of costs of the arbitration begins with the arbitrators' fees and expenses, (subs.(1)(a)). This heading, in turn, includes the fees and expenses of experts, legal advisers and assessors appointed by the tribunal for which the arbitrators are liable, (s.37(2)).

Also included within costs of the arbitration are the fees and expenses of any arbitral institution involved in the administration or conduct of the arbitration, (subs.(1)(b)).

The definition, of course, encompasses the normal legal costs of the parties, the costs of their experts, and other similar costs, (subs.(1)(c)). There will probably continue to be a debate as to whether arbitrators have the power to award the costs incurred by a party in taking proceedings in another jurisdiction in order to obtain security for his claim. Our view is that the section, read as a whole and read in conjunction with the other provisions dealing with costs, plainly covers only the costs incurred in the arbitration proceedings, not those of ancillary proceedings.

Finally, the definition includes the costs of proceedings by which the recoverable costs are themselves determined, (subs.(2)). These are known in the court context as proceedings for the 'taxation' of costs although the Act does not use the words 'tax' or 'taxation'. Such proceedings are dealt with in ss.63 and 64.

[59D]   *Model Law*
There is no comparable provision.

[59E]   *Rules*
CIMAR (somewhat ambiguously) defines costs (at Rule 13.6) by reference to s.59. In Rule 23 (Definitions) the ICE Arbitration Procedure substantially reproduces s.59. The CIArb Rules provide no definition. Of the international rules, the LCIA Rules do not formally define costs. The ICC Rules at Art.31 and the UNCITRAL Arbitration Rules at Art.38 each set out a comprehensive definition of costs, but they appear to have substantially similar effect to s.59.

Section 60 – Agreement to Pay Costs in Any Event

60. An agreement which has the effect that a party is to pay the whole or part of the costs of the arbitration in any event is only valid if made after the dispute in question has arisen.

Definitions

'agreement': s.5(1).
'costs of the arbitration': s.59.
'dispute': s.82(1).
'party': ss.82(2), 106(4).

[60A] ## Status

This section is derived from the Arbitration Act 1950, s.18(3). It is *mandatory*, (s.4(1) and Sched.1).

[60B] ## Summary

The section essentially preserves the law prior to this Act. A party may not agree to bear the whole or part of the costs of the arbitration in any event (that is to say, irrespective of the outcome of the arbitration) unless that agreement is made after the dispute has arisen.

The provision is based on considerations of public policy which the DAC felt were still valid. A party may only agree to commit itself to bearing costs irrespective of the outcome if it has at least knowledge of the dispute in question and, presumably, the opportunity to form a view as to its prospects.

Being mandatory, the parties may not avoid the effect of the section by agreement.

[60C] ### *Model Law*
The Model Law contains no comparable provision.

[60D] ### *Rules*
None of the rules we have examined covers this ground. In any event, a rule to contrary effect would be ineffective.

Section 61 – Award of Costs

61.–(1) The tribunal may make an award allocating the costs of the arbitration as between the parties, subject to any agreement of the parties.

(2) Unless the parties otherwise agree, the tribunal shall award costs on the general principle that costs should follow the event except where it appears to the tribunal that in the circumstances this is not appropriate in relation to the whole or part of the costs.

Definitions

'agreement', 'agree': s.5(1).
'costs of the arbitration': s.59.
'party': ss.82(2), 106(4).

[61A] Status

This section is derived from the Arbitration Act 1950, s.18(1). It is *non-mandatory*, (s.4(2)).

[61B] Summary

The section provides for how the costs of the arbitration are to be allocated and borne as between the parties. The costs of the arbitration are defined in s.59 and include both the arbitrators' fees and expenses and the costs of the parties. Note that unless the parties agree otherwise, only those costs that are determined as 'recoverable' under ss.63 and 64 will be affected by an agreement or award made under this section, (s.62).

For the question of costs as between the parties and the arbitrators, see s.28.

Points

[61C] *Agreement of the parties – Subs.(1)*

Subject to s.60, the parties may reach agreement as to how costs are to be allocated as between themselves at any time and such agreement may relate to the whole or any part of the costs. It may be made before or after any award to which it relates. (Of course, where the award has disposed of costs it is unlikely, though perfectly possible, that parties will make some different agreement as to the allocation of costs).

Thus for example, the parties might agree to bear their own costs of the arbitration proceedings and to bear the fees and expenses of the arbitrators in equal proportions; the parties might agree that the costs of a certain preliminary hearing be 'costs in the reference' (that is to say, that they will be allocated in accordance with the outcome of the main issues); or a party losing an award on liability might agree to pay all the costs of the other party and the arbitrators' fees and expenses.

Those formally drafting such agreements are reminded that they must deal with, or at least consider, the three different categories of costs set out in s.59.

[61D]   *Tribunal's power – Subs.(1)*
In the absence of agreement the tribunal makes the award of costs. We should draw attention to the fact that this is the only provision in the Act for the award of costs, and it thus covers interlocutory (or pre-liminary) matters as well as matters up to and including awards, whether interim or final.

The practice has long been immediately to dispose of related costs matters following interlocutory hearings. We are sure this will con-tinue. Arbitrators making 'orders' in respect of costs at preliminary hearings should ensure that these are later 'collected', or taken account of, in any more formal award subsequently made. Without an award of costs there is no way that a party can ask the court to make an assessment of them. Of course, it is possible to make awards related only to costs, but this will rarely be appropriate.

We imagine that, following the Woolf reforms, arbitrators may be more inclined than hitherto to make orders for costs of interlocutory applications to be borne by one party regardless of the outcome of the case (rather than saying 'costs in the reference') and may even wish to make orders for immediate payment of such costs. In the latter case, any such order will only be enforceable if it is embodied in an award.

[61E]   *Drafting*
In drafting their awards on costs, arbitrators now have to ensure that they consider the fees and expenses of any arbitral institution. S.63 appears to us to necessitate a more complex award than hitherto. An award on costs (omitting, for the sake of clarity, any provision for interest – as to which see paragraph 49F) might read:

'And it is ordered and directed that:
(1) Subject to paragraph 4 below, the respondent shall pay the claimant's costs of the arbitration which, unless agreed, shall be determined by me under s.63(3) of the Arbitration Act 1996 ('the 1996 Act') on the basis set out in s.63(5) of that Act;
(2) The respondent shall pay my fees and expenses in the sum of £[   ] plus VAT; provided that if, in the first instance, the claimant shall have paid any amount in respect of my fees and expenses, he shall be entitled to an immediate refund from the respondent of the sum so paid;

(3) The respondent shall pay to the claimant the fees and expenses of [such and such institution] which are hereby determined, under s.63(3) of the 1996 Act on the basis set out in s.63(5) of that Act, in the sum of £[   ], such fees and expenses being listed by item and amount in the attached Schedule;

(4) Provided that my orders of [dates] in which the claimant was ordered to pay the costs of the respondent arising out of the matters set out therein stand as part of this award. Such costs shall be determined by me under s.63(3) of the 1996 Act on the basis set out in s.63(5) of that Act unless agreed.'

We recommend the reference to s.63(5) in paragraphs 1, 3 and 4 of the draft, unless some other basis is ordered, for the reasons set out in paragraph 63H below. See paragraph 28E for commentary on paragraph 2 of the draft.

Even more complex drafting will be required where there has been a provisional order under s.39, particularly where a payment on account of costs requires adjustment, (s.39(2)).

[61F]   *Principle – Subs.(2)*
The parties are also at liberty to agree what principle the tribunal should adopt in awarding costs. Unless there is agreement to the contrary, the tribunal is obliged to adopt the principle that costs follow the event. However the tribunal is given a discretion to depart from that principle where it would appear not to be appropriate in relation to all or part of the costs.

The principle accords with practice generally both in court proceedings and in arbitration, although it was not formerly expressed statutorily in relation to arbitrations and is now expressed in terms of the successful and the unsuccessful party in the the CPR. Under it, generally speaking, a successful party may expect to be awarded the costs of the arbitration. So if a claimant claims a sum of money and recovers the whole or a substantial part of what he claims, he will be the winning party. 'The event' is the successful outcome for him, and costs will 'follow the event' by being awarded to him. On the other hand, if a respondent successfully defeats a claim, 'the event' is the successful outcome for him, and he will be entitled to the costs of the arbitration.

When might a tribunal use its discretion not to award a successful party all or part of their costs? One instance might be where that party had conducted itself in the course of the arbitration in a way that was unreasonable and oppressive, and thereby caused unnecessary delay and expense. Another might be where that party had misconducted itself in relation to the transaction out of which the arbitration arises.

A claimant may not be fully successful because he does not recover everything he wished, but he may yet recover a substantial part of what he originally claimed. In those circumstances, he will still normally be entitled to costs as a winning party. However, if there could have been a considerable saving in costs had the claim been restricted in the first place, or if the claimant has lost on a distinct issue or head of claim that significantly increased the costs, then the tribunal might depart from the general rule. It might also do so if the claimant only succeeded as to a small part of his claim.

In the first edition of this book we said that we doubted that the Act had changed the law as it stood previously in this respect, and we referred the reader to pages 396/7 of *Mustill and Boyd* (2nd Edition). Since then, the CPR has come into force. The tenor of the CPR is to move away from the position that some success is sufficient to obtain an order for all the costs. In future, in court proceedings there are likely to be many more orders awarding partial costs, thus more accurately reflecting the level of success achieved by the receiving party. Thus CPR Part 44 provides:

'44.3 (4) In deciding what order (if any) to make about costs, the court must have regard to all the circumstances, including –
(a) the conduct of all the parties;
(b) whether a party has succeeded on part of his case, even if he has not been wholly successful; and
(c) any payment into court or admissible offer to settle made by a party . . .

(5) The conduct of the parties includes –
(a) conduct before, as well as during, the proceedings, and . . .
(b) whether it was reasonable for a party to raise, pursue or contest a particular allegation or issue;
(c) the manner in which a party has pursued or defended his case or a particular allegation or issue; and
(d) whether a claimant who has succeeded in his claim, in whole or in part, exaggerated his claim.'

We think that orders for costs made by arbitrators are likely to move in a similar direction and that arbitrators who follow the guidelines of CPR Part 44 quoted are unlikely to fall foul of any appellate or review procedures.

We note that under Part 44 the court is given express power to make orders in respect of costs incurred before proceedings have begun. Costs incurred in contemplation of arbitration proceedings are generally allowable depending on relevance to the ultimate dispute. Provided arbitrators limit such costs to those conventionally permissible, to the extent that any practice develops in the court to make separate costs awards in respect of pre-litigation costs, we see no difficulty in arbitrators reflecting it.

In exercising his discretion on costs, an arbitrator must not take into account factors which have not been drawn to the attention of the parties who have thus not had an opportunity of dealing with them. That would not be to act fairly and impartially under s.33: *Gbangbola* v. *Smith & Sherriff Ltd* [1998] 3 All ER 730.

Of course, as is reflected in the example set out above in paragraph 61E, the successful party may have been unsuccessful in relation to individual applications or events in the course of the proceedings. If the costs of these have been agreed or decided against him, such costs will not follow the event.

[61G] *Offers*

A respondent, recognising that he has or may have some liability and is likely to 'lose' to some extent, may try to protect himself against an award of costs following the event. He may make a sealed offer which the tribunal will not open until it is considering the question of costs. Or he may make an offer that is expressed to be 'without prejudice except as to costs' (also known as a 'Calderbank' offer, after the case in which the practice was first approved). This, too, will not be revealed to the tribunal until the question of costs arises.

In each case the offer will represent a proposal to settle the proceedings on certain terms. If the claimant accepts the offer, the proceedings end. If he does not, and continues with the proceedings, he is at risk as to the costs arising after the date of the offer. If he eventually achieves a result that is better than that in the offer, for example by recovering a greater sum of money, then costs will follow the event in the usual way. But if he achieves a result that is less than that in the offer, for example by recovering a lesser sum of money, then the costs of the arbitration from the latest date the offer might have been accepted onwards will normally be awarded against him. To that extent, they will not 'follow the event'.

Two matters in particular arise from the introduction of CPR Part 36 which may be of interest to arbitrators. In the first place, a court may now take into account an offer to settle made before commencement of proceedings. We think it equally legitimate for an arbitrator to do so. Thus the whole of a respondent's costs might be paid by the claimant in an appropriate case. Secondly, Part 36 contains a procedure under which a claimant may make an offer to accept a certain sum. If the offer is rejected and the claimant ultimately recovers more, the court may order interest at up to 10% above base rate on money awarded to the claimant from the latest date when the defendant could have accepted the offer. It may also order indemnity costs and interest on them at the same rate and from the same date.

We do not think that the arbitrator can make penal awards of interest, but we do think that an arbitrator may be justified in awarding costs on an indemnity basis where the claimant has made an offer to the respondent to settle and has won a more advantageous award. Such indemnity costs would obviously run only from the latest date upon which the respondent might have accepted the offer.

[61H] *Model Law*

There is no comparable provision in the Model Law.

[61I]   *Rules*

At Rule 13 CIMAR covers the ground of s.61 and the account to be taken of offers in some detail. It reiterates that costs should in principle be borne by the losing party (avoiding, as does the CPR, the use of the expression 'the event'), but then gives the arbitrator 'the widest discretion' in awarding which party should bear what proportion of the costs. It requires him to have regard to all material circumstances as follows:

(1)   which of the claims has led to the incurring of substantial costs and whether they were successful;
(2)   whether any claim which has succeeded was unreasonably exaggerated;
(3)   the conduct of the party who succeeded on any claim and any concession made by the other party;
(4)   the degree of success of each party.

By Rule 13.9 the arbitrator should have 'regard to any offer of settlement or compromise from either party, whatever its description or form. The general principle which the arbitrator should follow is that a party who recovers less overall than was offered to him in settlement or compromise should recover the costs which he would otherwise have been entitled to recover only up to the date on which it was reasonable for him to have accepted the offer, and the offeror should recover his costs thereafter.' Although experience under the CPR is yet thin on the ground, the authors have often encountered the argument that sealed offers or offers expressed to be 'without prejudice except as to costs' in arbitration should mimic the payment into court procedure as closely as possible. CIMAR makes it quite clear that is not the case under these rules.

Art.10 of the CIArb Rules contains provisions similar to those in CIMAR whilst the ICE Arbitration Procedure simply provides (by Rule 19.7(a)) that the arbitrator has power to allocate the costs as he considers appropriate, thus leaving s.61(2) to apply as enacted.

Of the international rules, Arts.28.2 and 28.3 of the LCIA Rules provide that the tribunal may order the costs of one party to be paid by another, and shall determine in what proportion the parties shall bear the other costs of the arbitration. In both cases, by Art.28.4 the awards should reflect the parties' relative success and failure (unless this general approach is inappropriate). Any order for costs must be made with reasons in the award containing the order. Arts.28.2 to 28.4 thus replace s.61.

Art.31.1 of the ICC Rules provides that in the final award the tribunal should decide which of the parties should bear (all) the costs of the arbitration or in what proportion they should be borne by the parties. The article does not cover the principle to be adopted, so that, in our view, s.61(2) applies.

Art.40 of the UNCITRAL Arbitration Rules generally follows the regime contemplated by s.61, but by Art.40.2 it would appear that there is no presumption that the costs of legal representation follow the event. Art.40.4 specifically debars the tribunal from seeking additional fees where it has interpreted, corrected or completed its award.

Section 62 – Effect of Agreement or Award About Costs

62. Unless the parties otherwise agree, any obligation under an agreement between them as to how the costs of the arbitration are to be borne, or under an award allocating the costs of the arbitration, extends only to such costs as are recoverable.

Definitions

'agreement', 'agree': s.5(1).
'costs of the arbitration': s.59.
'party': ss.82(2), 106(4).
'recoverable costs': ss.63, 64.

[62A] Status

This is a new section. It is *non-mandatory*, (s.4(2)).

[62B] Summary

The allocation of costs may be agreed by the parties or determined in an award made under s.61. This section is concerned with how much of the costs falling within the scope of any such agreement or of an award will actually be met by the paying party and recovered by the receiving party. Such costs are termed the 'recoverable costs' of the arbitration.

Points

[62C] *Effect*

If there is no agreement to the contrary, then the recoverable costs of the arbitration will be determined in accordance with ss.63 and 64. S.63 deals with recoverable costs in general; and s.64 deals separately with that part of the recoverable costs comprising the fees and expenses of the arbitrators. The section should be read with s.65, where appropriate, in relation to 'capping' the amount of recoverable costs.

[62D] *Model Law*

There is no comparable provision.

[62E] *Rules*

As might be expected, all the rules we have considered are silent on the ground covered by s.62. In a sense, the section is a statement of the obvious. Note that the fees and expenses of the tribunal will, under the LCIA and ICC regimes, and to an extent under UNCITRAL, be fixed by the institution concerned.

Section 63 – The Recoverable Costs
of the Arbitration

63.–(1) The parties are free to agree what costs of the arbitration are recoverable.

(2) If or to the extent there is no such agreement, the following provisions apply.

(3) The tribunal may determine by award the recoverable costs of the arbitration on such basis as it thinks fit.

If it does so, it shall specify–

(a)   the basis on which it has acted, and
(b)   the items of recoverable costs and the amount referable to each.

(4) If the tribunal does not determine the recoverable costs of the arbitration, any party to the arbitral proceedings may apply to the court (upon notice to the other parties) which may–

(a)   determine the recoverable costs of the arbitration on such basis as it thinks fit, or
(b)   order that they shall be determined by such means and upon such terms as it may specify.

(5) Unless the tribunal or the court determines otherwise–

(a)   the recoverable costs of the arbitration shall be determined on the basis that there shall be allowed a reasonable amount in respect of all costs reasonably incurred, and
(b)   any doubt as to whether costs were reasonably incurred or were reasonable in amount shall be resolved in favour of the paying party.

(6) The above provisions have effect subject to section 64 (recoverable fees and expenses of arbitrators).

(7) Nothing in this section affects any right of the arbitrators, any expert, legal adviser or assessor appointed by the tribunal, or any arbitral institution, to payment of their fees and expenses.

Definitions

'agreement', 'agree': s.5(1).
'arbitrator': s.82(1).
'costs of the arbitration': s.59.
'the court': s.105.
'notice (or other document)': s.76(6).

'party': ss.82(2), 106(4).
'recoverable costs': ss.63, 64.
'upon notice' (to the parties or the tribunal): s.80.

[63A]   ## Status

This section derives from the Arbitration Act 1950, s.18(1) and (2). It is *non-mandatory*, (s.4(2)).

[63B] **Summary**

The section deals with the determination of the recoverable costs of the arbitration. The recoverable costs are such costs, falling within the scope of an agreement or award as to costs, as will actually be met by the paying party and recovered by the receiving party. Whilst these include the recoverable fees and expenses of the arbitrators (see s.59(1)(a)), the latter are also dealt with in s.64, because it would be wrong to allow arbitrators to determine the reasonableness of their own fees and expenses.

The section has had the important effect of changing the presumption as to who will assess the recoverable costs. Under the regime prior to this Act an arbitrator had specifically to reserve powers to do so; now it is the norm. As we note in paragraph 63H below, this is likely to have implications as to the methods used by the parties and their legal advisers to present their claims for costs and may result in higher awards being made. However if the 'capping' provisions in s.65 have been employed, the amount of recoverable costs may well be limited.

Points

[63C] *Recoverable costs*
As with court proceedings, so in relation to arbitrations, the fact that an agreement or award as to costs is made in favour of a party does not normally mean that that party will recover *all* its expenditure. If that were so, then a party engaging legal and expert assistance for every aspect of its case at extravagant expense would be able to impose an unreasonable costs burden upon the paying party, if successful. Pursuant to s.62, absent agreement to the contrary, only 'recoverable' costs are payable.

A determination is normally made of what proportion of the costs covered by an agreement or award should actually be paid or recovered. This section deals with the different methods of determination; and also with the different bases that may be adopted in reaching decisions on what costs may be recovered.

[63D] *Agreement – Subs.(1)*
The section essentially provides that the parties are at liberty to agree what costs of the arbitration are recoverable. In practice agreement is a very common method by which the amount of recoverable costs is determined. The receiving party or their representatives send to the paying party a bill of costs setting out their proposals and claim. The paying party or their representatives respond with comments, counter-proposals and arguments. Both parties try to reach agreement by negotiation.

Agreement is often attempted as a first step, with determination by the tribunal or the court in default. This is because agreement avoids incurring the further costs of proceedings by the tribunal or the court to determine the recoverable costs, which will themselves fall to be allocated as between the parties, (see s.59(2)).

It should be noted that agreement, in this context, may well involve the parties making their own ad hoc arrangements. For example, where there are commercial solicitors on both sides in an arbitration, they may agree to arrange for the recoverable costs to be determined by an experienced costs clerk in a third firm of commercial solicitors. In our view this would fall within s.63(1) since if parties are free to agree what costs are recoverable, it seems to us to follow that they must be free to agree how such a figure is to be determined. Similarly, rules of an institution which provide for that body to fix costs would, we believe, be covered by s.63(1).

[63E] *Default provisions – Subs.(2)*
The default provisions of subss.(3) to (6) (and, where appropriate, s.64(2) and (3)) apply if or to the extent that there is no agreement as to what costs are recoverable. Thus if the parties have only succeeded in making an agreement as to certain costs, for example in respect of certain categories or in relation to a specific hearing, then the default provisions will cover the remainder.

[63F] *Determination by tribunal – Subs.(3)*
If the parties do not agree, the determination of recoverable costs may be referred to the tribunal. This subsection preserves the power of the tribunal prior to this Act to determine recoverable costs (or to 'assess' costs, as it is known in the court context).

There are certain distinct advantages in referring such a determination to the tribunal. The tribunal is, of course, very familiar with the course the proceedings took, the issues that were raised and the eventual outcome. It is therefore well placed to decide arguments over whether, for example, preparation time claimed in relation to a particular issue was justified, or whether frequent reference to experts for assistance and advice on a particular point was really necessary. Determination by the tribunal may therefore be speedy and efficient.

The tribunal's resulting award must specify the items of recoverable costs and the amount referable to each. In this way the determination will reflect the items in the bill of costs submitted by the receiving party and commented upon by the paying party. Probably the simplest means of specifying the amount referable to each item will be by marking the figures allowed on a copy of the bill of costs which is then annexed to the award. The tribunal must also specify the basis on which it has acted, as to which see paragraph 63H below.

It should be noted that the tribunal is required to make its determination by an 'award', so that the formalities of s.52 have to be followed. In the normal case, the final award in the arbitration will be that which covers the recoverable costs. This award will also need to deal with the recoverable costs incidental to the proceedings to determine such recoverable costs (s.59(2)), otherwise there could be a never-ending sequence of such awards!

The determination of recoverable costs often involves detailed and difficult considerations. Tribunals may be reluctant to undertake the task through lack of familiarity with or experience of this kind of exercise, or simply because of its nature. In such circumstances, for the avoidance of doubt and out of courtesy, a prompt and categoric refusal will be appropriate.

[63G]   *Determination by the court – Subs.(4)*
Where the tribunal does not determine the recoverable costs an application may be made by any party for the court to determine them.

The court is defined in s.105 as the High Court or a county court. Under s.63(4)(b) the court may order costs to be determined by such means as it may specify. The assessment of costs is dealt with by costs judges in the High Court and by district judges in the County Courts. See generally CPR Parts 43, 44 and 47.

Note that where the arbitral tribunal consists of or includes a judge-arbitrator, the powers of the court under this subsection may only be exercised by the High Court, (Sched.2, para.9).

The application is an arbitration application within the terms of the Practice Direction on Arbitrations under CPR Part 49. We deal generally with arbitration applications in Part 3, Materials, Section H. The application must obviously be on notice to the other parties and the proceedings served on them.

In our view, the Court of Appeal has jurisdiction to hear appeals from a judge or a court under this section, and a party may apply to the judge or to the Court of Appeal for permission to appeal and such judge and the court have power to grant that permission: *Inco Europe Ltd* v. *First Choice Distribution* [1999] 1 WLR 270. Whilst the report refers to ss.64 and 65 we think it was intended to refer to ss.63 and 64.

[63H]   *Basis of determination – Subs.(5)*
Both subss.(3) and (4) give the tribunal or the court power to determine the recoverable costs on such basis as is thought fit. There are obviously a variety of such bases, but this subsection sets out the equivalent of the 'standard basis' in the court context, which is that the receiving party will be allowed a reasonable amount in respect of all costs reasonably incurred, and any doubts as to either of those aspects are resolved in favour of the paying party.

It should be noted that CPR Rule 44.4(2) has redefined the former standard basis to the effect that the court will:

(a) only allow costs which are proportionate to the matters in issue; and

(b) resolve any doubt which it may have as to whether costs were reasonably incurred or reasonable and proportionate in amount in favour of the paying party.

We see no difficulty in arbitrators applying the principle of pro-portionality where appropriate.

An alternative basis sometimes adopted is known in the court context as the 'indemnity basis': see CPR Rule 44.4(2) and (3). This differs from the standard basis in that the test of proportionality does not apply and any doubts as to whether costs were reasonably incurred or reasonable in amount is resolved in favour of the receiving party.

Plainly, the indemnity basis is far less favourable to the paying party, particularly since, under the CPR regime, the concept of proportion-ality bites on the standard basis, but not on the indemnity basis. Whilst the tribunal or the court has a discretion as to the basis it should adopt (subss.(3) and (4)(a)), it will only exceptionally depart from the usual standard basis, and then only with good reason. It may select the indemnity basis, for instance, where deliberate conduct on the part of the paying party has caused costs to be unnecessarily and wastefully incurred. We would not recommend arbitrators to order costs on an indemnity basis of their own volition; we think there should always be an application from a party for such an order.

Note that under CPR Rule 44.5(3) the court is required to have regard to a number of factors when assessing costs. These are:

(1) the conduct of all the parties, including in particular conduct before, as well as during the proceedings; and
(2) the efforts made, if any, before and during the proceedings in order to try to resolve the dispute;
(3) the amount or value of any money or property involved;
(4) the importance of the matter to all the parties;
(5) the particular complexity of the matter or the difficulty or novelty of the questions raised;
(6) the skill, effort, specialised knowledge and responsibility involved;
(7) the time spent on the case; and
(8) the place where and the circumstances in which work or any part of it was done.

Again, we think an arbitrator may specifically adopt these criteria, and an arbitrator who, say, adopts 'standard basis under the CPR' as that upon which he acts probably does so. In any event, these factors pro-vide useful guidance for tribunals.

The foregoing does not mean that the tribunal or the court is necessarily tied to a conventional basis of assessment. A solicitor's bill of costs presented in such an assessment is a complex document that bears little, if any, relationship to the account the solicitor has actually sent his client. Arbitrators, not surprisingly, have had difficulty in fairly determining what was previously the equivalent to recoverable costs from such documents. The Act now allows them, in our view, the latitude to assess such costs from an account containing those actually incurred by the client. Such assessment has, in recent years, come to be known as assessment on a 'commercial' basis. Provided there is due adherence to the principle that doubts are resolved in favour of the paying party, such assessment, even if it produces more for the receiving party than would result from a conventional bill of costs, will not produce an unfair consequence. If anything, it will be better attuned to commercial reality.

We note that it appears that now arbitrators (or the court, where appropriate) will be able to decide the basis on which the costs should be assessed at the time of the actual assessment, rather than at the earlier time of making the award in which liability for costs is allocated. However, we think arbitrators would be well advised to determine and express the basis on which assessment is to be made at that earlier stage. To do so will avoid surprises, help the parties to negotiate and assist the court, if it has to make the assessment.

[63I] *The bill of costs*

In the last edition we noted that neither the tribunal nor the court was necessarily tied to a conventional basis of assessment, and that the solicitor's bill of costs bore little relationship to the account sent to his client so that arbitrators generally had some difficulty in assessing an appropriate sum. We suggested arbitrators might legitimately seek documentation of a different nature.

Under the CPR the form of the bill of costs has changed. There is not space in this book to deal with it in detail. Suffice it to say that the reader will find it helpful to study the Practice Directions about costs supplementing Parts 43 to 48 of the CPR, in particular Section 2 which deals with the Form and Content of Bills of Costs and Form 2 of the Schedule of Costs Forms. The effect of these changes is, we believe, to bring the assessment of costs more into line with reality and to make the documentation more 'user friendly' and understandable. We think that parties will expect to use the CPR system and documentation in relation to arbitration and we anticipate that most arbitrators will become comfortable with it.

[63J] *Summary assessment*

In cases on the 'fast track' and in cases or applications lasting a day or less the court now makes a summary assessment of costs at the conclusion of the hearing on the basis of documentation provided in shortened form in advance of the hearing. Clearly, by agreement, arbitrators may do the same. When there is no agreement, provided the tribunal adheres to its s.33 duties with care, we see no objection to summary assessment in a suitable case. But arbitrators should note that they are required to issue an award, and that implies giving reasons as well. In court the authors have experienced a fairly abrupt oral process; this will not be possible in arbitration.

[63K] *Recoverable fees and expenses of arbitrators – Subs.(6)*
It should be noted that the foregoing provisions are made subject to s.64. Thus in relation to the fees and expenses of arbitrators, the basis for determining recoverability there set out will apply.

[63L] *Payment of fees and expenses of arbitrators – Subs.(7)*
The entitlement of the tribunal, those appointed by it and any arbitral institution to fees and expenses as against the parties is wholly unaffected by this section. In other words, s.63 deals only with the position as between the parties. For the question of costs as between the parties and the arbitrators, see s.28.

[63M] *Model Law*
The Model Law has no comparable provision.

[63N] *Rules*
CIMAR, by Rule 13.10, provides for the application of subss.(3) to (7). The ICE Arbitration Procedure, by Rule 19.7, provides that the arbitrator has power to direct the basis upon which the costs are to be determined and, in default of agreement, to determine the amount of recoverable costs. The balance of s.63 not specifically covered accordingly applies. The CIArb Rules do not cover this ground, so the default provisions apply.

   Of the international rules, the LCIA Rules (Art.28.3) require the tribunal to determine legal costs 'on such reasonable basis as it thinks fit'. We think, therefore, that s.63(5) continues to apply. The ICC Rules give no guidance save that the tribunal should fix the costs (Art.31.3). S.63 will therefore apply. By Arts.38 and 39 of the UNCITRAL Arbitration Rules the criterion is 'reasonable'. We think this removes the need to consider a formal basis and constitutes an agreement within s.63(1) as to the categories of costs recoverable.

Section 64 – Recoverable Fees and Expenses of Arbitrators

64.–(1) Unless otherwise agreed by the parties, the recoverable costs of the arbitration shall include in respect of the fees and expenses of the arbitrators only such reasonable fees and expenses as are appropriate in the circumstances.

(2) If there is any question as to what reasonable fees and expenses are appropriate in the circumstances, and the matter is not already before the court on an application under section 63(4), the court may on the application of any party (upon notice to the other parties)–

(a) determine the matter, or
(b) order that it be determined by such means and upon such terms as the court may specify.

(3) Subsection (1) has effect subject to any order of the court under section 24(4) or 25(3)(b) (order as to entitlement to fees or expenses in case of removal or resignation of arbitrator).

(4) Nothing in this section affects any right of the arbitrator to payment of his fees and expenses.

Definitions

'agreed': s.5(1).
'arbitrator': s.82(1).
'costs of the arbitration': s.59.
'the court': s.105.
'notice (or other document)': s.76(6).

'party': ss.82(2), 106(4).
'recoverable costs': ss.63, 64.
'upon notice' (to the parties or the tribunal): s.80.

[64A] **Status**

This is a new section although to some extent it derives from the Arbitration Act 1950, s.19. It is *non-mandatory*, (s.4(2)).

[64B] **Summary**

The section deals separately with the determination of the recoverable fees and expenses of arbitrators as between the parties. These form part of the costs of the arbitration, the determination of whose recoverability is generally dealt with in s.63. The fees and expenses of the arbitrators include the fees and expenses of experts, legal advisers and assessors appointed by the tribunal for which the arbitrators are liable, (s.37(3)).

The need for the section arises because it cannot be appropriate that arbitrators should be entitled finally to determine the reasonableness of their own charges. In addition, there is a possibility that one or more of the arbitrators may be paid excessively by agreement with one party, and any costs award made under s.61 would then result in the other party being required to pay unreasonable amounts without the protection of this section.

Note that, exceptionally, a judge-arbitrator may himself exercise the power of the court under this section, (Sched.2, para.10).

Points

[64C] *Basis of determination – Subss.(1), (3)*
A distinct basis of determination is provided for the assessment of the recoverability of arbitrators' fees and expenses. Unless the parties agree otherwise, only such reasonable fees and expenses as are appropriate in the circumstances are recoverable. This broad basis of recoverability appears to allow such considerations as the seniority and experience of the arbitrators, the length and complexity of the arbitration and any special factors such as the extraordinary size of the claim or the particular significance of the arbitration to the parties to be borne in mind.

However where, on the removal or resignation of an arbitrator, the court has had occasion to adjust his entitlement to fees or expenses, then recoverability must be assessed subject to that order of the court, (subs.(3)).

[64D] *Determination by the court – Subs.(2)*
Where the parties have not been able to agree as to the recoverable fees and expenses of the arbitrators, their only recourse is to the court. As we have noted, they may not seek a determination from the tribunal, since it would be inappropriate for the tribunal to have the responsibility of deciding the reasonableness of its own fees.

The question may come before the court in the context of an application relating to all the costs of the arbitration, (s.63(4)). Alternatively, if the recoverability of the costs of the arbitration apart from the arbitrators' fees and expenses is not before the court (for example, because it is being or has been agreed, or determined by the tribunal), an application may be made pursuant to this section.

The court need not carry out the determination itself. It may instead order the determination to be carried out by such means and upon such terms as it may specify. As we have indicated at paragraph 63G, the obvious means would be to pass the matter to a costs judge of the High Court or a district judge with suitable experience.

[64E] *The application*
The application under s.64(2) is an arbitration application falling within the terms of the Practice Direction on Arbitrations under CPR

Part 49. We deal generally with arbitration applications in Part 3, Materials, Section H. Specifically in respect of applications under s.64:

(i)   the application must be on notice to the other parties;

(ii)   the requirement for giving notice to the other parties is to be met by making those parties respondents to the application and serving on them the arbitration claim form and evidence in support;

(iii)   unless otherwise ordered, the hearing will be in private.

[64F]   *Payment of fees and expenses of arbitrators – Subs.(4)*

The entitlement of the tribunal and those appointed by it to fees and expenses as against the parties is wholly unaffected by this section. That is dealt with as a question of costs as between the parties and the arbitrators in s.28.

As we see it, if any challenge to the reasonableness of the fees and expenses of a tribunal under this section were to succeed, the precise consequences would depend upon a number of factors, including who had paid the costs in the first instance, who was liable for them (as a matter of allocation of responsibility for costs) and whether one or more of the arbitrators had a private agreement with one or more of the parties as to his fees. In our view, the court would take such factors – and the possible application of s.28(5) – into account when deciding the 'means' and 'terms' that it would direct in relation to an adjustment under s.28(2).

[64G]   *Appeals to Court of Appeal*

The Court of Appeal has jurisdiction to hear appeals from a judge or a court under this section, and a party may apply to the judge or to the Court of Appeal for permission to appeal and such judge and the court have power to grant that permission: *Inco Europe Ltd* v. *First Choice Distribution* [1999] 1 WLR 270.

[64H]   *Model Law*

There is no comparable provision.

[64I]   *Rules*

The domestic rules we have considered do not cover this ground so that s.64 will apply. Art.28.1 of the LCIA Rules provides for the LCIA Court to determine such fees and expenses in accordance with the Schedule of Costs and Art.29.1 provides that the decision of the LCIA Court is binding. We therefore think that s.64 is ousted, there being a contrary agreement. Note also our commentary on s.28 above. We are less convinced that the ICC Rules, the drafting of which was not specifically centred around this Act, have the same effect: see Art.31 and Appendix III. Nonetheless the parties have agreed to pay the fees of the arbitrator as set by the ICC Court and we certainly think it arguable that there is a contract with the arbitrators to pay their fees as so established and thus an agreement contrary to s.64.

Art.38 of the UNCITRAL Arbitration Rules provides that the fees of the arbitral tribunal shall be fixed '... by the tribunal itself in accordance with article 39'. Art.39(1) provides:

> 'The fees of the arbitral tribunal shall be reasonable in amount, taking into account the amount in dispute, the complexity of the subject-matter, the time spent by the arbitrators and any other relevant circumstances of the case'.

The balance of the article sets up a regime for consideration of any fee schedule published by any relevant institution and consultation with the institution itself.

We think the combination of Arts.38 and 39 constitute an agreement under s.64(1) that the recoverable costs are the fees fixed under Art.39, Art.40 in effect providing that the successful party recovers 'costs' as defined in Art.38. Thus s.64 would not apply to an arbitration subject to these Rules.

Section 65 – Power to Limit Recoverable Costs

65.–(1) Unless otherwise agreed by the parties, the tribunal may direct that the recoverable costs of the arbitration, or of any part of the arbitral proceedings, shall be limited to a specified amount.

(2) Any direction may be made or varied at any stage, but this must be done sufficiently in advance of the incurring of costs to which it relates, or the taking of any steps in the proceedings which may be affected by it, for the limit to be taken into account.

Definitions

'agreed': s.5(1).
'costs of the arbitration': s.59.
'party': ss.82(2), 106(4).
'recoverable costs': ss.63, 64.

[65A] Status

This is a new section. It is *non-mandatory*, (s.4(2)).

[65B] Summary

The section gives the tribunal a completely new power, which we believe to be unique, to limit in advance the amount of recoverable costs in relation to the whole arbitration or any part of the arbitral proceedings. The parties may agree that the tribunal should not have such a power, but if, for instance, they have adopted institutional rules which say nothing about this matter, it will be deemed included.

Points

[65C] *Purpose*

The power enables the tribunal, by imposing an upper limit on the recoverable costs, to encourage parties to exercise restraint and to avoid unnecessary expense. Whilst they may, of course, still incur costs beyond the limit imposed, the knowledge that they will not recover such costs from the other parties may deter them from so doing. A recoverable costs limit may also offer a measure of protection to financially weak parties facing financially stronger opponents. For the former, continuing with the arbitration once a limit has been imposed may be feasible; whereas without it, where the potential liability for costs might be catastrophic, it would not. As such, the power could be seen as a facet of the tribunal's s.33(b) duty to adopt procedures which, by avoiding unnecessary expense, provide a fair means for the resolution of the dispute.

Plainly, the power must also be exercised in accordance with the tribunal's s.33(a) duty to act fairly and impartially as between the parties. So, for example, a tribunal considering imposing a limit in relation to a part of the arbitral proceedings would have carefully to consider whether by so doing it might be introducing an imbalance in its treatment of the parties. A limit in relation to discovery, for example, might work to the advantage of a party who had very few documents to discover, and to the disadvantage of a party who had many documents to discover and placed considerable reliance upon them. Similarly, a limit in relation to expert evidence might work to the advantage of a party intending to call only a single expert, and to the disadvantage of a party that regarded it as vital to call more than one.

Where a party has limited means, it may well be that as part of its duty to act fairly and impartially as between the parties under s.33, a tribunal should be inclined to cap costs under this section if asked to do so. If no such request is made, then arguably the tribunal should, at the very least, raise the question.

It seems quite clear that any attempt to impose different financial caps on different parties, for example in an attempt to prevent a financially well-off party from abusing its position, would be improper.

[65D] *Drafting – Subs.(1)*

The power may be exercised in respect of the whole or 'any part' of the arbitral proceedings. We think 'any part' is wide enough to cover any stage of the proceedings, such as disclosure; any aspect, such as a preliminary issue; any specific issue; and possibly, even, a specific period which might be defined either by reference to a stage in the proceedings, by a date or by a period of time.

We have given some thought as to how suitable directions might be drafted. The following are examples:

'(1) I order and direct that recoverable costs falling within s.59(1)(c) of the Arbitration Act 1996, that is the legal and other costs of the parties, arising out of or in connection with my order of [date] requiring disclosure of documents shall not exceed £5,000;
(2) I order and direct that recoverable costs falling within s.59(1)(c) of the Arbitration Act 1996, that is the legal and other costs of the parties, arising out of or in connection with the determination of the following issue to be determined in an interim award specifically devoted thereto, namely ... shall not exceed £5,000;
(3) I order and direct that recoverable costs falling within s.59(1)(c) of the Arbitration Act 1996, that is the legal and other costs of the parties, arising out of or in connection with the arbi-

tration from the day immediately following this order shall not exceed £25,000.'

Our experience of drafting the above identified some potential pitfalls. In particular, it seems more satisfactory to define what is meant by recoverable costs than to leave it unclear. An order that simply read, 'recoverable costs shall not exceed £5,000' may be workable but it would be barely so, since the allocation of the sum between different types of costs is unknown. In particular, it is difficult to see how it could be safely used since it would include the arbitrators' fees and expenses (s.59(1)(a)) which are outwith the parties' control and knowledge.

In the first edition we suggested that whilst it might be feasible to cap all recoverable costs, we would recommend caps only to be placed on party costs, at least until there was wider experience as to how the section operated. We retain the view that it is better not to place a cap on the tribunal's costs, at least as part of an overall figure. It is, of course, open, and it may be appropriate in certain cases, for a tribunal to agree to limit its own costs to a specific amount, but unless so identified a figure inclusive of party and tribunal costs is difficult to monitor. Again, where the tribunal's costs are capped, we suggest it be made clear that a revision is possible.

Our experience to date has, in fact, demonstrated a wide disparity of views as to how the cap may be applied. Our own preference is that it is most easily applied to all recoverable costs in respect of the whole arbitration from the point of assessment, usually after the shape of the arbitration has been identified at the preliminary meeting. We also endorse the approach of those arbitrators who cap the amount that may be recovered in respect of expert evidence. Each of these orders gives little room for argument as to what falls within the cap and what is outside it. An arbitrator whose cap leaves room for argument may be creating problems for himself later on.

[65E] *Procedure – Subs.(2)*
The limit may be imposed or varied at any stage, so long as that is sufficiently in advance of the matters affected by it for the limit or its variation to be taken into account. Presumably, the possibility of a variation in the limit is available to enable the tribunal to deal with changing circumstances, or the perception that an initial limit imposed is likely to operate unfairly to one or more parties.

[65F] *Who may apply?*
The Act does not specify whether an application from a party is required, or whether the tribunal may act on its own initiative. We think both are possibilities. But in either case we think the tribunal should only act, so as to comply with its s.33 duty, on full information and after having given the parties adequate opportunity to make

submissions. In particular, the tribunal needs to have a very good idea of what it may be reasonable to spend on any particular matter. That involves fairly detailed information as to what the parties and their legal advisers may have to do at any stage, the amount of documentation, and so on. This in turn may well involve giving directions under s.34 so as to enable the tribunal to ascertain relevant information.

As we have indicated above, our own preference is to make these orders after the preliminary meeting has taken place. It is at that stage possible to ask the parties' representatives to make an assessment of the likely costs of the proposed arbitration. From this information, a figure may sensibly be established.

[65G]   *Model Law*
There is no comparable provision.

[65H]   *Rules*
Both CIMAR (Rule 13.4) and the ICE Arbitration Procedure (Rule 7.4) embrace s.65. CIMAR develops it: the arbitrator shall have regard primarily to the amounts in dispute and to any advisory or supplementary procedure under the contract to which the dispute relates. Claims which are not claims to money must be taken into account and, for the purposes of variation, the arbitrator may require the parties to submit at any time statements of costs incurred and foreseen. The CIArb Rules do not refer to the topic, so s.65 applies.

Each of the international rules we have considered are silent on this matter. Since s.65 applies unless agreed otherwise, the section is effective in the absence of any agreement to exclude it.

Powers of the Court in Relation to Award

Section 66 – Enforcement of the Award

66.–(1) An award made by the tribunal pursuant to an arbitration agreement may, by leave of the court, be enforced in the same manner as a judgment or order of the court to the same effect.

(2) Where leave is so given, judgment may be entered in terms of the award.

(3) Leave to enforce an award shall not be given where, or to the extent that, the person against whom it is sought to be enforced shows that the tribunal lacked substantive jurisdiction to make the award.

The right to raise such an objection may have been lost (see section 73).

(4) Nothing in this section affects the recognition or enforcement of an award under any other enactment or rule of law, in particular under Part II of the Arbitration Act 1950 (enforcement of awards under Geneva Convention) or the provisions of Part III of this Act relating to the recognition and enforcement of awards under the New York Convention or by an action on the award.

Definitions

'arbitration agreement': ss.6, 5(1).
'the court': s.105.
'enactment': s.82(1).
'substantive jurisdiction' (in relation to an arbitral tribunal): s.82(1),
 and see s.30(1)(a) to (c).

[66A] Status

This section is derived from the Arbitration Act 1950, s.26(1); and from the Model Law, Art.35. It is *mandatory*, (s.4(1) and Sched.1).

[66B] Summary

The section deals with the enforcement of an award by the court. It applies wherever the seat of the arbitration is (s.2(2)(b)), so that an arbitration award made in Switzerland, for example, may be enforced in England under this provision.

The section is a good example of the powers of the court being used to support arbitration. Because of its nature, the parties may not exclude this power by agreement.

Points

[66C] *Judgment in terms of award – Subs.(1), (2)*
Two methods of enforcement of an award are available under these subsections. They are cumulative, rather than alternative.

The first is an application directly to enforce the award in the same manner as a judgment or order to the same effect. If leave (now known, under the CPR, as 'permission') is given, the applicant may issue execution upon the award as if it were a judgment, without actually entering a judgment. Court procedures will then be available to enforce the award. Where the award is for the payment of a sum of money such procedures include the seizure and sale of the respondent's goods; the interception of a debt due to the respondent; and the charging and sale of the respondent's property. In the case of an order to do or refrain from doing something, the court's powers include that of committing the respondent to prison for contempt.

The second method, where permission has been given, is an application actually to enter judgment in terms of the award. There may be advantages in proceeding to this second stage rather than being content with the first. For example, the applicant may sue on the judgment or otherwise proceed by execution or registration or other method of enforcement in a foreign court; or he may obtain recognition of the judgment in a foreign court; or he may rely on the judgment as a judicial resolution of the issues that prevents any further action being brought in a foreign court.

A court cannot, when entering judgment in terms of an award, add on post-award interest if that has not been granted in the award: see *Walker* v. *Rome* referred to at paragraph 49E above.

[66D] *Discretion – Subs.(3)*
The section does not require the court to order enforcement in every case, but gives it a discretion to do so, the court exercising that discretion by the grant or refusal of leave.

Only in one case is the court bound to refuse permission, and that is where the person against whom enforcement is sought shows that the tribunal lacked substantive jurisdiction to make the award. Such a person may therefore raise a passive defence to the award at the enforcement stage, provided the right to do so has not been lost because the objection was not made in time, (s.73). (For the substantive jurisdiction of the tribunal, and preliminary objections made to the tribunal and the court on the ground of lack of substantive jurisdiction, see ss.30-32. For the challenge to an award on such a ground, see s.67.)

However, there are a number of other situations in which the person against whom enforcement is sought may successfully oppose the grant of permission to enforce the award. Consideration was initially given to enacting a non-exhaustive list of other situations in which the court would be required to refuse permission. These included instances where the award was so defective in form or substance that it was incapable of enforcement; where the award granted relief which (if enforced as a judgment or order of the court) would improperly affect the rights of persons other than parties to the arbitration agreement; where the award purported to decide matters which are not capable of resolution by arbitration; and where enforcement of the award would be contrary to public policy.

In the event, it was considered that a non-exhaustive list would be unsatisfactory since parties might think that matters not mentioned were not covered. As enacted, therefore, the section gives the court an unfettered discretion to grant or withhold permission to enforce in relation to objections made on a basis other than lack of substantive jurisdiction. Note, in this regard, the express saving of the common law rules relating to matters which are not capable of settlement by arbitration, and to the refusal of recognition and enforcement of an arbitral award on grounds of public policy, (s.81(1)(a) and (c) respectively).

[66E] *Judge-arbitrators*
Note that in the case of an award made by a judge-arbitrator, the judge-arbitrator may himself give permission to enforce under this section, (Sched.2, para.11).

[66F] *Other means of enforcement – Subs.(4)*
Although the section applies so as to permit enforcement of foreign awards, it is made clear that enforcement of an award as a judgment of the court under this section does not affect other regimes or methods of enforcement. In relation to foreign awards, the Geneva and New York Conventions (as to which see Part III) are notable examples.

We have had some difficulty assessing how ss.66 and 99 to 104 work together. It would appear to us that a New York Convention award may be enforced under either route, but it is likely that the s.66 route presents more obstacles. In particular, recognition and enforcement of a Convention award may only be refused in the cases set out in s.103(2), and it may prove the case that enforcement under s.66, which is discretionary, provides greater opportunity for possible refusal.

The words 'by an action on the award' refer to the possibility of bringing court proceedings founded on an arbitration award. Such proceedings need not be lengthy if there is no arguable defence, since the court may then enter summary judgment on the basis of documents and sworn depositions or witness statements. However such a course may also be appropriate where the objections to the award are such as cannot be resolved without trial.

[66G] *The application*
The procedure is set out in Part III of the Practice Direction on Arbitrations supplementary to CPR Part 49. The application may initially be made without notice on Form 8A, referred to for this purpose as 'the

enforcement form'. The application must be supported by an affidavit or witness statement exhibiting the arbitration agreement and the original award (or copies); stating the names and relevant addresses of the applicant and of the person against whom it is sought to enforce the award; and stating either that the award has not been complied with or the extent to which it has not been complied with at the date of the application. The court may direct service of the enforcement form on specified parties and, with permission, it may be served out of the jurisdiction. The matter then proceeds as if it were an arbitration application. The hearing is in private, unless the court orders otherwise. For the detailed requirements of the procedure, see para.31 of the Practice Direction on Arbitrations under CPR Part 49.

[66H] *Model Law*

Art.35 (Recognition and enforcement) provides:

> '(1) An arbitral award, irrespective of the country in which it was made, shall be recognised as binding and, upon application in writing to the competent court, shall be enforced subject to the provisions of this article and of article 36'.

The balance of the article deals with formalities; Art.36 repeats in substance the text of the New York Convention relevant to refusing recognition and enforcement. We note that Art.35 links the conditions for refusal of enforcement into those of the New York Convention, (as to which see ss.100 to 104). As with Art.35, enforcement under this section is not dependent upon the seat or place of arbitration.

Section 67 – Challenging the Award: Substantive Jurisdiction

67.–(1) A party to arbitral proceedings may (upon notice to the other parties and to the tribunal) apply to the court–

(a) challenging any award of the arbitral tribunal as to its substantive juris-diction; or

(b) for an order declaring an award made by the tribunal on the merits to be of no effect, in whole or in part, because the tribunal did not have sub-stantive jurisdiction.

A party may lose the right to object (see section 73) and the right to apply is subject to the restrictions in section 70(2) and (3).

(2) The arbitral tribunal may continue the arbitral proceedings and make a further award while an application to the court under this section is pending in relation to an award as to jurisdiction.

(3) On an application under this section challenging an award of the arbitral tribunal as to its substantive jurisdiction, the court may by order–

(a) confirm the award,
(b) vary the award, or
(c) set aside the award in whole or in part.

(4) The leave of the court is required for any appeal from a decision of the court under this section.

Definitions

'the court': s.105.
'notice (or other document)': s.76(6).
'party': ss.82(2), 106(4).
'substantive jurisdiction' (in relation to an arbitral tribunal): s.82(1), and see s.30(1)(a) to (c).
'upon notice' (to the parties or the tribunal): s.80.

[67A]   ## Status

This is a new section, deriving in part from the Model Law, Art.16 and 34. It is *mandatory*, (s.4(1) and Sched.1).

[67B]   ## Summary

The section provides for an application to court challenging an award insofar as it relates to the substantive jurisdiction of the tribunal. Because of its nature, the parties may not exclude the right to make such an application by agreement.

Points

[67C]   *Basis of the application – Subs.(1)*

Ss.31 and 32 provide for alternative ways of dealing with any question as to the tribunal's substantive jurisdiction. Which is appropriate depends to some extent upon whether the tribunal has power to rule on its own jurisdiction under s.30, that being a non-mandatory section where the scope of the tribunal's substantive jurisdiction is set out.

If the tribunal has that power to rule, then, pursuant to s.31, it may be called on to exercise it either by making a specific award as to jurisdiction, or by dealing with the jurisdiction objection in its eventual award on the merits. This latter choice is primarily one for the tribunal's discretion; however, the parties are able, by agreement, to require the tribunal to adopt one or other of these courses, (s.31(4)).

Whichever course it adopts, the tribunal's award is itself subject to challenge as to the jurisdiction aspect. Subject to the restrictions referred to in paragraph 67D below, an application to the court may be made by a party pursuant to this section.

Note the alternative procedure of an application to the court for the determination of a preliminary point of jurisdiction prior to any award being made, pursuant to s.32. Note also the possibility of opposing leave to enforce the award on the ground of lack of substantive jurisdiction, as to which see s.66(3).

[67D]   *Restrictions*

A party intending to make an application under this section must first exhaust available arbitral procedures in accordance with s.70(2)(a), (correction of the award or the making of an additional award under s.70(2)(b) being seemingly irrelevant). A party must also observe the 28 day time limit provided by s.70(3), subject to the court's power to extend the time limit pursuant to s.80(5). Note that there is no compliance with the 28 day time limit unless the arbitration claim form has been issued and all affidavits or witness statements in support have been filed by the expiry of the time limit: see para.22.1 of CPR Part 49G (Practice Direction). In addition, the applicant must not have lost the right to object through knowing delay by virtue of s.73.

With the purpose of avoiding unnecessary delay, the tribunal is able to proceed with the arbitration while an application to the court under the section is pending. We have thought arbitrators to have been timid in the past about proceeding in such circumstances. They will now have clear authority to do so if appropriate.

[67E]   *Remedies – Subs.(3)*

The relief available from the court varies according to whether the tribunal's ruling as to jurisdiction is contained in a specific award, or is included in the award on the merits. In the former case, covered by subs.(1)(a), the court has the option of confirming or varying the specific award, or setting it aside, in whole or in part, (subs.(3)). In the latter case, the appropriate relief is a declaration that the award on the merits is of no effect, either in whole or in part, for want of substantive jurisdiction, (subs.(1)(b)).

In hearing a challenge to an award as to an arbitration tribunal's substantive jurisdiction, the court may give directions enabling the challenging party to present its case and to challenge the other party's case with the full panoply of oral evidence and cross-examination so that, in effect, the challenge becomes a complete re-hearing of all that has already occurred before the arbitration tribunal. This is all the more so where a tribunal did not have the benefit of oral evidence and said that it came to its final decision with uncertainty.

In a case involving substantial issues of fact, including one as to whether a party had ever been a party to a contract, it may well be preferable to go straight to court rather than have the tribunal determine its own jurisdiction, either by agreement between the parties or upon an application and with the tribunal's consent: *Azov Shipping Co. v. Baltic Shipping Co.* [1999] 1 Lloyd's Rep. 68 and [1999] 2 Lloyd's Rep. 159.

On the other hand, in *Ranko Group* v. *Antarctic Maritime SA* (1998) LMLN 492 the court said it would be most unfortunate if parties could contest jurisdiction before an arbitrator and then, on a later application under this section, seek to introduce a raft of new evidence which could and should have been put before the arbitrator in the first instance.

[67F]   *The application*

The application under s.67 is an arbitration application falling within the terms of the Practice Direction on Arbitrations under CPR Part 49. We deal generally with arbitration applications in Part 3, Materials, Section H. Specifically in respect of an application under s.67:

(i)   the application must be on notice to the other parties and to the tribunal;

(ii)   the requirement for giving notice to the other parties is to be met by making those parties respondents to the application and serving on them the arbitration claim form and evidence in support;

(iii)   the requirement for giving notice to the tribunal is to be met by sending copies of the arbitration claim form to the arbitrators for their information at their last known address together with copies of the evidence in support;

(iv)   the arbitration claim form must show that any available arbitral process of appeal or review, or any available recourse under s.57 (correction or additional award) has been exhausted, and (subject to (v) below) the appeal is brought within 28 days of the award or, where there has been any arbitral process of appeal or review, within 28 days of the date when the applicant was notified of the result of that process;

(v) where the 28 day time limit has expired and the applicant seeks an extension of time, the arbitration claim form must also state why an order extending time should be made and the affidavit or witness statement in support must set out the evidence relied upon. (See further our commentary on s.70 below);

(vi) unless otherwise ordered, the hearing is in private.

[67G] *Prosecuting challenges*

Although decided under the 1950 Act the principles laid down in a recent case must apply to challenges to awards under the 1996 Act. They appear to be:

(1) The court has power to strike out for want of prosecution pro-ceedings challenging arbitral awards under the Arbitration Act.

(2) The discretion to strike out arises if there has been a failure to act with all deliberate speed to which attention is specifically drawn in the Commercial Court Guide.

(3) It is the duty of the party challenging the award to prosecute his application for leave to appeal and if leave is granted the appeal itself.

(4) It is not necessary for the party applying for the challenge to be struck out to show prejudice or a likelihood of prejudice caused by the delay.

(5) The discretion should be exercised in accordance with the policy of the law to promote speedy finality in the enforcement of arbi-tration awards, which policy involves wider interests than the private interests of the parties, namely the proper functioning of the arbitration system.

(6) Accordingly, although the exercise of the discretion to strike out will depend on the circumstances of each case, in general it should be exercised if there has been a failure to act with all deliberate speed unless there are good reasons for not doing so.

(7) Whilst prejudice is not a necessary condition for the exercise of the power to strike out, it may be relevant. The absence of prejudice is only one factor to take into consideration. The court should not place undue weight on this point because to do so would be to run the risk of introducing prejudice as a condition of the exercise of the court's discretion to strike out in the case of a failure to act with all deliberate speed.

(8) The foregoing principles apply with equal or greater force to an application to remit an award: *Huyton SA* v. *Jakil SpA* (unreported, Clarke J, 16 September 1997; affirmed by the Court of Appeal on other grounds: [1999] 2 Lloyd's Rep. 83). Security for the costs of a challenge may be ordered under s.70(6).

[67H] *Appeal – Subs.(4)*

Any appeal from a decision of the court may only be made with leave.

[67I]   ***Person who takes no part in proceedings***

It should be noted that a person alleged to be a party to arbitral proceedings, but who opts to take no part in them, may question the
substantive jurisdiction of the tribunal by court proceedings (s.72(1));
and may also challenge an award on the ground of lack of jurisdiction
by an application to court under this section (s.72(2)(a)). See s.72 for the
special considerations that apply in such a case.

[67J]   *Model Law*

Art.16(3) (set out at paragraph 31H) and Art.34 between them provide
a similar regime. The relevant part of Art.34 is as follows:

> '(2) An arbitral award may be set aside by the court ... only if:
> (a) the party making the application furnishes proof that:
> (i) ... the said [arbitration] agreement is not valid under the law to which the
> parties have subjected it or, failing any indication thereon, under the law of
> this State; or
>
> . . . . .
>
> (iii) the award deals with a dispute not contemplated by or not falling
> within the terms of the submission to arbitration, or contains decisions on
> matters beyond the scope of the submission to arbitration, provided that, if
> the decisions on matters submitted to arbitration can be separated from
> those not so submitted, only that part of the award which contains decisions
> on matters not submitted to arbitration may be set aside; or
> (iv) the composition of the arbitral tribunal ... was not in accordance with
> the agreement of the parties, unless such agreement was in conflict with a
> provision of this Law from which the parties cannot derogate, or, failing
> such agreement, was not in accordance with this Law ...'

[67K]   *Rules*

The section is mandatory so it will override any provision attempting
to oust the courts' jurisdiction. In any event the position that the court
has the last say on the matter is both logical and universal. The rules we
have considered contain nothing that attempts to change the position
under this section.

Section 68 – Challenging the Award: Serious Irregularity

68.–(1) A party to arbitral proceedings may (upon notice to the other parties and to the tribunal) apply to the court challenging an award in the proceedings on the ground of serious irregularity affecting the tribunal, the proceedings or the award.

A party may lose the right to object (see section 73) and the right to apply is subject to the restrictions in section 70(2) and (3).

(2) Serious irregularity means an irregularity of one or more of the following kinds which the court considers has caused or will cause substantial injustice to the applicant–

(a) failure by the tribunal to comply with section 33 (general duty of tribunal);
(b) the tribunal exceeding its powers (otherwise than by exceeding its substantive jurisdiction: see section 67);
(c) failure by the tribunal to conduct the proceedings in accordance with the procedure agreed by the parties;
(d) failure by the tribunal to deal with all the issues that were put to it;
(e) any arbitral or other institution or person vested by the parties with powers in relation to the proceedings or the award exceeding its powers;
(f) uncertainty or ambiguity as to the effect of the award;
(g) the award being obtained by fraud or the award or the way in which it was procured being contrary to public policy;
(h) failure to comply with the requirements as to the form of the award; or
(i) any irregularity in the conduct of the proceedings or in the award which is admitted by the tribunal or by any arbitral or other institution or person vested by the parties with powers in relation to the proceedings or the award.

(3) If there is shown to be serious irregularity affecting the tribunal, the proceedings or the award, the court may–

(a) remit the award to the tribunal, in whole or in part, for reconsideration,
(b) set the award aside in whole or in part, or
(c) declare the award to be of no effect, in whole or in part.

The court shall not exercise its power to set aside or to declare an award to be of no effect, in whole or in part, unless it is satisfied that it would be inappropriate to remit the matters in question to the tribunal for reconsideration.

(4) The leave of the court is required for any appeal from a decision of the court under this section.

Definitions

'the court': s.105.
'notice (or other document)': s.76(6).
'party': ss.82(2), 106(4).
'substantive jurisdiction' (in relation to an arbitral tribunal): s.82(1), and see s.30(1)(a) to (c).
'upon notice' (to the parties or the tribunal): s.80.

[68A] **Status**

This section derives from the Arbitration Act 1950, ss.22(1) and 23(2); and from the Model Law, Art.34. It is *mandatory*, (s.4(1) and Sched.1).

[68B] **Summary**

The section provides for an application to court to challenge an award on the ground of serious procedural irregularity affecting the tribunal, the proceedings or the award. Note that where the irregularity alleged is that of the tribunal exceeding its substantive jurisdiction, an application under s.67 is appropriate, (subs.(2)(b)).

Because of its nature the parties may not exclude an application under this section by agreement.

Points

[68C] *Basis of the application – Subs.(2)*
The application may only be founded on one of the grounds of serious irregularity listed in subs.2(a) to (i). Although a ground such as failure by the tribunal to comply with its s.33 duty is both broad and somewhat open-ended, most of the grounds are fairly specific. The list of grounds is exhaustive and deliberately so, since the DAC thought that the courts should not be free to invent more.

The attitude of the court to a number of instances of alleged misconduct in a recent case under the 1950 Act would suggest that a robust approach is likely to be taken to any allegations of serious irregularity under this section: *GEC Alsthom Metro Cammell Ltd* v. *Firema Consortium* (unreported, Longmore J, 25 September 1997).

As to what does constitute a serious irregularity for the purposes of this section, it has been held that an arbitrator influenced in the exercise of his discretion on costs by factors which had not been drawn to the attention of the parties, who had thus not had an opportunity of dealing with them, did not act fairly and impartially under s.33, and thus his conduct amounted to a serious irregularity under the present section: *Gbangbola* v. *Smith & Sherriff Ltd* [1998] 3 All ER 730.

On the other hand, there was held to be no serious irregularity in a case where trade arbitrators refused to hear expert evidence in a trade arbitration, the governing rules of which gave them the widest discretion to determine procedure and entitled them to allow, refuse or limit the appearance of witnesses, whether factual or expert: *Egmatra AG* v. *Marco Trading Corporation* [1999] Lloyd's Rep. 862. Similarly in another trade arbitration where an arbitrator and an associate of a party had earlier been in dispute: *Rustal Trading Ltd* v. *Gill & Duffus SA* [2000] 1 Lloyd's Rep. 14. It has also been held that an arbitrator's decision not to order production of a ship's log in a case under a Small Claims

Procedure fell far short of anything which could properly be regarded as a serious irregularity: *Ranko Group* v. *Antarctic Maritime SA* (1998) LMLN 492.

In another recent case under the 1950 Act (ss.2 and 23) the Court of Appeal remitted to arbitrators a costs award made because the arbitrators thought that a party had not beaten a 'Calderbank' offer when in fact that offer had been beaten and the arbitrators' view was based on a simple mathematical error, although such error was made after 'endless argument' and was justified by lengthy written reasons, and was not accidental or admitted by the arbitrators. The court commented that if the arbitration had been governed by the 1996 Act, broadly similar issues would have arisen. This suggests that the court took the view that the arbitrators' error was such as to put them in breach of their s.33 duty. On the particular facts of the case, there would have been a 'substantial injustice' if there had been no remission (*Danae Air Transport SA* v. *Air Canada* [1999] 2 All ER (Comm.) 943).

In addition to considering whether there has been 'serious irregularity' within the rather restricted list set out in subs.2, the court must also consider whether the irregularity relied upon in each case has caused or will cause substantial injustice to the applicant. This is a clear indication that a challenge on the ground of serious irregularity is only intended to be available in cases where the arbitral procedure has so far departed from what might reasonably have been expected as to justify the corrective intervention of the court (see, e.g., *Conder Structures* v. *Kvaerner* [1999] ADRLJ 305).

The court's power is essentially designed to be supportive of arbitration in the sense of maintaining its good reputation by being available to rectify glaring and indefensible irregularities that would occasion injustice. The limitations on that power restrain the court from interfering with the arbitral process on any lesser occasion. For example, it will not now be able to remit an award for further consideration of an argument not pursued before arbitrators, as it did in *Indian Oil Corporation* v. *Coastal (Bermuda) Ltd* [1990] 2 Lloyd's Rep. 407.

By contrast with the previous position, where the court could remit an award for reconsideration whenever it thought fit, and where the setting aside of the award was linked to the difficult concept of the arbitrator's 'misconduct', this section gives considerably more definition and structure to the court's power to intervene on the basis of irregularity.

[68D] *Restrictions – Subs.(1)*

A party intending to make an application under this section must first exhaust available arbitral procedures and available recourse to correction of the award or the making of an additional award, in accordance with s.70(2). Any ambiguities or uncertainties in an award must first be dealt with by the tribunal in the light of s.57 and s.70(2)(b) before any application or appeal relating to them is taken to the court, but any appeal or application in relation to a part of the award unaffected by any decisions on ambiguities or uncertainties should not be held up pending resolution of those: *Gbangbola* v. *Smith & Sherriff Ltd* (above). A party must also observe the 28 day time limit provided by s.70(3), subject to the court's power to extend the time limit pursuant to s.80(5). Note that there is no compliance with the 28-day time limit unless the arbitration claim form has been issued and all affidavits or witness statements in support have been filed by the expiry of the time limit: see para.22.1 of CPR Part 49G (Practice Direction). In addition, the applicant must not have lost the right to object through knowing delay by virtue of s.73.

[68E]   *Remedies – Subs.(3)*
In addition to the remedies previously available to the court of remitting the award for reconsideration or setting aside the award (which are now exercisable as to whole or part of the award), a new remedy is introduced of declaring the award to be of no effect, in whole or in part. The court may only set aside an award or declare it to be of no effect, in whole or in part, where remission would be inappropriate.

In our view, remission is likely to be the usual remedy under subs.(2)(d), (f) and (h). Setting aside may be appropriate under the other subsections. Other than in respect of lack of substantive jurisdiction, which is dealt with in s.67, we have not been able to think of any case where declaring the award to be of no effect would be appropriate in preference to setting aside.

[68F]   *The application*
Our commentary in paragraph 67F (relating to an application under s.67) applies equally to s.68 and we refer the reader to it. For more detailed commentary on an application for extension of time, see s.70 below.

[68G]   *Prosecuting applications*
As to the duty to prosecute applications under this section expeditiously see paragraph 67G.

[68H]   *Appeal – Subs.(4)*
Any appeal from a decision of the court may only be made with leave.

[68I]   *Person who takes no part in proceedings*
It should be noted that a person alleged to be a party to arbitral proceedings, but who opts to take no part in them, may challenge an award on the ground of serious irregularity affecting him by making an application to court under this section, (s.72(2)(b)). See s.72(2) for the special considerations that apply in such a case.

[68J]   *Model Law*

This section is intended to reflect what the DAC referred to as 'the internationally accepted view that the Court should be able to correct serious failure to comply with the 'due process' of arbitral proceedings'.

Art.34 of the Model Law provides a remedy under which an award may only be set aside, but the setting aside proceedings may be suspended, on application by a party, to give the tribunal an opportunity to eliminate the grounds for setting aside. The scope of the Article overlaps with s.67 since it also provides a remedy for lack of substantive jurisdiction. Time limits are more generous than under this section, being three months from receipt of the award. The list of grounds is not wholly comparable except that ground (ii) reflects s.33 (see subs.(2)(a)); and part of ground (iv) corresponds to subs.(2)(c).

> '(2) An arbitral award may be set aside by the court ... only if:
> (a) the party making the application furnishes proof that:
> .....
> (ii) the party making the application was not given proper notice of the appointment of an arbitrator or of the arbitral proceedings or was otherwise unable to present his case...
> .....
> (iv) ... the arbitral procedure was not in accordance with the agreement of the parties, unless such agreement was in conflict with a provision of this Law from which the parties cannot derogate, or, failing such agreement, was not in accordance with this Law...
> .....
> (4) The court, when asked to set aside an award, may, where appropriate and so requested by a party, suspend the setting aside proceedings for a period of time determined by it in order to give the arbitral tribunal an opportunity to resume the arbitral proceedings or to take such other action as in the arbitral tribunal's opinion will eliminate the grounds for setting aside.'

[68K]   *Rules*

Since this is a mandatory section parties adopting rules will nonetheless find themselves subject to it. Such rules may contain arbitral processes of appeal or review which will have to be exhausted before the section is invoked. See, for instance, Rule 12.9 of CIMAR (correction of award and additional award) and Rule 21.3 of the ICE Arbitration Procedure to similar effect. The CIArb Rules are silent on the point. The international rules also generally provide for correction or additional awards; see Art.27 of the LCIA Rules, Art.29 of the ICC Rules and Arts.35 to 37 of the UNCITRAL Arbitration Rules. In particular Art.29 of the ICC Rules and Art.35 of the UNCITRAL Arbitration Rules contain provisions under which the tribunal may give an interpretation of its award, thus removing uncertainty or ambiguity (see subs.(2)(f)).

Section 69 – Appeal on Point of Law

69.–(1) Unless otherwise agreed by the parties, a party to arbitral proceedings may (upon notice to the other parties and to the tribunal) appeal to the court on a question of law arising out of an award made in the proceedings.

An agreement to dispense with reasons for the tribunal's award shall be considered an agreement to exclude the court's jurisdiction under this section.

(2) An appeal shall not be brought under this section except–

(a) with the agreement of all the other parties to the proceedings, or
(b) with the leave of the court.

The right to appeal is also subject to the restrictions in section 70(2) and (3).

(3) Leave to appeal shall be given only if the court is satisfied–

(a) that the determination of the question will substantially affect the rights of one or more of the parties,
(b) that the question is one which the tribunal was asked to determine,
(c) that, on the basis of the findings of fact in the award–

    (i)   the decision of the tribunal on the question is obviously wrong, or
    (ii) the question is one of general public importance and the decision of the tribunal is at least open to serious doubt, and

(d) that, despite the agreement of the parties to resolve the matter by arbitration, it is just and proper in all the circumstances for the court to determine the question.

(4) An application for leave to appeal under this section shall identify the question of law to be determined and state the grounds on which it is alleged that leave to appeal should be granted.

(5) The court shall determine an application for leave to appeal under this section without a hearing unless it appears to the court that a hearing is required.

(6) The leave of the court is required for any appeal from a decision of the court under this section to grant or refuse leave to appeal.

(7) On an appeal under this section the court may by order–

(a) confirm the award,
(b) vary the award,
(c) remit the award to the tribunal, in whole or in part, for reconsideration in the light of the court's determination, or
(d) set aside the award in whole or in part.

The court shall not exercise its power to set aside an award, in whole or in part, unless it is satisfied that it would be inappropriate to remit the matters in question to the tribunal for reconsideration.

(8) The decision of the court on an appeal under this section shall be treated as a judgment of the court for the purposes of a further appeal.

But no such appeal lies without the leave of the court which shall not be given unless the court considers that the question is one of general importance or is one which for some other special reason should be considered by the Court of Appeal.

Definitions

'agreement', 'agreed': s.5(1).
'the court': s.105.
'notice (or other document)': s.76(6).
'party': ss.82(2), 106(4).

'question of law': s.82(1).
'upon notice' (to the parties or the tribunal): s.80.

[69A]   ## Status

This section is derived from the Arbitration Act 1979, ss.1 and 3. It is *non-mandatory*, (s.4(2)).

[69B]   ## Summary

Unless there is agreement to the contrary, the section provides for an appeal to the court on a question of law arising out of an award. By virtue of the definition of 'question of law' in s.82(1), this is a reference to the law of England and Wales or Northern Ireland as appropriate, questions relating to any other law being matters of fact.

Points

[69C]   *Should there be a right of appeal?*
Anxious consideration was given to the question of whether, given the rare opportunity to recast the principles of arbitration law represented by the Act, a right of appeal on the substantive issues in the arbitration should be preserved at all. The Model Law envisages no such appeal, and it is absent in the arbitration law of many jurisdictions. The principle that the parties are free to agree how their disputes are resolved with the minimum of court intervention would, prima facie, indicate against a substantive appeal. From a commercial point of view, the prospect of long drawn-out court proceedings following on from an award would be likely to make England a less attractive option to parties choosing a forum for international arbitrations.

   In the event, the Act provides for a right of appeal on a point of law arising out of an award. The rationale would seem to be that the parties must not be taken, in the ordinary case, to have agreed that the tribunal would obviously misapply the relevant law. In cases of wider significance, there is a general interest in enabling a seriously doubtful decision to be reviewed. These are instances of safeguards, necessary in the public interest, that delimit the freedom of the parties to choose their tribunal and abide by its decision.

   The right of appeal is limited and restricted in a number of ways, thus ensuring that only rarely will the award not constitute the final decision on the substantive issues in the arbitration.

[69D]   *Exclusion agreements – Subs.(1)*
The section is non-mandatory, from which it follows that the parties may exclude the right of appeal by agreement. An agreement to dispense with reasons for the tribunal's award is deemed to be such an exclusion agreement. It should be noted that the restriction on exclusion agreements in certain categories of case (the 'special categories') provided by the Arbitration Act 1979, s.4, has not been perpetuated.

S.87 not having been brought into effect, the position in respect of both domestic and international arbitrations is the same. It is now open to the parties to exclude the right of appeal in any category of case, and at any time, that is to say before or after the commencement of the arbitration.

[69E]   *Restrictions – Subs.(2)*
A party intending to appeal under this section must first exhaust available arbitral procedures, such as the agreed appeal procedures provided for in commodities associations' arbitration rules, and available recourse to correction of the award or the making of an additional award (although these may only rarely be relevant), in accordance with s.70(2). A party must also observe the 28 day time limit provided by s.70(3), subject to the court's power to extend the time limit pursuant to s.80(5). Note that there is no compliance with the 28-day limit unless the arbitration claim form has been issued and all affidavits or witness statements in support have been filed by the expiry of the time limit: see para.22.1 of CPR Part 49G (Practice Direction).

In addition, a party intending to appeal must obtain *either* the agreement of all the other parties to the proceedings or the permission of the court. An agreement consenting to the right of appeal made in advance of the proceedings, for example in the arbitration clause or the substantive contract, would continue to be valid for these purposes. This is unfortunate. Under the law prior to this Act, consents of this nature have been included in printed standard forms (particularly in the construction industry, in forms produced by the Joint Contracts Tribunal) and they have led to unnecessary and wasteful appeals. This practice can continue under the Act, although at least one commentator is making strenuous efforts to persuade the relevant bodies against it.

The question what constitutes an agreement consenting to a right of appeal arose in a 1996 Act arbitration started under an agreement in which the parties agreed and consented 'pursuant to s.1(2)(a) and 2(1)(b) of the Arbitration Act 1979 that either party' might appeal to the High Court on any question of law arising out of an award. It was held that leave to appeal was not required, notwithstanding that the reference was to the 1979 Act and that there was no reference to the 1996 Act. The parties' essential agreement was that either party would have the right to appeal to the court on a question of law: *Taylor Woodrow Civil Engineering Ltd* v. *Hutchison IDH Development Ltd* [1999] ADRLJ 83.

[69F]   ***Permission to appeal – Subs.(3)***
The most important limitations on the right of appeal comprise a number of conditions which must be satisfied before the appellant (who does not have the agreement of the other parties) will be granted permission. The court is likely carefully to scrutinise intended appeals in the light of the criteria laid down.

The court must first be satisfied that the point of law substantially affects the rights of one of the parties. An insignificant or inconsequential point will not qualify. This is an existing requirement, see *Pioneer Shipping Ltd* v. *BTP Tioxide Ltd (The 'Nema')* [1982] AC 724.

Next, the court must be satisfied that the point of law was one which the tribunal was asked to determine. This is a new requirement which reverses existing case law. It has been introduced to preclude appeals on points of law not raised or considered in the course of the arbitration, but subsequently perceived by parties scanning the reasons for the award. To qualify for appeal, the question of law must have been fairly in issue before the tribunal. In this context comments made by Judge Lloyd QC when granting permission to appeal in *Gbangbola* v. *Smith & Sherriff Ltd* [1998] 3 All ER 730 (not reported on this point) may be rather sweeping. The judge rejected a submission that points on which permission to appeal was sought had to have been argued before the arbitrator, and preferred a submission that 'the question of law has to arise out of the award and until the findings of fact are made it may not be possible to see why the decision was wrong and how the question of law arises. Accordingly the question of law is admissible provided it was integral to the resolution of the dispute which was argued before the arbitrator'. In many cases such an approach will sit somewhat strangely with the requirement of subs.3(b) that 'the question [of law] is one which the tribunal was asked to determine', and in our view a narrower approach should be adopted.

Further, the point of law may only be considered against the findings of fact made in the award. This clarifies the intention that questions of fact decided by the award should remain settled, and should not be subjected to review by being disguised as questions of law. We note that although the subsection appears to contemplate the findings of fact already made in the award, s.70(4) provides for the court to order the tribunal to state the *reasons* for its award in sufficient detail to enable the court to consider an application for permission or an appeal. Under the law prior to this Act, the courts treated *reasons* as including *findings of fact*, and we see no reason why the position should be different under s.70. Thus we think that a court faced with an application for permission under s.69 which it is unable to determine in the absence of certain findings of fact will be able to refer the matter back to the arbitrators for further findings.

Against the background of facts as established, the court must then be satisfied, in an ordinary case, that the decision of the tribunal on the question of law is obviously wrong. Less stringently, in a case of general public importance the decision of the tribunal on the question of law must at least be open to serious doubt. These are the existing criteria already applied by the courts following the decision of the House of Lords in 'The Nema'. They represent the demarcation line between the parties' freedom to arbitrate, and the public interest that justifies a very limited degree of court intervention. In *India Steamship Co. Ltd* v. *Arab Potash Co. Ltd* (unreported, Colman J, 12 December 1997) the court commented that in a case of a specialist tribunal, where there is more than one arguable construction of words used in a contract and the tribunal arrived at a construction other than that which the court considered might be right, the court should not interfere by virtue of s.69(3)(d). However, if a case arose where the courts had determined that words which were in substance indistinguishable from those used in the clause in question had a particular meaning, and the tribunal had taken the view that the words in the contract in question had quite another meaning, it would be hard to see how it could be just and proper in all the circumstances for the court to refuse leave to appeal.

Finally, there is an overriding obligation on the court to consider, in all the circumstances, whether its intervention is appropriate and justified despite the parties' agreement to arbitrate. This involves the court considering not only the issues which are raised on the application for permission to appeal, but also the circumstances in which the arbitration was set up and in which the reference came about: *India Steamship Co. Ltd* v. *Arab Potash Co. Ltd* (above). For example, in circumstances such as an arbitration conducted against a very tight timetable, where the parties plainly intend to arrange their affairs on the basis of a speedy and final decision, the court has a discretion to refrain from intervening, even if other elements of the criteria for permission may be satisfied.

[69G] *Procedure for obtaining permission – Subss.(4), (5)*
There are formal requirements in these subsections that the application for permission should identify the point of law which is the subject of the intended appeal, and should state the grounds on which permission is sought.

Beyond these, the court is required to determine the application for permission on written materials only and without a hearing, unless it appears to the court that a hearing is required. Because the criteria for permission are relatively difficult to satisfy, in most cases the court should be able to decide on documents alone whether permission is appropriate. In any event, lengthy and expensive permission applications are plainly to be avoided.

The formal requirements of the Act are now supplemented by the Practice Direction on Arbitrations issued under CPR Part 49. The application is an arbitration application within the meaning of that Practice Direction. This requires the issue of an arbitration claim form (see our commentary in paragraph 69H below). Specifically in respect of the application for permission:

(1)   the arbitration claim form must identify the question of law and state the grounds on which the applicant alleges that permission should be granted;

(2)   the affidavit or witness statement in support of the application must set out any evidence relied on by the applicant for the purpose of satisfying the court of the matters mentioned in subs.(3) and for satisfying the court that permission should be granted;

(3)   the affidavit or witness statement filed by the respondent must:

    (i)   state the grounds on which the respondent opposes the grant of permission;

    (ii)   set out any evidence relied on by him relating to the matters mentioned in subs.(3); and

    (iii)   specify whether the respondent wishes to contend that the award should be upheld for reasons not expressed (or not fully expressed) in the award and, if so, state these reasons;

(4)   as soon as practicable after the filing of the affidavits and witness statements, the court will determine the application for permission in accordance with subs.(5).

Note that in *Matthew Hall Ortech Ltd* v. *Tarmac Roadstone Ltd* (1998) 87 BLR 96 it was held that a judge hearing an appeal should not refer to the reasons given by the judge who grants permission to appeal.

[69H]   *The application*

The application under s.69 is an arbitration application falling within the terms of the Practice Direction on Arbitrations under CPR Part 49. We deal generally with arbitration applications in Part 3, Materials, Section H. In paragraph 69G above we have dealt with the application for permission to appeal, and we deal with applications for extension of time under s.70 below. Otherwise in respect of applications under s.69:

(i)   the application must be on notice to the other parties and to the tribunal;

(ii)   the requirement for giving notice to the other parties is to be met by making those parties respondents to the application and serving on them the arbitration claim form and evidence in support;

(iii)   the requirement of giving notice to the tribunal is to be met by sending copies of the arbitration claim form to the arbitrators for their information at their last known addresses together with copies of the evidence in support;

(iv) where applicable the arbitration claim form must state that the appeal is brought with the agreement of all the other parties. In such a case the affidavit or witness statement in support must give details of the agreement and exhibit any document evidencing that agreement. In all other cases the arbitration claim form will need to state that permission is required. (See further paragraph 69G above);

(v) the arbitration claim form must show that any available arbitral process of appeal or review, or any available recourse under s.57 (correction or additional award) has been exhausted, and (subject to (vi) below) that the appeal is brought within 28 days of the award or, where there has been any arbitral process of appeal or review, within 28 days of the date when the applicant was notified of the result of that process;

(vi) where the 28 day time limit has expired and the applicant seeks an extension of time, the arbitration claim form must also state why an order extending time should be made and the affidavit and witness statement in support must set out the evidence relied upon. (See further our commentary on s.70 below);

(vii) unless otherwise ordered, the substantive hearing is in public.

[69I] *Remedies – Subs.(7)*

The court may confirm or vary an award, or remit for reconsideration or set aside an award, in whole or in part. However it may only set aside an award, in whole or in part, where remission would be inappropriate.

[69J] *Prosecuting applications*

As to the duty to prosecute applications under this section expeditiously see paragraph 67G.

[69K] *Appeals – Subss.(6), (8)*

An appeal to the Court of Appeal against the grant or refusal of permission would itself require permission.

If the substantive appeal proceeds, then an appeal to the Court of Appeal against the decision of the court would also require permission, with the added requirement that such permission should not be given unless the question is one of general importance or, for some other special reason, is appropriate for consideration by the Court of Appeal.

In both situations, only the court (and not the Court of Appeal) has the power to grant permission.

Note that in relation to judge-arbitrators, references to the court will mean the Court of Appeal; and references to the Court of Appeal will mean the House of Lords, (Sched.2, para.2).

[69L]   *Model Law*

As we have stated above, the Model Law contains no provision for appeal on a point of law. There was considerable debate as to whether the position under the Act should be the same, but the arguments in favour of a right of appeal prevailed.

[69M]   *Rules*

Unless institutional rules exclude the right to appeal on a point of law, s.69 will apply. CIMAR, the ICE Arbitration Procedure and the CIArb Rules do not attempt to prevent appeals, merely requiring the arbitrator to be informed. In contrast, Art.26.9 of the LCIA Rules provides:

> '...the parties also waive irrevocably their right to any form of appeal, review or recourse to any state court or other judicial authority, insofar as such waiver may be validly made.'

The article bites on s.69, but not on ss.67 and 68, which are mandatory. Art 28.6 of the ICC Rules has similar effect:

> '...the parties ... shall be deemed to have waived their right to any form of recourse insofar as waiver can validly be made.'

The UNCITRAL Arbitration Rules, on the other hand, contain no such exclusion.

Section 70 – Challenge or Appeal:
Supplementary Provisions

70.–(1) The following provisions apply to an application or appeal under section 67, 68 or 69.

(2) An application or appeal may not be brought if the applicant or appellant has not first exhausted–

(a) any available arbitral process of appeal or review, and
(b) any available recourse under section 57 (correction of award or additional award).

(3) Any application or appeal must be brought within 28 days of the date of the award or, if there has been any arbitral process of appeal or review, of the date when the applicant or appellant was notified of the result of that process.

(4) If on an application or appeal it appears to the court that the award–

(a) does not contain the tribunal's reasons, or
(b) does not set out the tribunal's reasons in sufficient detail to enable the court properly to consider the application or appeal,

the court may order the tribunal to state the reasons for its award in sufficient detail for that purpose.

(5) Where the court makes an order under subsection (4), it may make such further order as it thinks fit with respect to any additional costs of the arbitration resulting from its order.

(6) The court may order the applicant or appellant to provide security for the costs of the application or appeal, and may direct that the application or appeal be dismissed if the order is not complied with.

The power to order security for costs shall not be exercised on the ground that the applicant or appellant is–

(a) an individual ordinarily resident outside the United Kingdom, or
(b) a corporation or association incorporated or formed under the law of a country outside the United Kingdom, or whose central management and control is exercised outside the United Kingdom.

(7) The court may order that any money payable under the award shall be brought into court or otherwise secured pending the determination of the application or appeal, and may direct that the application or appeal be dismissed if the order is not complied with.

(8) The court may grant leave to appeal subject to conditions to the same or similar effect as an order under subsection (6) or (7).

This does not affect the general discretion of the court to grant leave subject to conditions.

Definitions

'available arbitral process': s.82(1).
'costs of the arbitration': s.59.
'the court': s.105.

[70A] ## Status

This section derives from the Arbitration Act 1950, s.23(3); and the Arbitration Act 1979, ss.1(5), (6). It is *mandatory*, (s.4(1) and Sched.1).

[70B] ## Summary

The section brings together a number of procedural provisions and applies them uniformly to applications to challenge the award on the grounds of lack of substantive jurisdiction in the tribunal (s.67) and of serious irregularity (s.68), as well as to appeals on a point of law (s.69). Because of its nature, the section may not be excluded by agreement of the parties.

Points

[70C] *Restrictions – Subs.(2)*
Applications and appeals require the prior exhaustion of available arbitral procedures, and available recourse to correction of the award or the making of an additional award, (see s.57). Obviously, this requirement will only apply insofar as appropriate. As we have indicated in our commentaries on the individual applications, it will be necessary for the arbitration claim forms and evidence, where applicable, to deal with compliance with this requirement.

[70D] *Time limit – Subs.(3)*
Applications and appeals must be brought within a time limit of 28 days of the date of the award, or, where applicable, the date of notification of the result of any arbitral process of appeal or review. For the date of the award, see s.54. It is important to note that there is no compliance with the 28-day time limit unless the arbitration claim form has been issued and all affidavits or witness statements in support have been filed by the expiry of the time limit: see para.22.1 of the CPR Part 49G (Practice Direction).

The DAC, in paragraph 294 of its Report, suggested that the court has power to extend the time limit pursuant to s.79. We believe that the court's power in fact arises under s.80(5). This would be the means by which difficulties arising from the tribunal exercising its power to withhold the award until payment, pursuant to s.56, might be overcome. If the award is not released until the time limit for challenge or appeal has expired, then an application to the court for an extension of time under s.80(5) would be appropriate. We do, however, feel that where the difficulties have arisen because of the applicant's failure to pay for and collect the award promptly, the applicant will have a heavy burden placed upon him to justify his conduct and thus obtain an extension.

[70E]   *Applications for extension of time*
An applicant who wishes to proceed under ss.67, 68 or 69 may, where the time limit of 28 days has not yet expired, apply without notice on affidavit or witness statement for an order extending the time limit. It will be good practice to exhibit a draft of the proposed claim form to the application wherever possible.

Where the time limit of 28 days has expired, the arbitration claim form should be issued. The following applies:

(1)   in addition to the other matters required by ss.67, 68 or 69 (as applicable) the applicant must state in his arbitration claim form the grounds why an order extending time should be made and his affidavit or witness statement in support must set out any evidence on which he relies;

(2)   a respondent who wishes to oppose the making of an order extending time must file an affidavit or witness statement within seven days after service of the applicant's evidence;

(3)   the court decides whether or not to extend time without a hearing unless it considers that a hearing is required;

(4)   where time is extended the respondent must file his evidence in response to the arbitration application within 21 days of the order.

[70F]   *Reasons – Subs.(4), (5)*
The court is empowered to order the tribunal to state the reasons for its award in sufficient detail to enable the court properly to deal with an application to challenge or an appeal.

The award is bound to contain reasons, (s.52(4)), unless it is an agreed award (in which case it need not contain reasons, and in any event an application to challenge or an appeal seems unlikely), or the parties have agreed to dispense with reasons (in which case an appeal is precluded, see s.69(1)). It therefore appears that the court's power to order the tribunal to state reasons would be most likely to be exercised to compensate for an omission on the part of the tribunal to state the reasons when it should have done so. The court may also order the tribunal to elaborate on its reasons. The original deficiency may, of course, be the fault of the tribunal, but may equally result from the failure of an advocate to address a point or emphasise its importance.

As indicated above at paragraph 31C, we think it advisable that any award deciding a tribunal's substantive jurisdiction should contain the reasons for the conclusion on that topic, even if the parties have agreed that there should be no reasons. S.67 is mandatory and thus there is an automatic right of challenge, and under s.70(4) the court can order a tribunal to state reasons sufficiently to enable it to consider any challenge.

The court has the additional power of making an order with respect to any additional costs of the arbitration resulting from a direction to the tribunal to state reasons. Thus the court might, we think, order the applicant to bear such costs, or the party whose fault it was – if fault can be shown – that the reasons were inadequate. If such inadequacy arose because of the tribunal's failure, we suppose the court could direct that no further payment be due to the tribunal for additional reasons. In an extreme case it might even order the tribunal to bear the parties' further costs, though we think that unlikely.

[70G] *Security for costs – Subs.(6)*

The court has the power to order security for the costs of an application to challenge or an appeal. As with the tribunal's power to order security for the costs of the arbitration (s.38(3)), the power is subject to a negative stipulation, that security is not to be ordered on the basis that the applicant or appellant is foreign. For further comment on this aspect, see paragraph 38F. No other indication is given as to how the power is to be exercised. In particular, any reference to the rules and case-law relating to security for costs in the court context has been avoided. The inference would seem to be that the court may exercise its discretion very flexibly, and in a manner that may well diverge from that which it would adopt if dealing with a court action.

In *Azov Shipping Co. v. Baltic Shipping Co. (No. 2)* [1999] 1 All ER (Comm.) 716 the court commented, on an application for security for the costs of a challenge to jurisdiction under s.67, that in the light of s.1(a) 'the cases will be rare in which a court or indeed an arbitrator would think it right to order security for costs if an applicant for relief has sufficient assets to meet any order for costs and if those assets are available for satisfaction of any such order for costs'.

The sanction for failure to comply with an order for security is the dismissal of the application or appeal.

[70H] *Bringing money into court – Subs.(7)*

The court also has the power to order the bringing into court or securing by other means of money payable under the award pending the determination of the application to challenge or the appeal. The sanction for failure to comply, as with an order for security, is the dismissal of the application or appeal.

[70I] *Permission subject to conditions – Subs.(8)*

The grant of permission to appeal may be made conditional on requirements to the same or similar effect as orders for security for costs and the bringing into court of money payable under the award. The court may also grant permission to appeal subject to such conditions as it may, in its general discretion, impose.

[70J]   *Prosecuting applications*

As to the duty to prosecute applications under this section expeditiously, see paragraph 67G.

[70K]   *Model Law*

As does subs.(3), Art.34(3) of the Model Law sets a time limit for making applications to set aside an award. Such an application '... may not be made after three months have elapsed from the date on which the party making that application had received the award or, if a request had been made under article 33 [Correction and interpretation of award; additional award], from the date on which that request had been disposed of by the arbitral tribunal'.

It is noteworthy that the DAC had some difficulty in determining the date from which the time limit should start. The Model Law does not, of course, contain provision for the arbitrators to withhold the award, so the article has a simpler formula than the Act.

The Model Law does not cover the balance of the ground covered by s.70.

[70L]   *Rules*

This is a mandatory section relating to applications to court. It is not, therefore, an appropriate area for the application of rules.

Section 71 – Challenge or Appeal: Effect of Order of Court

71.–(1) The following provisions have effect where the court makes an order under section 67, 68 or 69 with respect to an award.

(2) Where the award is varied, the variation has effect as part of the tribunal's award.

(3) Where the award is remitted to the tribunal, in whole or in part, for reconsideration, the tribunal shall make a fresh award in respect of the matters remitted within three months of the date of the order for remission or such longer or shorter period as the court may direct.

(4) Where the award is set aside or declared to be of no effect, in whole or in part, the court may also order that any provision that an award is a condition precedent to the bringing of legal proceedings in respect of a matter to which the arbitration agreement applies, is of no effect as regards the subject matter of the award or, as the case may be, the relevant part of the award.

Definitions

'arbitration agreement': ss.6, 5(1).
'the court': s.105.
'legal proceedings': s.82(1).

[71A] ## Status

This section derives in part from the Arbitration Act 1950, s.22(2) and the Arbitration Act 1979, s.1(8). It is *mandatory*, (s.4(1) and Sched.1).

[71B] ## Summary

The section deals with the effect of certain orders of the court with respect to an award. It applies uniformly to orders made on applications to challenge the award on the grounds of lack of substantive jurisdiction in the tribunal (s.67) and of serious irregularity (s.68), as well as on appeals on a point of law (s.69). Because of its nature, the section may not be excluded by agreement of the parties.

Points

[71C] *Variation – Subs.(2)*
A variation made by the court is deemed to take effect as if it were part of the tribunal's award. This preserves the position prior to this Act.

[71D] *Remission – Subs.(3)*

The tribunal is given three months within which to make a new award after remission of the original award, in whole or in part, although the court may change this period. This preserves the position prior to this Act. The new award deals *only* with matters remitted. Where not everything is, the original award will stand to that extent, and need not be re-dealt with in the new one.

It appears that the court has no power to extend the time for making a new award under s.79, since that section applies only to time limits agreed by the parties and those which have effect under this Part of the Act in default of agreement. Subs.(3) does not fit into either category. On the other hand, no sanction is specified for non-compliance by an arbitrator with the three-month limit. These are minor shortcomings that the legislature ought to address in due course.

As to the status of a remitted award (or part of an award) it appears likely that the remission 'suspends' but does not annul that which is remitted, but that once the second award is published, the first falls away and becomes null: see the discussion by the Court of Appeal in *Huyton SA* v. *Jakil SpA* [1999] 2 Lloyd's Rep. 83.

[71E] *Setting aside/declaration of no effect – Subs.(4)*

This subsection deals with the effect of a clause in the arbitration agreement which makes an award a condition precedent to legal proceedings to enforce the contract (known as a *Scott* v. *Avery* clause, after the case of that name). The court now has power to order that such a clause is of no effect when it sets the award aside or declares it to be of no effect. A party may not then defend subsequent legal proceedings on the grounds that no award has been made.

Note that this subsection does not apply in relation to a statutory arbitration, s.97(c).

[71F] *Model Law*

There being no procedure under the Model Law resulting in variation or remission, there is no article to similar effect. However, we would again draw attention to Art.34(4) (set out at paragraph 68J above) which allows the court to suspend proceedings to set aside an award whilst the tribunal has the opportunity to eliminate the grounds upon which it would otherwise be set aside.

Miscellaneous

Section 72 – Saving for Rights of Person Who Takes No Part in Proceedings

72.-(1) A person alleged to be a party to arbitral proceedings but who takes no part in the proceedings may question–

(a) whether there is a valid arbitration agreement,
(b) whether the tribunal is properly constituted, or
(c) what matters have been submitted to arbitration in accordance with the arbitration agreement,

by proceedings in the court for a declaration or injunction or other appropriate relief.

(2) He also has the same right as a party to the arbitral proceedings to challenge an award–

(a) by an application under section 67 on the ground of lack of substantive jurisdiction in relation to him, or
(b) by an application under section 68 on the ground of serious irregularity (within the meaning of that section) affecting him;

and section 70(2) (duty to exhaust arbitral procedures) does not apply in his case.

Definitions

> 'arbitration agreement': ss.6, 5(1).
> 'the court': s.105.
> 'party': ss.82(2), 106(4).
> 'substantive jurisdiction' (in relation to an arbitral tribunal): s.82(1), and
> see s.30(1)(a) to (c).

[72A] ## Status

This is a new provision. It is *mandatory*, (s.4(1) and Sched.1).

[72B] ## Summary

The section provides certain rights of challenge for a person who does not participate in arbitral proceedings to which he is alleged to be a party. Such rights may not be excluded by agreement.

Points

[72C]   *Passive non-participation*
The section contemplates the situation of a person who is alleged to be a party to arbitral proceedings, which contention he wishes to dispute without, however, participating in the proceedings. The procedures available to *parties* to raise a preliminary objection to the tribunal's jurisdiction pursuant to ss.30 to 32 would not be appropriate to such a person.

One course that is open to a person who denies that the tribunal has jurisdiction over him is simply to ignore the arbitral proceedings entirely. The danger with this course is that the tribunal may make an enforceable award against him. If leave were sought to enforce the award as a judgment, however, it would still be open to such a person to raise a passive defence by opposing leave on the ground that the tribunal lacked substantive jurisdiction to make the award, see s.66(3).

This section gives an alleged party active means of objecting to jurisdiction without him having to take part in the proceedings.

For a case in which a party was held not to have taken part in arbitral proceedings by advising the ICC that it was not a party to the contract and thus not a party to the arbitration agreement, see *Caparo Group Ltd v. Fagor Arrasate Soc. Cooperativa* (unreported, Clarke J, 7 August 1998).

[72D]   *Proceedings for declaration – Subs.(1)*
Two possible courses are envisaged for the person alleged to be a party. The first is to bring proceedings in court to challenge the substantive jurisdiction of the tribunal. Subs.(1)(a) to (c) sets out the same defining elements of substantive jurisdiction as are set out in s.30(1)(a) to (c).

These proceedings are quite distinct from the arbitral proceedings, the court granting a declaration, injunction or other relief as appropriate. The alleged party is thereby permitted to preserve his independence from and attitude of non-participation towards the arbitration.

The application is nonetheless an arbitration application within the terms of the Practice Direction on Arbitrations under CPR Part 49: see para.2.1. It will be necessary for the applicant to complete and file an arbitration claim form together with appropriate evidence. This section does not say to whom notice should be given. In our view it will be appropriate to serve the proceedings on the other parties to the arbitral proceedings. It may be appropriate to send copies of the proceedings to the arbitrators for their information. Unless otherwise ordered, any hearing will be in private. See our General Note on Arbitration Applications in Part 3, Materials, Section H.

[72E]   *Challenges to jurisdiction – Subs.(2)*
Alternative courses are available if an award purportedly affecting

324

such a person has been made. The alleged party then has the same rights as a party to challenge jurisdiction. He may challenge a specific award as to jurisdiction or an award on the merits by an application to court pursuant to s.67. He may challenge an award on the ground of serious irregularity affecting him (as to which see, particularly, s.68(2)(b) or (e)) by making an application to court pursuant to s.68.

To the extent that these applications involve participation in the challenge or appeal process, the alleged party would have to abide by the time limits prescribed by s.70(3), see s.67(1) and s.68(1). However, the alleged party need not comply with the s.70(2) duty to exhaust arbitral procedures that would otherwise apply. S.73, concerning the loss of right to object, would also not apply in such a case since this section is concerned with an alleged party who takes no part in the arbitral proceedings whilst s.73(1) applies to 'a party to arbitral proceedings'.

[72F]   *Appeals to Court of Appeal*
The Court of Appeal has jurisdiction to hear appeals from a judge or a court under this section, and a party may apply to the judge or to the Court of Appeal for permission to appeal and such judge and the court have power to grant that permission: *Inco Europe Ltd* v. *First Choice Distribution* [1999] 1 All ER 820.

Section 73 – Loss of Right to Object

73.–(1) If a party to arbitral proceedings takes part, or continues to take part, in the proceedings without making, either forthwith or within such time as is allowed by the arbitration agreement or the tribunal or by any provision of this Part, any objection–

(a) that the tribunal lacks substantive jurisdiction,
(b) that the proceedings have been improperly conducted,
(c) that there has been a failure to comply with the arbitration agreement or with any provision of this Part, or
(d) that there has been any other irregularity affecting the tribunal or the proceedings,

he may not raise that objection later, before the tribunal or the court, unless he shows that, at the time he took part or continued to take part in the proceedings, he did not know and could not with reasonable diligence have discovered the grounds for the objection.

(2) Where the arbitral tribunal rules that it has substantive jurisdiction and a party to arbitral proceedings who could have questioned that ruling–

(a) by any available arbitral process of appeal or review, or
(b) by challenging the award,

does not do so, or does not do so within the time allowed by the arbitration agreement or any provision of this Part, he may not object later to the tribunal's substantive jurisdiction on any ground which was the subject of that ruling.

Definitions

'arbitration agreement': ss.6, 5(1).
'available arbitral process': s.82(1).
'the court': s.105.
'party': ss.82(2), 106(4).

[73A]   ## Status

This section is derived from the Model Law, Art.4. It is *mandatory*, (s.4(1) and Sched.1).

[73B]   ## Summary

The section is intended to prevent parties from delaying proceedings or avoiding the effect of an award by raising late objections as to lack of jurisdiction and irregularity.

Points

[73C] *Knowing participation – Subs.(1)*

The section places parties under an obligation to make objections as to jurisdiction or irregularity very promptly. They must do so either immediately or within the prescribed period of time. Parties failing to do so and knowingly participating, or continuing to participate in the proceedings, lose their right to object. They will only retain that right if they can show (the onus being on them) that they did not know or could not with reasonable diligence have discovered the grounds for the objection.

In *Rustal Trading Ltd* v. *Gill & Duffus SA* [2000] 1 Lloyd's Rep. 14 it was held that any objection to the arbitrator in question had existed from the moment of his appointment and although the complaining party had thereafter taken no positive step in the arbitration, it had nonetheless continued to take part in the proceedings, even after the conclusion of the hearing and until the publication of the award, and had therefore lost its right to object.

[73D] *Failure to challenge award as to jurisdiction – Subs.(2)*

Similarly, where the tribunal has made a positive ruling as to its own jurisdiction in an award made pursuant to s.31(4), the objecting party is under an obligation to challenge the award within the appropriate period of time either by any available process of appeal or review or by an application to the court made pursuant to s.67. A party failing to make such a challenge within the appropriate time limit or at all may not later object to the tribunal's substantive jurisdiction on any ground dealt with in its ruling.

In particular, therefore, if permission is sought to enforce the award on the merits as a judgment pursuant to s.66, the right to oppose the grant of permission on the ground of the tribunal's lack of substantive jurisdiction (s.66(3)) may be lost in this way.

[73E] *Model Law*

Art.4 provides:

> 'A party who knows that any provision of this Law from which the parties may derogate or any requirement under the arbitration agreement has not been complied with and yet proceeds with the arbitration without stating his objection to such non-compliance without undue delay or, if a time-limit is provided therefor, within such period of time, shall be deemed to have waived his right to object.'

There are two important differences between this provision and the Act. Under the Model Law actual knowledge is the relevant criterion for debarring the objection, whereas under the Act it is sufficient that the party seeking to raise the objection might with reasonable diligence have discovered the grounds. Under the Model Law the party facing the objection has to prove knowledge on the part of the party making the objection, whereas under the Act it is for the party making the objection to prove that he neither knew of the grounds nor might with reasonable diligence have discovered them.

[73F]   *Rules*

Those domestic rules we considered do not cover this ground, so that s.73, which is mandatory, applies. Art.32 of the LCIA Rules is in general terms (which are not inconsistent with s.73):

> 'A party who knows that any provision of the Arbitration Agreement (including these Rules) has not been complied with and yet proceeds with the arbitration without promptly stating its objection to such non-compliance, shall be treated as having irrevocably waived its right to object.'

Art.33 of the ICC Rules and Art.30 of the UNCITRAL Arbitration Rules are in similar terms. Delay in taking a point of jurisdiction is generally dealt with by requiring the point to be raised no later than the defence.

Section 74 – Immunity of Arbitral Institutions etc.

74.–(1) An arbitral or other institution or person designated or requested by the parties to appoint or nominate an arbitrator is not liable for anything done or omitted in the discharge or purported discharge of that function unless the act or omission is shown to have been in bad faith.

(2) An arbitral or other institution or person by whom an arbitrator is appointed or nominated is not liable, by reason of having appointed or nominated him, for anything done or omitted by the arbitrator (or his employees or agents) in the discharge or purported discharge of his functions as arbitrator.

(3) The above provisions apply to an employee or agent of an arbitral or other institution or person as they apply to the institution or person himself.

Definitions

'arbitrator': s.82(1).
'party': ss.82(2), 106(4).

[74A] ## Status

This section is a new provision. It is *mandatory*, (s.4(1) and Sched.1).

[74B] ## Summary

The section introduces a limited degree of immunity for appointing bodies and persons as regards their appointing functions. It complements s.29, which confers immunity from suit on arbitrators for acts or omissions in the course of their functions as arbitrators, unless these are shown to have been in bad faith.

The effect is that parties will not be able to allege negligence on the part of an appointing body or person with regard to its appointing or nominating function, subject to an exception of bad faith in the discharge of that function. Nor will parties be able to allege vicarious liability solely by reason of the appointment or nomination.

The section has mandatory status, so that the parties are not able, by agreement between themselves, to deprive appointing bodies or persons of this protection.

Points

[74C] *Immunity for acts of institution – Subs.(1)*
Immunity from suit is extended to institutions and persons who appoint arbitrators in respect of the exercise of their appointing or

nominating function, unless the act or omission is shown to have been in bad faith.

The term 'bad faith' is not further defined, and may have a variety of meanings in different contexts. See, for example, the views of Mr Justice Lightman in *Melton Medes* v. *Securities and Investments Board* [1995] 3 All ER 880 at 889j to 890b. It remains to be seen whether, in the context of the Act, the court will decide that bad faith has a moral ingredient, and connotes, for example, malice or dishonesty (in the sense of knowing of the absence of a basis for making a particular decision); or whether it will bear a wider interpretation.

[74D]   *Immunity for acts of arbitrator – Subs.(2)*

Immunity from suit is also extended to institutions and persons who appoint arbitrators in respect of vicarious liability that may be alleged for any negligent acts or omissions of the arbitrator appointed or nominated by them.

Note that the immunity only protects the institution from the consequences of anything done or not done by the arbitrator simply because it had appointed or nominated him. It is not intended to afford protection in other circumstances. Thus, if such a body or person has administrative functions in relation to an arbitration and acts negligently in the discharge of those, this section will afford no protection.

[74E]   *Employees and agents – Subs.(3)*

It is expressly provided that employees or agents of institutions and persons who appoint arbitrators enjoy the protection of immunity as it applies to their employers or principals.

[74F]   *Model Law*

There is no comparable provision.

[74G]   *Rules*

CIMAR is not an institution and need not seek immunity in relation to its acts. On the other hand, the CIArb Rules say nothing (so the section applies) but Rule 25.1 of the ICE Arbitration Procedure confirms the position of the Institution of Civil Engineers under s.74, Art.31.1 of the LCIA Rules does so for the LCIA and its Court and Art.34 of the ICC Rules does so for the ICC.

Section 75 – Charge to Secure Payment of Solicitors' Costs

75. The powers of the court to make declarations and orders under section 73 of the Solicitors Act 1974 or Article 71H of the Solicitors (Northern Ireland) Order 1976 (power to charge property recovered in the proceedings with the payment of solicitors' costs) may be exercised in relation to arbitral proceedings as if those proceedings were proceedings in the court.

Definitions

'the court': s.105.

[75A] ## Status

This section is derived from the Arbitration Act 1950, s.18(5). It is *mandatory*, (s.4(1) and Sched.1).

[75B] ## Summary

Where a solicitor engaged in court proceedings successfully recovers or preserves property for his client, the court may declare him entitled to a charge over that property for his recoverable costs in relation to those proceedings, and may make orders for the payment of those costs out of the property.

This section allows the court to exercise the same powers in relation to arbitrations. It preserves the position prior to this Act; and may not be excluded by agreement of the parties.

Points

[75C] *Solicitors Act 1974, s.73*

S.73 of the Solicitors Act 1974 provides as follows:

'(1) Subject to subsection (2) any court in which a solicitor has been employed to prosecute or defend any suit, matter or proceeding may at any time –
(a) declare the solicitor entitled to a charge on any property recovered or preserved through his instrumentality for his taxed costs in relation to that suit, matter or proceeding; and
(b) make such orders for the taxation of those costs and for raising money to pay or for paying them out of the property recovered or preserved as the court thinks fit; and all conveyances and acts done to defeat, or operating to defeat, that charge shall, except in the case of a conveyance to a bona fide purchaser for value without notice, be void as against the solicitor.
(2) No order shall be made under subsection (1) if the right to recover the costs is barred by any statute of limitations.'

331

[75D]   *Judge-arbitrators*
Note that a judge-arbitrator may exercise the powers of the court under this section himself, (Sched.2, para.12).

[75E]   *Appeals to Court of Appeal*
In our view, the Court of Appeal has jurisdiction to hear appeals from a judge or a court under this section, and a party may apply to the judge or to the Court of Appeal for permission to appeal and such judge and the court have power to grant that permission: *Inco Europe Ltd* v. *First Choice Distribution* [1999] 1 WLR 270. Whilst the report refers to s.74, we think it was intended to refer to this section.

[75F]   *Model Law*
This power is, of course, particular to the law of England and Wales, or (under the corresponding provision) Northern Ireland. There is no equivalent in the Model Law.

Supplementary

Section 76 – Service of Notices etc.

76.–(1) The parties are free to agree on the manner of service of any notice or other document required or authorised to be given or served in pursuance of the arbitration agreement or for the purposes of the arbitral proceedings.

(2) If or to the extent that there is no such agreement the following provisions apply.

(3) A notice or other document may be served on a person by any effective means.

(4) If a notice or other document is addressed, pre-paid and delivered by post–

(a)   to the addressee's last known principal residence or, if he is or has been carrying on a trade, profession or business, his last known principal business address, or
(b)   where the addressee is a body corporate, to the body's registered or principal office,

it shall be treated as effectively served.

(5) This section does not apply to the service of documents for the purposes of legal proceedings, for which provision is made by rules of court.

(6) References in this Part to a notice or other document include any form of communication in writing and references to giving or serving a notice or other document shall be construed accordingly.

Definitions

'agreement', 'agree': s.5(1).
'arbitration agreement': ss.6, 5(1).
'the court': s.105.
'legal proceedings': s.82(1).
'notice (or other document)': s.76(6).

'party': ss.82(2), 106(4).
'serve', 'service' (of notice or other document): s.76(6).
'in writing': s.5(6).

[76A]   ## Status

This section is derived from the Model Law, Art 3. It is *non-mandatory*, (s.4(2)).

[76B]   ## Summary

The section essentially provides that the parties are at liberty to agree how notices and other documents connected with the arbitration should be served. In default of agreement, the section provides certain options for service.

Note the court's powers to assist where service in accordance with this section is impracticable, (s.77).

Points

[76C] *Any effective means – Subss.(3), (6)*
The first option is for service of notices or other documents to be performed by any effective means. The serving party has virtually unlimited scope to as to the means adopted. This is particularly appropriate in view of the very wide definition of a notice or document as any form of communication in writing (subs.(6)), which in turn includes a notice or document that is recorded by any means (s.5(6)).

The only apparent requirement is that the means of service achieves the result that the notice or document is actually received. In the event of a dispute as to service, the onus would seem to be on the serving party to show that the means adopted was in fact effective.

[76D] *Deemed effective service – Subs.(4)*
If one of the means set out here is adopted, that is sending a notice or document by pre-paid post to the last known principal residence or business address of an individual, or to the registered or principal office of a corporation, service will be treated as effective, even if it is not. Delivery must still be demonstrated, but receipt by the addressee need not be shown. It will therefore be wise to use recorded delivery, registered post, or some other method whereby a record of delivery is made.

We do, however, think that clear evidence of non-receipt in fact (such as the return of the notice or document to the sender unopened; or evidence to the effect that the intended recipient did not in fact receive it at the material time) would displace the presumption in favour of service.

[76E] *Legal proceedings – Subs.(5)*
Service of documents connected with legal proceedings, as to which rules of court apply, must comply with the requirements of those rules of court, and thus does not fall within this section. See now, however, paras.7 and 8 of the Practice Direction on Arbitrations supplementing CPR Part 49 which deal with the service of arbitration claim forms within and out of the jurisdiction.

[76F] *Model Law*
Art.3 provides:

> '(1) Unless otherwise agreed by the parties:
> (a) any written communication is deemed to have been received if it is delivered to the addressee personally or if it is delivered at his place of

business, habitual residence or mailing address; if none of these can be found after making a reasonable inquiry, a written communication is deemed to have been received if it is sent to the addressee's last-known place of business, habitual residence or mailing address by registered letter or any other means which provides a record of the attempt to deliver it;
c) the communication is deemed to have been received on the day it is so delivered.
(2) The provisions of this article do not apply to communications in court proceedings.'

We note that this provision additionally permits service at a 'mailing address'; and it specifically provides for service by registered letter or any other means which provides a record of the attempt to deliver it. Subs.(4) equally has the effect that service/receipt is achieved on the date of delivery.

[76G] *Rules*
Rule 14.2 of CIMAR provides for the application of subss.76(3) to (6) preceding this with:

> 'The arbitrator shall establish and record postal addresses and other means, including facsimile or telex, by which communication in writing may be effected for the purposes of the arbitration.'

Whilst containing a similar provision (at Art.11.2) the CIArb Rules do not otherwise refer to s.76. The ICE Arbitration Rules are silent on the matter.

In contrast Art.4 of the LCIA Rules sets out a detailed regime. Note Art.4.5:

> 'Notwithstanding the above, any notice in communication by one party may be addressed to another party in the manner agreed in writing between them or, failing such agreement, according to the practice followed in the course of their previous dealings or in whatever manner ordered by the Arbitral Tribunal.'

In contrast, the ICC Rules contemplate that communications will generally pass through the secretariat – see Art.3. Art.2 of the UNCITRAL Arbitration Rules has a provision of similar effect to s.76, additionally permitting service at a 'mailing address'.

Section 77 – Powers of Court in Relation to Service of Documents

77.–(1) This section applies where service of a document on a person in the manner agreed by the parties, or in accordance with provisions of section 76 having effect in default of agreement, is not reasonably practicable.

(2) Unless otherwise agreed by the parties, the court may make such order as it thinks fit–

(a)   for service in such manner as the court may direct, or
(b)   dispensing with service of the document.

(3) Any party to the arbitration agreement may apply for an order, but only after exhausting any available arbitral process for resolving the matter.

(4) The leave of the court is required for any appeal from a decision of the court under this section.

Definitions

'agreement', 'agreed': s.5(1).
'arbitration agreement': ss.6, 5(1).
'available arbitral process': s.82(1).
'the court': s.105.

'notice (or other document)': s.76(6).
'party': ss.82(2), 106(4).
'service' (of notice or other document): s.76(6).

[77A]   ## Status

This is a new section. It is *non-mandatory*, (s.4(2)).

[77B]   ## Summary

Where service of notices and other documents connected with the arbitration in accordance with any agreement between the parties or with s.76 is impracticable, the court has powers to assist. These powers are designed to be used where the intended recipients of notices or documents are apparently evading service, or their whereabouts are unknown, so that the normal methods of service become difficult or impossible to use effectively.

Since the section is non-mandatory, the parties may exclude such powers by agreement.

Points

[77C]   *Service as the court directs – Subs.(2)(a)*
The court may make an order for service in such manner as it may direct. Such an order might be appropriate where the intended

336

recipient is evading service, but it is nevertheless believed that service in a particular manner may be effective, or would probably come to his attention. Examples are where service is directed to be effected on a solicitor who has recently acted for the intended recipient; or where it is directed to be effected on the intended recipient's insurers; or upon an habitual agent or broker; or upon an individual who is the alter ego of a corporation that seems to have no locus.

Another case where such an order might assist is if the whereabouts of the intended recipient are not known other than very generally. Then, service may be directed to be effected by advertisement in an appropriate newspaper or journal if there is a reasonable expectation that the advertisement would come to his attention.

[77D] *Dispensing with service – Subs.(2)(b)*
The court may dispense with service altogether. It would presumably only do so in extreme cases, where it was persuaded that the intended recipient was successfully evading service of proceedings of which he was aware or if his whereabouts were wholly unknown, so that service by any means would be impracticable or impossible.

[77E] *Judge-arbitrators*
Note that the power of the court under subs.(2) is exercisable by a judge-arbitrator, with any appeal from his decision going to the Court of Appeal with the leave of that court, (Sched.2, para.13).

[77F] *Restriction – Subs.(3)*
Any available arbitral process for dealing with service must be exhausted before the court may be approached.

[77G] *The application*
The application under s.77(3) is an arbitration application falling within the terms of the Practice Direction on Arbitrations under CPR Part 49. We deal generally with arbitration applications in Part 3, Materials, Section H. Specifically in respect of an application under s.77:

(i)   the application may be made without notice;
(ii) the arbitration claim form must show that any available arbitral process for resolving the matter has been exhausted;
(iii) subject to contrary order, the hearing will be in private.

[77H] *Appeal – Subs.(4)*
Any appeal from a decision of the court may only be made with leave.

[77I] *Model Law*
There is no comparable provision.

[77J]   ***Rules***

The section applies 'unless otherwise agreed by the parties'. It would thus technically apply to all the rules we have considered, but (particularly in the case of the international rules) there is little or no scope for its application in practice.

Section 78 – Reckoning Periods of Time

78.–(1) The parties are free to agree on the method of reckoning periods of time for the purposes of any provision agreed by them or any provision of this Part having effect in default of such agreement.

(2) If or to the extent there is no such agreement, periods of time shall be reckoned in accordance with the following provisions.

(3) Where the act is required to be done within a specified period after or from a specified date, the period begins immediately after that date.

(4) Where the act is required to be done a specified number of clear days after a specified date, at least that number of days must intervene between the day on which the act is done and that date.

(5) Where the period is a period of seven days or less which would include a Saturday, Sunday or a public holiday in the place where anything which has to be done within the period falls to be done, that day shall be excluded.

In relation to England and Wales or Northern Ireland, a 'public holiday' means Christmas Day, Good Friday or a day which under the Banking and Financial Dealings Act 1971 is a bank holiday.

Definitions

'agreement', 'agree', 'agreed': s.5(1).
'party': ss.82(2), 106(4).

[78A]   Status

This is a new section. It is *non-mandatory*, (s.4(2)).

[78B]   Summary

The section deals with how time should be counted with regard to the periods of time for doing things agreed by the parties or required by Part I of the Act.

Its purpose is primarily to avoid the need to refer to other sources (such as rules of court or other statutes) when reckoning periods of time. However note the effect of s.80(5), which is that in relation to time limits for applications or appeals to the court, time must be reckoned in accordance with the relevant rules of court.

The parties are essentially at liberty to agree how time should be counted, with the section providing rules that will apply to the extent that there is no such agreement.

Points

[78C]   *Periods from a specified date – Subs.(3)*
The principal rule is that where a period of time is stated to run after or from a specified date, that day is not included in the reckoning of the period, and the period begins to run on the following day.

For example, s.16(3) provides that, 'If the tribunal is to consist of a sole arbitrator, the parties shall jointly appoint the arbitrator not later than 28 days after service of a request in writing by either party to do so.' If the date of service of the request is 1 June, that day is disregarded, and 2 June is the first day of the 28 day period. It follows that the joint appointment must be made on or before 29 June.

[78D]   *Clear days – Subs.(4)*
A further rule is that where the period of time is expressed in clear days after a specified date, neither that day nor the day on which the act is subsequently done are included in the reckoning of the period. The reckoning of time in clear days therefore provides for a minimum period of time that must elapse between dates or events.

For example, s.17(2) deals with party A's entitlement to appoint his arbitrator as sole arbitrator where party B is in default in not having appointed an arbitrator. If party B does not take certain steps 'within 7 clear days' of notice given by party A, party A may make such an appointment. In this example, the day of notice is excluded from the reckoning, as is the day on which party A may make his appointment. At least seven whole days must intervene between the two. In addition, because the period is seven days, weekends and public holidays would not be counted, see paragraph 78E below.

[78E]   *Weekends and public holidays – Subs.(5)*
In relation to periods of seven days or less, weekends and local public holidays are excluded in reckoning the period. For England, Wales and Northern Ireland, public holidays are defined as Christmas Day, Good Friday and bank holidays.

Returning to the example in paragraph 78D above, suppose, in relation to an arbitration in England, notice were given to party B on 16 December; 21 and 22 December are a weekend – Saturday and Sunday; 25 and 26 December are public holidays – Christmas Day and Boxing Day; none of these days would be included in the reckoning. The seven clear days would therefore be 17, 18, 19, 20, 23, 24 and 27 December. The first day on which party A could appoint his arbitrator as sole arbitrator would be 28 December.

[78F]   *Specific time limits*
We thought it helpful to set out here a table of specific time limits in the Act:

Section	Provision	Time limit	Subsection
16(3)	Appointment of sole arbitrator	Not later than 28 days after service of a request in writing by either party	78(3)
16(4)	Appointment of two arbitrators	Each party must appoint theirs within 14 days after service of request by either party	78(3)
16(5)	Appointment of three arbitrators	Each party must appoint theirs within 14 days after service of request by either party, and third appointed forthwith	78(3)
17(2)	Default appointment of sole arbitrator	Seven clear days after notice being given	78(4) and (5)
57(4)	Application for correction of award	Within 28 days of date of award	78(3)
57(5)	Correction to award	Within 28 days of receipt of application or date of award	78(3)
57(6)	Additional award	Within 56 days of date of award	78(3)
70(3)	Application for challenge or appeal under ss.67, 68, 69	Within 28 days of date of award or result of arbitral process	78(3)

[78G] *Model Law*

There is no comparable provision.

[78H] *Rules*

By Rule 14.3 CIMAR adopts s.78(3) to (5), and the subsections apply by default under the ICE Arbitration Procedures. Art.11.3 of the CIArb Rules provides that periods of time shall be reckoned as provided in s.78. Art.4.6 of the LCIA Rules, Art.3.4 of the ICC Rules and Art.2.2 of the UNCITRAL Arbitration Rules all make broadly similar provision which differs from that under the Act by including weekends and holidays, but extending time if the last day falls on a non-business day. The concept of 'clear' days is not usually embodied into those rules, so there is no scope for application of any part of s.78.

Section 79 – Power of Court to Extend Time Limits Relating to Arbitral Proceedings

79.–(1) Unless the parties otherwise agree, the court may by order extend any time limit agreed by them in relation to any matter relating to the arbitral proceedings or specified in any provision of this Part having effect in default of such agreement.

This section does not apply to a time limit to which section 12 applies (power of court to extend time for beginning arbitral proceedings, &c.).

(2) An application for an order may be made–

(a) by any party to the arbitral proceedings (upon notice to the other parties and to the tribunal), or
(b) by the arbitral tribunal (upon notice to the parties).

(3) The court shall not exercise its power to extend a time limit unless it is satisfied–

(a) that any available recourse to the tribunal, or to any arbitral or other institution or person vested by the parties with power in that regard, has first been exhausted, and
(b) that a substantial injustice would otherwise be done.

(4) The court's power under this section may be exercised whether or not the time has already expired.

(5) An order under this section may be made on such terms as the court thinks fit.

(6) The leave of the court is required for any appeal from a decision of the court under this section.

Definitions

'agreement', 'agree', 'agreed': s.5(1).   'party': ss.82(2), 106(4).
'the court': s.105.   'upon notice' (to the parties or the
'notice (or other document)': s.76(6).   tribunal): s.80.

[79A] Status

This is a new section. It is *non-mandatory*, (s.4(2)).

[79B] Summary

The court is given a general power to extend time limits relating to arbitral proceedings, where these have either been agreed by the parties or are specified by Part I of the Act as having effect in default of agreement.

Note the separate powers of the court to extend time for beginning arbitral proceedings provided for by s.12, and to extend time for making an award provided for by s.50. Note also the effect of s.80(5), which is that in relation to time limits for applications or appeals to the court, time may only be extended in accordance with the relevant rules of court. In *Ranko Group* v. *Antarctic Maritime SA* (1998) LMLN 492 the court rejected an argument that it had no jurisdiction to extend time to permit an application for permission to challenge an award to be made. It held that it could so extend time under s.80(5) and the then Order 73, rule 22, the latter not being ultra vires.

Since the section is non-mandatory, the parties may exclude this general power of the court to extend time by agreement.

Points

[79C] *Who may apply – Subs.(2)*
The arbitral tribunal may apply for an order under this section as well as the parties. It may wish to do so, for example, to obtain an extension of time within which to correct an award. However, in most cases any desire the tribunal might have to extend time will probably be founded on a wish of one of the parties for an extension, and the tribunal will probably prefer to leave it to that party to go to the trouble and expense of applying.

[79D] *Court's discretion – Subss.(3), (4), (5)*
The court's power to extend a time limit is discretionary, and an extension may be granted even though the relevant time limit has passed. It may be granted on terms. In deciding whether to extend time, it is relevant for the court to consider to what extent injustice would be done by a refusal, and that may involve a consideration of the merits of the case which the applicants wish to argue: *Ranko Group* v. *Antarctic Maritime SA* (above).

However, the power is subject to certain qualifications. Thus the court may not act unless it is satisfied that other available arbitral procedures for obtaining an extension (for example, under the rules of any administering body) have been exhausted and that a substantial injustice would result if it did not act.

In short, the court's intervention is intended to be exceptional, and only justified if it supports arbitration in the sense of preventing the process from operating unfairly by causing an injustice.

[79E] *The application*
The application under s.79(2) is an arbitration application falling within the terms of the Practice Direction on Arbitrations under CPR Part 49. We deal generally with arbitration applications in Part 3, Materials, Section H. Specifically in respect of an application under s.79:

(i)   the application may be made by any party or by the tribunal;

(ii)   as appropriate to the application, notice must be given to the (other) parties and to the tribunal;

(iii)   the requirement for giving notice to the other parties is to be met by making those parties respondents to the application and serving on them the arbitration claim form and evidence in support;

(iv)   the requirement for giving notice to the tribunal is to be met by sending copies of the arbitration claim form together with copies of the evidence in support to the arbitrators for their information to their last known addresses;

(v)   the arbitration claim form must show that any available recourse to the tribunal, or anyone vested by the parties with power in that regard, has first been exhausted and that a substantial injustice would be done if the court did not exercise its power;

(vi)   unless otherwise ordered, the hearing is in private.

[79F]   *Judge-arbitrators*

Note that a judge-arbitrator may exercise the power of the court conferred by this section, with any appeal from his decision going to the Court of Appeal with the leave of that court, (Sched.2, para.14).

[79G]   *Appeal – Subs.(6)*

Any appeal from a decision of the court may only be made with leave.

[79H]   *Model Law*

There is no comparable provision.

[79I]   *Rules*

CIMAR, the ICE Arbitration Procedure and the CIArb Rules do not provide to the contrary so that s.79 will apply. The LCIA Rules at least in part provide to the contrary via Art.22.1(b) and Art.22.2. Art.32 of the ICC Rules provides that the ICC Court may extend time limits making the burden under s.79(3) harder to discharge, but otherwise it would seem s.79 would apply. The UNCITRAL Arbitration Rules do not cover the ground at all, so that s.79 applies.

Section 80 – Notice and Other Requirements in Connection with Legal Proceedings

80.-(1) References in this Part to an application, appeal or other step in relation to legal proceedings being taken 'upon notice' to the other parties to the arbitral proceedings, or to the tribunal, are to such notice of the originating process as is required by rules of court and do not impose any separate requirement.

(2) Rules of court shall be made–

(a) requiring such notice to be given as indicated by any provision of this Part, and
(b) as to the manner, form and content of any such notice.

(3) Subject to any provision made by rules of court, a requirement to give notice to the tribunal of legal proceedings shall be construed–

(a) if there is more than one arbitrator, as a requirement to give notice to each of them; and
(b) if the tribunal is not fully constituted, as a requirement to give notice to any arbitrator who has been appointed.

(4) References in this Part to making an application or appeal to the court within a specified period are to the issue within that period of the appropriate originating process in accordance with rules of court.

(5) Where any provision of this Part requires an application or appeal to be made to the court within a specified time, the rules of court relating to the reckoning of periods, the extending or abridging of periods, and the consequences of not taking a step within the period prescribed by the rules, apply in relation to that requirement.

(6) Provision may be made by rules of court amending the provisions of this Part–

(a) with respect to the time within which any application or appeal to the court must be made,
(b) so as to keep any provision made by this Part in relation to arbitral proceedings in step with the corresponding provision of rules of court applying in relation to proceedings in the court, or
(c) so as to keep any provision made by this Part in relation to legal proceedings in step with the corresponding provision of rules of court applying generally in relation to proceedings in the court.

(7) Nothing in this section affects the generality of the power to make rules of court.

Definitions

'arbitrator': s.82(1).
'the court': s.105.
'legal proceedings': s.82(1).

'notice (or other document)': s.76(6).
'upon notice' (to the parties or the tribunal): s.80.

[80A]   **Status**

This is a new section. Whilst it is not expressly made mandatory by s.4(1) and Sched.1 and is therefore, in theory at least, capable of being supplanted by agreement of the parties in accordance with s.4(2), we think that by reason of its nature, its application could not be affected by contrary agreement.

[80B]   **Summary**

The section ties together the rules of court and those sections of this Part of the Act which refer to proceedings in court. Since such proceedings must be subject to the rules of court concerned, the section gives the rules of court precedence where they run parallel to or overlap with the provisions of this Part.

Points

[80C]   *Notice – Subss.(1), (2)*

As an example of the precedence of the rules of court, an application to the court to challenge an award (ss.67, 68) or an appeal to the court on a question of law (s.69) requires the giving of notice to the other parties and the tribunal. Subs.(1) provides that it is the court's requirement for notifying the originating process that must be adhered to for this purpose, and not any separate or additional requirement.

Subs.(2) provides for rules of court to be made establishing the requirement for notice to be given where appropriate, and setting the manner, form and content.

Such rules of court are presently to be found in the Practice Direction on Arbitrations supplementary to CPR Part 49. Para.10 of the Practice Direction provides:

'10.1   Where the Arbitration Act requires that an application to the Court is to be made upon notice to other parties notice shall be given by making those parties respondents to the application and serving on them the arbitration claim form and any affidavit or witness statement in support.

10.2   Where an arbitration application is made under section 24, 28 or 56 of the Arbitration Act, the arbitrators or, in the case of an application under section 24, the arbitrator concerned shall be made respondents to the application and notice shall be given by serving on them the arbitration claim form and any affidavit or witness statement in support.

10.3   In cases where paragraph 10.2 does not apply, an applicant shall be taken as having complied with any requirement to give notice to the arbitrator if he sends a copy of the arbitration claim form to the arbitrator for his information at his last known address with a copy of any affidavit or witness statement in support.

10.4   This paragraph does not apply to applications under section 9 of the Arbitration Act to stay legal proceedings.'

We have dealt with the application of these requirements under each relevant section in the text above.

[80D]   *Notice to tribunal – Subs.(3)*

Notice of legal proceedings must be given to each member of the tribunal, or each member so far appointed. However this requirement, also, is subject to any rules of court. Again we have quoted the relevant rule of court immediately above. Note that in the case of applications under s.24 it is only necessary to serve the proceedings on the arbitrator concerned.

[80E]   *Application/appeal – Subss.(4), (5)*

Rules of court also dictate questions relating to time periods within which applications to court and appeals must be made. The reckoning of time, its extension and abridgment and the consequences of not complying with the limits are governed by such rules, rather than the provisions of this Part of the Act, (subs.(5)). Moreover making an effective application or appeal will depend upon issuing the appropriate originating process in accordance with the rules of court, (subs.(4)).

The time limit to which subs.(4) primarily applies is the 28-day time limit for applications relating to challenges or appeals under ss.67, 68 and 69. In this respect, the relevant rule of court is again the Practice Direction on Arbitrations supplementary to CPR Part 49. Para.22.1 of this provides:

> 'An applicant shall not be taken as having complied with the time limit of 28 days ... unless the arbitration claim form has been issued, and all the affidavits or witness statements in support have been filed, by the expiry of that time limit.'

The extension of the 28-day time limit is governed by paras.22.2 and 22.3 of the Practice Direction. We have dealt with this at paragraph 70E above. The reckoning of time in relation to court proceedings is now governed by Rule 2.8 of the CPR.

In *Ranko Group* v. *Antarctic Maritime SA* (1998) LMLN 492 the court rejected an argument that the then Order 73, rule 22 was *ultra vires*, and held that it did have power to extend the time for an application for permission to challenge an award.

[80F]   *Amendment of this Part – Subss.(6), (7)*

The general power to make rules of court is maintained. Specifically, rules of court may amend this Part of the Act to keep it in step with the court as regards time limits for appeals and corresponding rules of court as they apply there.

[80G] *Sections affected by 'notice'*

9(1)	Application for stay of legal proceedings is 'upon notice'.
12(2)	Application to extend time for beginning arbitral proceedings.
17(3)	Application to set aside default appointment of sole arbitrator.
18(2)	Application where failure of appointment procedure.
21(5)	Application to court in respect of replacement of arbitrators by umpire.
24(1)	Removal of arbitrator.
25(3)	Application by arbitrator on resignation.
28	Application in respect of arbitrator's fees.
32(1)	Determination of preliminary point of jurisdiction.
42(2)	Application to enforce a peremptory order.
44(4)	Application to court to exercise powers in support of arbitral proceedings where no urgency.
45(1)	Determination of preliminary point of law.
50(2)	Application for extension of time for making an award.
56(2)	Court intervention where tribunal refuses to deliver award except on full payment of fees.
63(4)	Application to determine recoverable costs.
64(2)	Application to determine recoverable fees and expenses of arbitrators as between parties.
67(1)	Challenging the award: substantive jurisdiction.
68(1)	Challenging the award: serious irregularity.
69(1)	Appeal on a point of law.
79(2)	Extension of time.

[80H] *Model Law*

There is no comparable provision.

Section 81 – Saving for Certain Matters Governed by Common Law

81.–(1) Nothing in this Part shall be construed as excluding the operation of any rule of law consistent with the provisions of this Part, in particular, any rule of law as to–

(a) matters which are not capable of settlement by arbitration;
(b) the effect of an oral arbitration agreement; or
(c) the refusal of recognition or enforcement of an arbitral award on grounds of public policy.

(2) Nothing in this Act shall be construed as reviving any jurisdiction of the court to set aside or remit an award on the ground of errors of fact or law on the face of the award.

Definitions

'arbitration agreement': ss.6, 5(1).
'the court': s.105.

[81A] Status

This is a new section. Whilst it is not expressly made mandatory by s.4(1) and Sched.1 and is therefore, in theory at least, capable of being supplanted by agreement of the parties in accordance with s.4(2), we think that by reason of its nature, its application could not be affected by contrary agreement.

[81B] Summary

It is clear from this section that whilst the Act is a fundamental restatement of the principles and rules relating to arbitration, it is not an exhaustive code. Accordingly, other rules of law that are consistent with the Act will continue to apply. These rules will most often be found in decided cases (when they are known as 'common law' rules – a term which the Act avoids). However they may also be found in other legislation.

Points

[81C] *Unaffected areas – Subs.(1)*
As an example of the section's operation, the Act does not deprive a purely oral arbitration agreement of any effect because it is not in writing, but allows it to continue to have such effect as the common law will give to it, (subs.(1)(b)).

As a further example, when the court is asked for permission to enforce an award as a judgment (s.66), permission may be refused if the award deals with matters which are not capable of resolution by arbitration (subs.(1)(a)). The category of such matters is left open, and may therefore evolve and change as the need arises. Similarly, permission to enforce an award may be refused on public policy grounds (subs.(1)(c)), which are also open to development.

A significant advantage of this approach is that areas of law that would, for various reasons, be difficult to codify have simply not been touched by the Act. The sensitive area of confidentiality and privacy in relation to arbitrations is one of these. Whilst its importance was clearly recognised by the DAC, this area was thought to be too controversial and difficult to codify. They considered that to attempt to do so would lead to rules (and indeed litigation) that would impede rather than facilitate arbitration. The solution was to allow commercial good sense and case law to continue to govern this area of the law. It may therefore continue to be developed and applied to individual cases in a flexible way.

[81D]   *No revival – Subs.(2)*
Common law rules relating to errors on the face of the award are not revived by the repeal of existing statutes. The procedure for setting aside or remitting an award on such grounds was abolished by the Arbitration Act 1979, s.1(1), and it will not revive on the repeal of the 1979 Act by this Act.

Section 82 – Minor Definitions

82.–(1) In this Part–

'arbitrator', unless the context otherwise requires, includes an umpire;

'available arbitral process', in relation to any matter, includes any process of appeal to or review by an arbitral or other institution or person vested by the parties with powers in relation to that matter;

'claimant', unless the context otherwise requires, includes a counterclaimant, and related expressions shall be construed accordingly;

'dispute' includes any difference;

'enactment' includes an enactment contained in Northern Ireland legislation;

'legal proceedings' means civil proceedings in the High Court or a county court;

'peremptory order' means an order made under section 41(5) or made in exercise of any corresponding power conferred by the parties;

'premises' includes land, buildings, moveable structures, vehicles, vessels, aircraft and hovercraft;

'question of law' means–

(a)   for a court in England and Wales, a question of the law of England and Wales, and

(b)   for a court in Northern Ireland, a question of the law of Northern Ireland;

'substantive jurisdiction', in relation to an arbitral tribunal, refers to the matters specified in section 30(1)(a) to (c), and references to the tribunal exceeding its substantive jurisdiction shall be construed accordingly.

(2) References in this Part to a party to an arbitration agreement include any person claiming under or through a party to the agreement.

[82A]   ## Status

This section is new. Whilst it is not expressly made mandatory by s.4(1) and Sched.1 and is therefore, in theory at least, capable of being supplanted by agreement of the parties in accordance with s.4(2), we think that by reason of its nature, its application could not be affected by contrary agreement.

[82B]   ## Summary

The section sets out the definitions of a number of expressions frequently used in this Part. See s.83 for an index both to these and to other important expressions which are explained elsewhere in the Act.

Under the Contracts (Rights of Third Parties) Act 1999 where a contract provides that a third party may enforce an arbitration agreement or purports to confer a benefit on him, that party is to be treated as a party to the arbitration agreement.

Section 83 – Index of Defined Expressions: Part I

83. In this Part the expressions listed below are defined or otherwise explained by the provisions indicated–

agreement, agree and agreed	section 5(1)
agreement in writing	section 5(2) to (5)
arbitration agreement	sections 6 and 5(1)
arbitrator	section 82(1)
available arbitral process	section 82(1)
claimant	section 82(1)
commencement (in relation to arbitral proceedings)	section 14
costs of the arbitration	section 59
the court	section 105
dispute	section 82(1)
enactment	section 82(1)
legal proceedings	section 82(1)
Limitation Acts	section 13(4)
notice (or other document)	section 76(6)
party–	
– in relation to an arbitration agreement	section 82(2)
– where section 106(2) or (3) applies	section 106(4)
peremptory order	section 82(1) (and see section 41(5))
premises	section 82(1)
question of law	section 82(1)
recoverable costs	sections 63 and 64
seat of the arbitration	section 3
serve and service (of notice or other document)	section 76(6)
substantive jurisdiction (in relation to an arbitral tribunal)	section 82(1) (and see section 30(1)(a) to (c))
upon notice (to the parties or the tribunal)	section 80
written and in writing	section 5(6)

[83A] Status

This section is new. Whilst it is not expressly made mandatory by s.4(1) and Sched.1 and is therefore, in theory at least, capable of being supplanted by agreement of the parties in accordance with s.4(2), we think that by reason of its nature, its application could not be affected by contrary agreement.

[83B] Summary

This section provides an index to a number of expressions frequently used in this Part and defined in s.82, and to other important expressions which are explained elsewhere in the Act.

Section 84 – Transitional Provisions

84.-(1) The provisions of this Part do not apply to arbitral proceedings commenced before the date on which this Part comes into force.

(2) They apply to arbitral proceedings commenced on or after that date under an arbitration agreement whenever made.

(3) The above provisions have effect subject to any transitional provision made by an order under section 109(2) (power to include transitional provisions in commencement order).

Definitions

'arbitration agreement': ss.6, 5(1).
'commencement' (in relation to arbitral proceedings): s.14.

[84A] ## Status

This new section corresponds to the Arbitration Act 1950, s.33. Whilst it is not expressly made mandatory by s.4(1) and Sched.1 and is therefore, in theory at least, capable of being supplanted by agreement of the parties in accordance with s.4(2), we think that by reason of its nature, its application could not be affected by contrary agreement.

[84B] ## Summary

The section determines which arbitrations were to be governed by this Part of the Act when it came into force.

Points

[84C] *Existing arbitrations – Subs.(1)*
This Part of the Act came into force on 31 January 1997 pursuant to the Arbitration Act (Commencement No.1) Order 1966 (SI 1996 No.3146), (s.109(1)). Arbitral proceedings previously in existence continue under the law prior to this Act, unaffected by this Part.

[84D] *New arbitrations – Subs.(2)*
Arbitral proceedings commenced on 31 January 1997 or at any later time are governed by it.
    The critical question is when the arbitral proceedings commence, as to which see s.14. The fact that the relevant arbitration agreement may have been made prior to this Part coming into force does *not* affect the application of this Part.

[84E] ***Transitional provisions – Subs.(3)***

The Commencement Order referred to in paragraph 84C brought into effect as of 17 December 1996 certain provisions of the Act as set out in Sched.1 thereto, and provided for the rest of the Act (with the exception of ss.85 to 87) to come into force on 31 January 1997. It also contained further transitional provisions in Sched.2. The Order itself is set out in the Appendices of this book.

Essentially the 'old law' (i.e. the enactments specified in s.107 as they stood before their amendment or repeal by the Act) continues to apply to arbitration proceedings started before 31 January 1997, to arbitration applications started or made before that date and to arbitration applications started or made on or after that date but relating to arbitration proceedings started before 31 January 1997, whereas the provisions of the present Act brought into force by the Order apply to any other arbitration application.

Further, in the case of an agreement to decide a dispute in accordance with provisions other than law under s.46(1)(b), the Order provided that the agreement is to have effect in accordance with the rules of law (including any conflict of laws rules) as they stood immediately before 31 January 1997.

Part II

Other Provisions Relating to Arbitration

Domestic Arbitration Agreements

General Note on Sections 85 to 88

IMPORTANT NOTE: as explained below, ss.85 to 87 were NOT brought into effect

Background

Arbitration law prior to this Act distinguished between domestic and other arbitrations for two main purposes. These concern differences in the power of the court to stay court proceedings when there is an arbitration agreement; and differences in the rights of the parties to exclude the court's jurisdiction to determine a preliminary point of law or to entertain an appeal on a point of law.

This distinction began with the Arbitration Act 1975, which enacted the Convention on the Recognition and Enforcement of Foreign Arbitral Awards Done at New York on 10 June 1958 (the New York Convention). Prior to that, and pursuant to the Arbitration Act 1950, s.4, a stay of court proceedings in respect of a matter agreed to be referred to arbitration was discretionary in all cases. The New York Convention introduced, in respect of international cases, the requirement for a stay of court proceedings to be mandatory, unless the arbitration agreement was null and void, inoperative or incapable of being performed, see Art.III.

S.1(1) of the 1975 Act gave effect to the New York Convention. However, domestic arbitration agreements were excluded from its operation, and continued to be governed by s.4 of the 1950 Act. Thus a stay of court proceedings in international cases was mandatory unless the stated exceptions applied; whereas in domestic cases it remained discretionary.

The distinction continued with the Arbitration Act 1979. S.3 of that Act permitted parties to enter into agreements excluding their rights to seek the determination of a preliminary point of law from the court or to appeal on a point of law. In international cases, such agreements could be entered into without restriction. However in domestic cases, they could only be entered into after the commencement of the arbitration, (s.3(6) of the 1979 Act).

In shipping, insurance and commodity cases, whether domestic or international (and known as the 'special categories'), particular restrictions on exclusion agreements also applied (s.4 of the 1979 Act).

The reasoning behind the development of different international and domestic rules appears to have been a concern to preserve, where possible, what was then seen as the courts' supervisory role over arbitrations. International treaty obligations might require a mandatory stay of legal proceedings, but where, in the domestic context, they did not, the court retained its discretionary powers. International opinion and expectation might require an unfettered right to exclude the court's jurisdiction on points of law but where, in the domestic context, it did not, the right of exclusion was restricted so that reference to the court would be more possible.

Abolish the distinction?

Given the opportunity of a fundamental reappraisal of arbitration law, both the DAC and Parliament considered whether the distinction between international and domestic arbitration rules should be abolished.

It was apparent that measures designed to preserve a supervisory role for the courts at the expense of party autonomy were widely at variance with the objects and tenor of the present Act. S.1(b) declares that the parties should be 'free to agree how their disputes are resolved, subject only to such safeguards as are necessary in the public interest'. It seemed difficult to justify, for domestic purposes, measures that are not regarded as necessary in an international context.

In addition to the philosophical objection to the distinction, powerful arguments in favour of its abolition may be noted. It seemed logically inconsistent, for example, for an arbitration between two English-based companies to be subject to rules different from those that would apply to an arbitration between an English-based company and one based overseas. The same observation applies to an arbitration between two English people and an arbitration between an English person and one of a different nationality, who may yet spend all his time in England.

Moreover, concern was expressed that such different treatment discriminated against European Community nationals who are not English, and that it might be drawn to the attention of the European Court with an invitation to rule it contrary to European law.

On a practical level, the DAC's Report, focusing on the operation of the discretionary power to stay court proceedings, found that the refusal of a stay in domestic cases often amounted to an unwarranted intervention in the rights of contracting parties. It took the view that a consensual choice of arbitration was not compatible with a discretionary power in the court to stay court proceedings.

In relation both to stays of court proceedings and exclusion agreements, abolition of the distinction would be taking nothing from the parties. If a stay of court proceedings became mandatory in all cases, then the freedom to arbitrate would be more meaningful. Moreover, the removal of restrictions on exclusion agreements would make little practical difference since the parties could simply refrain from entering into an exclusion agreement in any event, if that was their choice.

The DAC concluded in its February 1996 Report that both aspects of the distinction should be abolished. However, it was felt both by the DAC and by Parliament that notwithstanding the powerful arguments in favour of abolishing the distinction, insufficient consultation had taken place (particularly amongst small firms) to enable it to be done at that stage.

Subject to certain changes, therefore, the differences between international and domestic arbitrations were retained in ss.85 to 87 of the Act, but with specific provision for the relevant sections to be repealed in s.88.

In July 1996 there were two developments of considerable importance in relation to this question. The Court of Appeal held, in *Philip Alexander Securities and Futures Limited* v. *Bamberger* [1996] CLC 1757 that the exclusion from protection under the Consumer Arbitration Agreements Act 1988 of arbitration clauses in non-domestic arbitration agreements amounted to unlawful discrimination contrary to the EC Treaty. At the same time, the Department of Trade and Industry published a consultation document on the commencement of the Act in which, amongst other matters, views were sought on the question of the possible domestic/international distinction. The majority of respondents were in favour of the abolition of it and the application of the international regime throughout. Accordingly, it was decided not to bring ss.85 to 87 into effect.

In the circumstances, whilst we have set out the sections in question, we have not commented upon them in this edition.

Section 85 – Modification of Part I in Relation to Domestic Arbitration Agreement

(NOT brought into effect)

[85.–(1) In the case of a domestic arbitration agreement the provisions of Part I are modified in accordance with the following sections.

(2) For this purpose a 'domestic arbitration agreement' means an arbitration agreement to which none of the parties is–

(a) an individual who is a national of, or habitually resident in, a state other than the United Kingdom, or

(b) a body corporate which is incorporated in, or whose central control and management is exercised in, a state other than the United Kingdom,

and under which the seat of the arbitration (if the seat has been designated or determined) is in the United Kingdom.

(3) In subsection (2) 'arbitration agreement' and 'seat of the arbitration' have the same meaning as in Part I (see sections 3, 5(1) and 6).]

Section 86 – Staying of Legal Proceedings

(NOT brought into effect)

[86.–(1) In section 9 (stay of legal proceedings), subsection (4) (stay unless the arbitration agreement is null and void, inoperative, or incapable of being performed) does not apply to a domestic arbitration agreement.

(2) On an application under that section in relation to a domestic arbitration agreement the court shall grant a stay unless satisfied–

(a) that the arbitration agreement is null and void, inoperative, or incapable of being performed, or

(b) that there are other sufficient grounds for not requiring the parties to abide by the arbitration agreement.

(3) The court may treat as a sufficient ground under subsection (2)(b) the fact that the applicant is or was at any material time not ready and willing to do all things necessary for the proper conduct of the arbitration or of any other dispute resolution procedures required to be exhausted before resorting to arbitration.

(4) For the purposes of this section the question whether an arbitration agreement is a domestic arbitration agreement shall be determined by reference to the facts at the time the legal proceedings are commenced.]

Section 87 – Effectiveness of Agreement to Exclude Court's Jurisdiction

(NOT brought into effect)

[**87.**–(1) In the case of a domestic arbitration agreement any agreement to exclude the jurisdiction of the court under–

(a) section 45 (determination of preliminary point of law), or
(b) section 69 (challenging the award: appeal on point of law),

is not effective unless entered into after the commencement of the arbitral proceedings in which the question arises or the award is made.

(2) For this purpose the commencement of the arbitral proceedings has the same meaning as in Part I (see section 14).

(3) For the purposes of this section the question whether an arbitration agreement is a domestic arbitration agreement shall be determined by reference to the facts at the time the agreement is entered into.]

Section 88 – Power to Repeal or Amend
Sections 85 to 87

88.–(1) The Secretary of State may by order repeal or amend the provisions of sections 85 to 87.

(2) An order under this section may contain such supplementary, incidental and transitional provisions as appear to the Secretary of State to be appropriate.

(3) An order under this section shall be made by statutory instrument and no such order shall be made unless a draft of it has been laid before and approved by a resolution of each House of Parliament.

[88A] ## Status

This is a new section.

[88B] ## Summary

As mentioned above in the General Note on ss.85 to 88 (paragraph 85GN) pursuant to this section, ss.85 to 87 were not brought into effect.

Consumer Arbitration Agreements

General Note on Sections 89 to 91

Background

A measure of protection for consumers had, prior to the Act, been effected by the Consumer Arbitration Agreements Act 1988. Under that Act, arbitration agreements could not be enforced against consumers in respect of causes of action falling within the jurisdiction of the county court unless they consented in writing after the dispute had arisen; or they had submitted to arbitration; or the court disapplied the restriction in respect of a particular cause of action. The 1988 Act applied to domestic arbitration agreements within the United Kingdom.

Consideration was given simply to re-enacting the 1988 Act in this Act. However since 1 July 1995, a parallel regime for protecting consumer interests has been in force, namely the Unfair Terms in Consumer Contracts Regulations 1994 (SI 1994 No.3159), and their successor, the Unfair Terms in Consumer Contracts Regulations 1999 (SI 1999 No.2083), which implement Council Directive 93/13. These Regulations avoid any distinction between domestic and international arbitration for consumers. They accord the same rights to consumers throughout the European Economic Area. As such, they overcome any objection that UK consumers are treated more favourably than their counterparts in other member states of the European Community.

In the event it was decided to repeal the 1988 Act in its entirety and to base the protection of consumer interests solely on the Regulations. This avoided two concurrent and overlapping regimes, which would have made for complexity and confusion, and permitted both the simplification and clarification of consumer rights.

The new regime

The 1999 Regulations, which are set out in full in the Appendix, broadly apply to terms in a contract concluded between a seller or supplier and a consumer which have not been individually negotiated, (Regs.4(1), 5(1)). Terms included in standard conditions put forward by the seller or supplier are therefore plainly caught. Terms drafted in advance and in respect of which the consumer has not been able to influence the substance must always be regarded as not having been individually negotiated, (Reg.5(2)) . The onus is on the seller or supplier to prove that a term was individually negotiated, (Reg.5(4)). For the question of who is a consumer, see the commentary on s.90 below.

A term which has not been individually negotiated and which, contrary to the requirement of good faith, causes a significant imbalance in the parties' rights and obligations under the contract to the detriment of the consumer is an unfair term, (Reg.5(1)). The assessment of unfairness must take into account certain factors relevant to the contract in question, together with the terms of any other related contract, (Reg.6(1)).

If a term is unfair, it is not binding on the consumer, (Reg.8(1)). However the remainder of the contract will remain in force if it is capable of so doing without the unfair term, (Reg.8(2)).

Sched.2 contains an indicative and non-exhaustive list of terms which may be regarded as unfair. It includes, at para.1(q), terms which have the object or effect of 'excluding or hindering the consumer's right to take legal action or exercise any other legal remedy, particularly by requiring the consumer to take disputes exclusively to arbitration not covered by legal provisions...'. It is noteworthy that in the course of the debate in the House of Lords, that paragraph was partially quoted as referring to terms 'excluding ... the consumer's right to take legal action ... by requiring [him] to take disputes exclusively to arbitration'. The last five words '*not covered by legal provisions...*' were omitted.

Those words were not in the original draft of the Council Directive implemented by the Regulations, but were inserted later to take account of the particular situations in the Netherlands, Portugal and, it is believed, also Spain, which have special statutory arbitration schemes to allow consumers easy access to justice. They do not cover non-statutory schemes or schemes where the arbitration process is offered to the consumer as a voluntary option (such as those offered by ABTA or BT). The precise effect of the paragraph is therefore to render unfair any clause providing exclusively for arbitration to the exclusion of the courts other than a special statutory scheme specifically designed to assist consumers.

S.91 provides, in addition, that a consumer arbitration agreement falling within the Regulations will be deemed to be unfair, and therefore non-binding, to the extent that the monetary value of the claim brought in respect of it does not exceed a certain amount. This amount was specified for England and Wales as £5,000 by the Unfair Arbitration Agreements (Specified Amount) Order 1999, SI 1999 No.2167, which came into force on 1 January 2000.

Effect

During the course of the debate on this new regime in the House of Lords the government spokesman insisted that not only would it not detract from the rights of consumers, but it would strengthen those rights in certain respects. We wonder whether that intention has been fulfilled.

Whilst s.91 renders consumer arbitration agreements automatically non-binding where the claims are small, in all other cases it is for the consumer to show that the arbitration clause is unfair. In order to do this, he must show that it causes a significant imbalance in his rights, to his detriment. Bearing in mind the first principle enunciated by the Act, which declares the object of arbitration to be 'the fair resolution of disputes by an impartial tribunal without unnecessary delay or expense' (s.1(a)), it is, as a matter of principle, difficult to see what that imbalance might be. Perhaps it will be held that the cost of the tribunal itself, however reasonable, is capable of comprising the relevant detriment, bearing in mind that courts and judges do not charge for their services.

The Regulations have not been in force very long. It remains to be seen whether, as applied by these sections, they have indeed strengthened consumer protection, as predicted.

Application

It should be noted that whereas the Act generally applies only to England and Wales and Northern Ireland, but not to Scotland (which has its own arbitration law), these consumer protection provisions do apply to Scotland, (s.108(3)). Since the Regulations apply throughout the United Kingdom, it was thought sensible to make common provisions for consumer protection for all parts of the United Kingdom.

Section 89 – Application of Unfair Terms Regulations to Consumer Arbitration Agreements

89.–(1) The following sections extend the application of the Unfair Terms in Consumer Contracts Regulations 1994 in relation to a term which constitutes an arbitration agreement.

For this purpose 'arbitration agreement' means an agreement to submit to arbitration present or future disputes or differences (whether or not contractual).

(2) In those sections 'the Regulations' means those regulations and includes any regulations amending or replacing those regulations.

(3) Those sections apply whatever the law applicable to the arbitration agreement.

[89A] ## Status

This is a new section.

[89B] ## Summary

This section, and ss.90 and 91, provide for the protection of consumer rights by extending the application of the Unfair Terms in Consumer Contracts Regulations 1994, superseded by the Unfair Terms in Consumer Contracts Regulations 1999, to arbitration agreements. For a summary of the operation and effect of the Regulations, see the general note to ss.89 to 91 at paragraph 89GN above.

Points

[89C] *Defined expression – Subs.(1)*
'Arbitration agreement' is here separately defined, and the definition includes the extended meaning of 'disputes' as including 'differences', because the index of defined expressions in s.83 only applies to Part I of the Act.

[89D] *Foreign law – Subs.(3)*
Reg.9 of the Regulations provides for their application notwithstanding a choice of law clause purporting to apply the law of a non-member state of the European Economic Area, provided the contract has a close connection with the territory of the member states.

This subsection continues the effect of Reg.9 by applying the relevant sections of the Act, whatever the law applicable to the arbitration agreement.

Section 90 – Regulations Apply Where Consumer is a Legal Person

90. The Regulations apply where the consumer is a legal person as they apply where the consumer is a natural person.

[90A]   ## Status

This is a new section.

[90B]   ## Summary

The Regulations define 'consumer' as any natural person who, in contracts covered by the Regulations, is acting for purposes outside his trade, business or profession (Reg.3(1)). In short, they apply to individuals, acting in their private capacity.

This section extends the definition of 'consumer' to legal persons, as well. Thus corporations would be included, if they are acting for purposes outside their business.

A company purchasing a computer for use in its office, for example, would not come within the definition of consumer. By contrast, the same company purchasing a computer for the private use, at home, of one of its directors, would qualify as a consumer.

Section 91 – Arbitration Agreement Unfair
Where Modest Amount Sought

91.–(1) A term which constitutes an arbitration agreement is unfair for the purposes of the Regulations so far as it relates to a claim for a pecuniary remedy which does not exceed the amount specified by order for the purposes of this section.

(2) Orders under this section may make different provision for different cases and for different purposes.

(3) The power to make orders under this section is exercisable–

(a)   for England and Wales, by the Secretary of State with the concurrence of the Lord Chancellor,

(b)   for Scotland, by the Secretary of State with the concurrence of the Lord Advocate, and

(c)   for Northern Ireland, by the Department of Economic Development for Northern Ireland with the concurrence of the Lord Chancellor.

(4) Any such order for England and Wales or Scotland shall be made by statutory instrument which shall be subject to annulment in pursuance of a resolution of either House of Parliament.

(5) Any such order for Northern Ireland shall be a statutory rule for the purposes of the Statutory Rules (Northern Ireland) Order 1979 and shall be subject to negative resolution, within the meaning of section 41(6) of the Interpretation Act (Northern Ireland) 1954.

[91A]   ## Status

This is a new section.

[91B]   ## Summary

This section provides that a consumer arbitration agreement falling within the Regulations will be deemed to be unfair, and therefore non-binding, to the extent that the monetary value of the claim brought in respect of it does not exceed a certain amount.

For a summary of the operation and effect of the Regulations, see the general note to ss.89 to 91 at paragraph 89GN above.

Points

[91C]   *Specified amount – Subss.(2) to (5)*
The level of monetary claims below which consumer arbitration agreements are deemed unfair was specified for England and Wales as £5,000 by the Unfair Arbitration Agreements (Specified Amount)

Order 1999, SI 1999 No.2167, which came into force on 7 January 2000.

The level may be changed from time to time by virtue of the order-making power provided in these subsections. They constitute a flexible means of adjusting the level when and where necessary.

Small Claims Arbitration in the County Court

Section 92 – Exclusion of Part I in Relation to Small Claims Arbitration in the County Court

92. Nothing in Part I of this Act applies to arbitration under section 64 of the County Courts Act 1984.

[92A] ### Status

This section derives from the Arbitration Act 1979, s.7(3).

[92B] ### Summary

The section provides that nothing in Part I of the Act applies to small claims arbitration in the county court.

Points

[92C] *Small claims arbitration*
The County Courts Act 1984, s.64, provided for a statutory form of arbitration known as small claims arbitration. This comprised a distinct and self-contained regime which was different from non-court based arbitration (although it borrowed some of the features of the latter). With the introduction of the CPR, small claims arbitration was replaced by the small claims track. This is a fully integrated court procedure that does not use the term 'arbitration' or purport to share its features. With that change, this section became redundant.

Appointment of Judges as Arbitrators

Section 93 – Appointment of Judges as Arbitrators

93.–(1) A judge of the Commercial Court or an official referee may, if in all the circumstances he thinks fit, accept appointment as a sole arbitrator or as umpire by or by virtue of an arbitration agreement.

(2) A judge of the Commercial Court shall not do so unless the Lord Chief Justice has informed him that, having regard to the state of business in the High Court and the Crown Court, he can be made available.

(3) An official referee shall not do so unless the Lord Chief Justice has informed him that, having regard to the state of official referees' business, he can be made available.

(4) The fees payable for the services of a judge of the Commercial Court or official referee as arbitrator or umpire shall be taken in the High Court.

(5) In this section–

'arbitration agreement' has the same meaning as in Part I; and
'official referee' means a person nominated under section 68(1)(a) of the Supreme Court Act 1981 to deal with official referees' business.

(6) The provisions of Part I of this Act apply to arbitration before a person appointed under this section with the modifications specified in Schedule 2.

Definitions

'arbitration agreement': ss.6, 5(1).
'official referee': s.93(5).

[93A] ## Status

This section derives from the Administration of Justice Act 1970, s.4 and Sched.3, and from the Arbitration Act 1950, s.11.

[93B] ## Summary

The section brings together and continues aspects of the law prior to this Act by which judges of the Commercial Court and official referees are able to be appointed arbitrators.

Points

[93C]   *Accepting appointments – Subss.(1) to (3)*

It is noteworthy that a commercial judge or official referee may only accept appointment as a sole arbitrator or umpire, but not in another capacity. Thus he may not simply become a member of a tribunal, or the chairman of a tribunal. For the functions of an umpire, see s.21.

It is also clear that a commercial judge or official referee may only accept an appointment when court commitments permit him to do so. His availability has first to be cleared with the Lord Chief Justice.

[93D]   *Defined expressions – Subs.(5)*

The expressions 'arbitration agreement' and 'official referee' are here separately defined, although in the first case, by reference to Part I. The index of defined expressions in s.83 only applies to Part I of the Act.

[93E]   *Schedule 2 – Subs.(6)*

Sched.2 sets out certain modifications to Part I of the Act as it applies to judge-arbitrators. The modifications are individually noted in the commentary to the sections which they affect. Essentially, they change various sections of Part I to reflect the particular status of the judge-arbitrator.

Thus a number of powers which would normally be exercisable by the court – such as extending time for making an award under s.50, or giving permission to enforce an award as a judgment under s.66 – may be exercised by the judge-arbitrator himself, (Sched.2, paras.5, 11).

However, instances where reference is made in Part I to the court are, except where Sched.2 provides differently, to be construed as references to the Court of Appeal, (Sched.2, para.2(1)). Thus an application to remove a judge-arbitrator under s.24 and an application by a judge-arbitrator for relief from liability after his resignation under s.25(3) would both have to be made to the Court of Appeal. This avoids the difficulty of a single judge exercising jurisdiction over a commercial judge or official referee.

[93F]   *Extent*

Note that the provisions of this section and of Sched.2 do not extend to Northern Ireland, (s.108(2)).

[93G]   *Other kinds of judge-arbitrator*

Consideration was given by the DAC to providing for any judge to be appointed as an arbitrator, rather than just the two kinds of judge presently included. Whilst such a change was felt to be desirable, there was insufficient time for the necessary agreements to this proposal to be obtained to enable it to be included in the Act.

Statutory Arbitrations

Section 94 – Application of Part I to Statutory Arbitrations

94.–(1) The provisions of Part I apply to every arbitration under an enactment (a 'statutory arbitration'), whether the enactment was passed or made before or after the commencement of this Act, subject to the adaptations and exclusions specified in sections 95 to 98.

(2) The provisions of Part I do not apply to a statutory arbitration if or to the extent that their application–

(a) is inconsistent with the provisions of the enactment concerned, with any rules or procedure authorised or recognised by it, or
(b) is excluded by any other enactment.

(3) In this section and the following provisions of this Part 'enactment'–

(a) in England and Wales, includes an enactment contained in subordinate legislation within the meaning of the Interpretation Act 1978;
(b) in Northern Ireland, means a statutory provision within the meaning of section 1(f) of the Interpretation Act (Northern Ireland) 1954.

Definitions

> 'enactment': s.94(3).
> 'statutory arbitration': s.94(1).

[94A] ## Status

Ss.94 to 98 derive from the Arbitration Act, 1950, s.31.

[94B] ## Summary

The section operates to apply Part I of the Act to statutory arbitrations, subject to various adaptations and exclusions.

Points

[94C] *Statutory arbitration – Subss.(1), (3)*
A statutory arbitration is one that arises not where the parties have agreed to arbitrate their disputes (private consensual arbitration), but where an Act or other piece of legislation provides that certain types of dispute arising within its remit should be referred to arbitration.

Such arbitration provisions are comparatively rare. Statutes concerning agricultural tenancies have stipulated arbitration as a means of dispute resolution for a number of aspects of the landlord/tenant relationship, see the Agricultural Holdings Act 1986. More recently, the Agricultural Tenancies Act 1995, s.28(1), provides, '...any dispute between the landlord and the tenant under a farm business tenancy, being a dispute concerning their rights and obligations under this Act, under the terms of the tenancy or under any custom, shall be determined by arbitration'.

Other examples are more obscure. The Companies Clauses Consolidation Act 1985, for instance, was passed to comprise in one Act the clauses or provisions relating to the constitution and management of companies incorporated to carry on undertakings of a public nature (e.g. railway companies) that were at that time usually introduced into the special Acts of Parliament incorporating such companies. Ss.128 to 134 of that Act provide a scheme of arbitration for settling disputes authorised by the Act or the special Act. Under the Highways Act 1980, s.115, the reasonableness of adjoining occupiers withholding consent to works proposed by a council to a highway is to be referred to arbitration.

The legislation setting up such statutory arbitration schemes need not be an Act of Parliament, but may be subordinate legislation, (subs. (3)). The Interpretation Act 1978, s.21(1) defines subordinate legislation as 'Orders in Council, orders, rules, regulations, schemes, warrants, byelaws and other instruments made or to be made under any Act'. The enactment may have been made before or after this Act was passed.

[94D] *Exclusions – Subs.(2)*

Part I will not apply to the extent that it is inconsistent with the statutory arbitration scheme. It will also not apply if it is excluded by another enactment.

Section 95 – General Adaptation of Provisions in Relation to Statutory Arbitrations

95.–(1) The provisions of Part I apply to a statutory arbitration–

(a) as if the arbitration were pursuant to an arbitration agreement and as if the enactment were that agreement, and
(b) as if the persons by and against whom a claim subject to arbitration in pursuance of the enactment may be or has been made were parties to that agreement.

(2) Every statutory arbitration shall be taken to have its seat in England and Wales or, as the case may be, in Northern Ireland.

Definitions

'enactment': s.94(3).
'statutory arbitration': s.94(1).

[95A] ## Status

Ss.94 to 98 derive from the Arbitration Act, 1950, s.31.

[95B] ## Summary

Where, and to the extent that, Part I applies to a statutory arbitration, certain general adaptations of its provisions are made by this section. More specific adaptations and exclusions are made by ss.96 and 97 respectively.

Points

[95C] ### Agreement and parties – Subs.(1)
In order to make up for the absence of an agreement and consensual parties in the statutory context, Part I is made to apply as if the enactment providing for arbitration were the agreement and the persons on either side of the claim which is referred to arbitration were the parties.

[95D] ### Seat of the arbitration – Subs.(2)
Whereas for the purposes of a private consensual arbitration, the seat of the arbitration may be designated or determined in accordance with a number of different factors (as to which, see s.3), statutory arbitrations are deemed to have their seat in England and Wales or Northern Ireland, as appropriate. The effect is to bring them firmly within s.2(1),

which provides that Part I applies where the seat of the arbitration is in England and Wales or Northern Ireland. The overall consequence is that for statutory arbitrations, this subsection eliminates the possibility of the provisions of Part I not applying because of a designation or determination of the seat of the arbitration outside England and Wales or Northern Ireland.

Section 96 – Specific Adaptations of Provisions in Relation to Statutory Arbitrations

96.–(1) The following provisions of Part I apply to a statutory arbitration with the following adaptations.

(2) In section 30(1) (competence of tribunal to rule on its own jurisdiction), the reference in paragraph (a) to whether there is a valid arbitration agreement shall be construed as a reference to whether the enactment applies to the dispute or difference in question.

(3) Section 35 (consolidation of proceedings and concurrent hearings) applies only so as to authorise the consolidation of proceedings, or concurrent hearings in proceedings, under the same enactment.

(4) Section 46 (rules applicable to substance of dispute) applies with the omission of subsection (1)(b) (determination in accordance with considerations agreed by parties).

Definitions

'enactment': s.94(3).
'statutory arbitration': s.94(1).

[96A] Status

Ss.94 to 98 derive from the Arbitration Act, 1950, s.31.

[96B] Summary

Where, and to the extent that, Part I applies to a statutory arbitration, certain specific adaptations of its provisions are made by this section. More general adaptations and exclusions are made by ss.95 and 97 respectively.

Points

[96C] *Adaptations – Subss.(2) to (4)*
Since, for the purposes of statutory arbitrations, the enactment providing for arbitration is treated as the agreement (s.95(a)), a tribunal ruling on its substantive jurisdiction under s.30(1)(a) could find itself questioning the validity of the enactment. As this would hardly be appropriate, sub.(2) adapts the tribunal's function to one of considering whether the enactment applies to the dispute referred to it. The phrase 'dispute or difference' is used here because the definition of 'dispute' in s.82(1) (as including any 'difference') only applies to Part I of the Act.

In relation to statutory arbitrations, the power to consolidate proceedings or order concurrent hearings (which in any event is dependent on the parties' agreement, s.35) only arises in relation to proceedings under the same enactment, (subs.(3)). This obviates inconsistencies that might arise on consolidation of proceedings under different enactments, or even mixed consolidations between statutory and private consensual arbitrations.

For the purposes of statutory arbitrations, the tribunal is only permitted to decide the dispute in accordance with the relevant law, (s.46(1)(a)). It may not do so in accordance with the broad justice of the case, on the principles of equity and good conscience, or on some other non-legal basis. S.46(1)(b) is expressly excluded.

[96D] *Impartial and independent tribunal*

During the course of debates in the House of Lords, consideration was given to expressly adapting s.1(a) so that for the purposes of statutory arbitrations, 'an impartial tribunal' should be construed as requiring the tribunal to be independent of the executive. The object would have been to comply with Art.VI of the European Convention on Human Rights which provides that everyone is entitled, in the determination of his civil rights obligation, to an *independent* as well as an impartial tribunal established by law.

Whilst case law has established that the Convention does not apply to private consensual arbitration, where the state is engaged in an arbitration (which it would be in the case of statutory arbitrations, either because a government body was a party or simply because it had established the statutory framework) the Convention does apply.

The matter was dealt with by Lord Fraser, on behalf of the government, saying on 28 February 1996, 'It is clear that the Bill must comply with the Convention in relation to statutory arbitrations. The fact that the Bill does not expressly require tribunals in statutory arbitrations to be independent as well as impartial should not be taken to mean that independence is not a necessary attribute of such tribunals. The provisions of the Bill are consistent with a need for the tribunal to be independent. There is no provision for the Executive to overturn a decision of the tribunal. There are various avenues of appeal to the court for an aggrieved party. If there is any doubt about a particular tribunal, the proper way of dealing with the matter is in relation to that tribunal. If necessary, the power in [s.98(1)] would be available. I am confident that the Bill is not inconsistent with the requirements of the Convention, and that no change is necessary to the text.'

Section 97 – Provisions Excluded from Applying to Statutory Arbitrations

97. The following provisions of Part I do not apply in relation to a statutory arbitration–

(a) section 8 (whether agreement discharged by death of a party);
(b) section 12 (power of court to extend agreed time limits);
(c) sections 9(5), 10(2) and 71(4) (restrictions on effect of provision that award condition precedent to right to bring legal proceedings).

Definitions

'statutory arbitration': s.94(1).

[97A] Status

Ss.94 to 98 derive from the Arbitration Act, 1950, s.31.

[97B] Summary

Where, and to the extent that, Part I applies to a statutory arbitration, certain exclusions are made by this section. General and more specific adaptations are made by ss.95 and 96 respectively.

Points

[97C] *The exclusions*

The reason for excluding the specified sections would seem to be because there is no arbitration agreement (as such) in the context of a statutory arbitration, see s.95(1)(a).

Thus it is not necessary to have a section dealing with the effect of the death of the parties on an arbitration agreement since there is no agreement, but only an enactment. Hence the exclusion of s.8. Similarly, restrictions on *Scott* v. *Avery* clauses (as to which, see paragraph 9G) need not be included since without an arbitration agreement, there can be no such clauses. Hence the exclusion of ss.9(5), 10(2) and 71(4).

S.12, dealing with the court's power to extend time for beginning arbitral proceedings, cannot apply to a statutory arbitration because the enactment in question will provide for the commencement of the arbitration, and clauses barring a claim if some step is not taken in time simply will not arise.

Section 98 – Power to Make Further Provision by Regulations

98.–(1) The Secretary of State may make provision by regulations for adapting or excluding any provision of Part I in relation to statutory arbitrations in general or statutory arbitrations of any particular description.

(2) The power is exercisable whether the enactment concerned is passed or made before or after the commencement of this Act.

(3) Regulations under this section shall be made by statutory instrument which shall be subject to annulment in pursuance of a resolution of either House of Parliament.

Definitions

'enactment': s.94(3).
'statutory arbitration': s.94(1).

Status

[98A]

Ss.94 to 98 derive from the Arbitration Act, 1950, s.31.

Summary

[98B]

The section permits further adaptations or exclusions affecting Part I to be made by statutory instrument, so as to suit its operation to statutory arbitrations.

Points

[98C] *Future statutory schemes – Subs.(2)*
Note that this power may be exercised not only in relation to existing statutory schemes, but also in relation to those that may come into being after this Act.

Recognition and Enforcement of Certain Foreign Awards

Enforcement of Geneva Convention Awards

Section 99 – Continuation of Part II of the Arbitration Act 1950

99. Part II of the Arbitration Act 1950 (enforcement of certain foreign awards) continues to apply in relation to foreign awards within the meaning of that Part which are not also New York Convention awards.

[99A] **Status**

This section derives from the Arbitration Act 1950, Part II (ss.35 to 42) and from the Arbitration Act 1975, s.2.

[99B] **Summary**

Part III of the Act continues the law relating to the recognition and enforcement of foreign arbitral awards contained in Part II of the Arbitration Act 1950 and in the Arbitration Act 1975. These give effect to the United Kingdom's obligations under international treaties to which it is a party. Such treaties require enactment in order for them to become part of United Kingdom law.

This section maintains the effect of the Geneva Convention, (see paragraph 99C). Note that s.36 of the 1950 Act is amended by Sched.3 of the Act to substitute s.66 as the enforcement mechanism.

Points

[99C] *Conventions*
Part II of the Arbitration Act 1950 gives effect to treaty obligations under the Convention on the Execution of Foreign Arbitral Awards

signed at Geneva on behalf of His Majesty on 26 September 1927 (the Geneva Convention). Awards falling within the Geneva Convention are termed 'foreign awards' by the 1950 Act and by this section.

Ss.3 to 7 of the Arbitration Act 1975 gave effect to treaty obligations under the Convention on the Recognition and Enforcement of Foreign Arbitral Awards adopted by the United Nations Conference on International Commercial Arbitration on 10 June 1958 (the New York Convention). The substance of those sections is re-enacted in ss.100 to 104 of this Act. Awards falling within the New York Convention are termed 'New York Convention awards' by the new sections.

[99D]   *Parallel procedures*

There are therefore two parallel procedures for enforcing awards made outside the United Kingdom. Many states are a party to both Conventions. However the Geneva Convention only remains in force as between state parties which have not subsequently become parties to the New York Convention, (see Art.VII(2) of the New York Convention).

Accordingly, this section provides that Part II of the Arbitration Act 1950, giving effect to the Geneva Convention, only continues to apply in relation to foreign awards that are not also New York Convention awards. The latter are now governed by ss.100 to 104.

For example, an award made in Denmark would be enforced as a New York Convention award since Denmark has become a party to that Convention, even though it is also a party to the Geneva Convention. By contrast, an award made in Malta could only be enforced under the Geneva Convention since Malta has not become a party to the New York Convention.

[99E]   *Part II of the Arbitration Act 1950*

Since there are only a small number of states that, whilst parties to the Geneva Convention, have not become parties to the New York Convention, Part II of the 1950 Act has relatively little application. Accordingly it was not thought necessary to restate its provisions in this Act (as has been done in relation to ss.3 to 7 of the Arbitration Act 1975). This section has merely provided for its continuing application to foreign awards that are not New York Convention awards.

For the sake of convenience, the provisions of Part II of the Arbitration Act 1950 (as amended by this Act) together with the Schedules to that Act are reproduced at Appendix II. It should be noted that by s.41 of the 1950 Act, these provisions apply also to Scotland.

[99F]   *The application*

The application for enforcement is made under s.66. The procedure is governed by Part III of the Practice Direction on Arbitrations supplementary to CPR Part 49. For further detail, we refer the reader to our commentary on s.66 above at paragraph 66G.

[99G] ***Geneva Convention States***

By virtue of the Arbitration (Foreign Awards) Order 1984 (SI 1984 No. 1168) the territories to which the Geneva Convention applies are as follows:

The United Kingdom of Great Britain and Northern Ireland
Anguilla
British Virgin Islands
Cayman Islands
Falkland Islands
Falkland Islands Dependencies
Gibraltar
Hong Kong
Montserrat
Turks and Caicos Islands
Antigua and Barbuda
Austria
Bahamas
Bangladesh
Belgium
Belize
Czechoslovakia
Denmark
Dominica
Finland
Federal Republic of Germany
France
German Democratic Republic (now part of the Federal Repubic of Germany)
Greece
Grenada
Guyana
India
Republic of Ireland
Israel
Italy
Japan
Kenya
Luxembourg
Malta
Mauritius
Netherlands (including Curacao)
New Zealand
Pakistan
Portugal
Romania
Saint Christopher and Nevis
St. Lucia
Spain
Sweden
Switzerland
Tanzania
Thailand
Western Samoa
Former Yugoslavia (present position uncertain)
Zambia

Recognition and Enforcement of New York Convention Awards

Section 100 – New York Convention Awards

100.–(1) In this Part a 'New York Convention award' means an award made, in pursuance of an arbitration agreement, in the territory of a state (other than the United Kingdom) which is a party to the New York Convention.

(2) For the purposes of subsection (1) and of the provisions of this Part relating to such awards–

(a) 'arbitration agreement' means an arbitration agreement in writing, and
(b) an award shall be treated as made at the seat of the arbitration, regardless of where it was signed, despatched or delivered to any of the parties.

In this subsection 'agreement in writing' and 'seat of the arbitration' have the same meaning as in Part I.

(3) If Her Majesty by Order in Council declares that a state specified in the Order is a party to the New York Convention, or is a party in respect of any territory so specified, the Order shall, while in force, be conclusive evidence of that fact.

(4) In this section 'the New York Convention' means the Convention on the Recognition and Enforcement of Foreign Arbitral Awards adopted by the United Nations Conference on International Commercial Arbitration on 10th June 1958.

[100A] **Status**

This section derives from the Arbitration Act 1975, s.7.

[100B] **Summary**

The section sets out certain definitions relating to the recognition and enforcement of awards made outside the United Kingdom under the New York Convention. For the background to and general scheme for the enforcement of these awards, and foreign awards under the Geneva Convention, see paragraphs 99B to 99G above.

Points

[100C] *New York Convention – Subs.(4)*
The New York Convention – here given its full title – is an essential element of international arbitration. It permits parties from different

trading nations to arbitrate in a neutral country with confidence. This is because an award made in a New York Convention state will be recognised in and enforced by any other New York Convention state in accordance with a consistent and predictable scheme laid down by the Convention.

[100D]   *Agreement in writing – Subss.(1), (2)*

There are two essential elements that make a New York Convention award capable of recognition or enforcement as such. The first is that it should be made pursuant to an arbitration agreement in writing. Art.II(2) of the New York Convention provides that an 'agreement in writing' includes (but is not limited to) an 'arbitral clause in a contract or an arbitration agreement, signed by the parties or contained in an exchange of letters or telegrams'.

This is enacted by providing, in subs.(1), that a New York Convention award means an award made 'in pursuance of an arbitration agreement...'. Subs.(2)(a) goes on to provide that 'arbitration agreement' means an arbitration agreement in writing, and that 'agreement in writing' has the same meaning as in Part I. S.83 provides that 'agreement in writing', for the purposes of Part I, is defined or explained by ss.5(2) to (5).

As the commentary on s.5 indicates, 'agreement in writing' is there given a very broad definition. It extends beyond that which the New York Convention allows for. However the DAC took the view that the words 'shall include' in the English text of Art.II(2) of the Convention (which is an authentic text, see Art.XVI) comprise a non-exhaustive definition, leaving scope for expansion in the manner the Act has adopted.

The DAC therefore concluded that the s.5 definition is consonant with the New York Convention. It also noted that Parts I and III of the Act being consistent in this regard would have the further advantage of ensuring that the enforcement of foreign awards under s.66 and enforcement under the New York Convention are, in this respect, in line with each other.

[100E]   *Made in a Convention state – Subss.(1), (2)*

The other essential element to a New York Convention award is that it should be 'made in the territory of a State other than the State...' where recognition and enforcement of the award is sought.

This is enacted by providing, in subs.(1), that a New York Convention award means an award 'made ... in the territory of a state (other than the United Kingdom) which is a party to the New York Convention'.

Subs.(2)(b) goes on to provide that such an award is treated as made at the seat of the arbitration, regardless of where it was signed, despatched or delivered to the parties. This subsection corresponds, for international purposes, with s.53, which applies to England and Wales and Northern Ireland. The 'seat of the arbitration' is also defined as having the same meaning as in Part I, as to which see s.3.

The effect of these provisions is to reverse the judgment of the House of Lords in the case of *Hiscox* v. *Outhwaite* [1992] 1 AC 562, in which an award was held to have been made in Paris, France, purely because it was signed there, even though the arbitration (which had been conducted in London) had no connection with France.

In short, an award will be made in the territory of a New York Convention state if the seat of the arbitration was in that state. The vagaries of where the award happened to be signed, despatched or received will not affect the situation.

[100F] *States which are parties – Subs.(3)*

States which are parties to the New York Convention are declared by Order in Council. At present the following states have been specified as being parties:

Algeria
Antigua and Barbuda
Argentina
Australia (including external territories for whose international relations Australia is responsible)
Austria
Bahrain
Belgium
Belize
Benin
Botswana
Bulgaria
Burkina Faso
Byelorussian Soviet Socialist Republic
Cambodia
Cameroon
Canada
Central African Republic
Chile
China
Colombia
Costa Rica
Cuba
Cyprus
Czechoslovakia
Denmark (including Greenland and the Faroe Islands)
Djibouti
Dominica
Ecuador
Egypt
Finland
Federal Republic of Germany

France (including all territories of the French Republic)
German Democratic Republic (now part of the Federal Republic of Germany)
Ghana
Greece
Guatemala
Haiti
Holy See
Hungary
India
Indonesia
Republic of Ireland
Israel
Italy
Japan
Jordan
Kenya
Korea
Kuwait
Luxembourg
Madagascar
Malaysia
Mexico
Monaco
Morocco
Netherlands (including Netherlands Antilles)
New Zealand
Niger
Nigeria
Norway
Panama
Peru

Philippines
Poland
Romania
San Marino
Singapore
South Africa
Spain
Sri Lanka
Sweden
Switzerland
Syria
Tanzania
Thailand
Trinidad and Tobago

Tunisia
Ukrainian Soviet Socialist Republic
(now the Ukraine)
Former Union of Soviet Socialist
Republics (now dissolved and
position uncertain)
United States of America (including
all territories for whose
international relations the USA is
responsible)
Uruguay
Former Yugoslavia (present position
uncertain)

Section 101 – Recognition and Enforcement of Awards

101.–(1) A New York Convention award shall be recognised as binding on the persons as between whom it was made, and may accordingly be relied on by those persons by way of defence, set-off or otherwise in any legal proceedings in England and Wales or Northern Ireland.

(2) A New York Convention award may, by leave of the court, be enforced in the same manner as a judgment or order of the court to the same effect.

As to the meaning of 'the court' see section 105.

(3) Where leave is so given, judgment may be entered in terms of the award.

[101A] Status

This section derives from the Arbitration Act 1975, s.3.

[101B] Summary

The section provides for the recognition and enforcement of New York Convention awards. For the definition of a New York Convention award, see s.100.

Points

[101C] *Recognition – Subs.(1)*

A New York Convention award is recognised by the court acknowledging its binding effect as between the parties to it. In legal proceedings within those jurisdictions to which the Act extends (and see the commentary on s.108(4) for the position in Scotland), such an award may be relied upon 'by way of defence, set-off or otherwise' without further formality.

'Legal proceedings' are defined in s.82(1), for the purposes of Part I, as 'civil proceedings in the High Court or a county court'. That definition does not apply to this section; nor were such proceedings defined in the 1975 Act. We therefore presume they are not limited to the Part I definition and might, for instance, include an arbitration.

A 'set-off' is a monetary cross-claim which is raised as a defence to the claim made in the proceedings. The defendant 'sets off' his own claim against that of the plaintiff as a way of cancelling out the latter.

The defendant may wish to go further, and make a positive counterclaim in the proceedings in reliance on the award. That would be a means of effectively enforcing, rather than merely recognising, the award. It would amount to an action on the award, which s.104 would permit.

Recognition of the award 'otherwise' than by defence or set-off might occur if one of the persons bound by the award tried to re-open the issues decided by it in court proceedings. Then the award might be relied upon in a pre-trial application designed to stay or strike out the court proceedings on the ground that the issues had already been judicially resolved as between those parties.

[101D]   *Enforcement – Subss.(2), (3)*

Two methods of enforcement of a New York Convention award are open to an applicant under these subsections. They are cumulative rather than alternative.

The first is an application directly to enforce the award in the same manner as a judgment or order to the same effect. If permission is given, the applicant may issue execution upon the award as if it were a judgment, without actually entering a judgment. Court procedures will then be available to enforce the award. Where the award is for the payment of a sum of money such procedures include the seizure and sale of the respondent's goods; the interception of a debt due to the respondent; and the charging and sale of the respondent's property. In the case of an order to do or refrain from doing something, the court's powers include that of committing the respondent to prison for contempt.

The second method, where permission has been given, is an application actually to enter a judgment in terms of the award. There may be advantages in proceeding to this second stage rather than being content with the first. For example, the applicant may sue on the judgment or otherwise proceed by execution or registration or other method of enforcement in a foreign court; or he may obtain recognition of the judgment in a foreign court; or he may rely on the judgment as a judicial resolution of the issues that prevents any further action being brought in a foreign court.

Note that these methods of enforcement are not exhaustive. S.104 leaves other possibilities available.

[101E]   *The court – Subs.(2)*

For the purposes of enforcement, reference is expressly made to s.105 for the purposes of identifying the relevant court.

The practice prior to this Act was that New York Convention awards were enforced only in the High Court. The former practice of High Court enforcement will continue, with the additional possibility of enforcement in the Mercantile Courts and the Central London County Court. This is the combined effect of para.4 of the High Court and County Courts (Allocation of Arbitration Proceedings) Order 1996, SI 1996 No. 3215, and para.5 of the Practice Direction on Arbitrations under CPR Part 49.

[101F]   ***The application***

The procedure is set out in Part III of the Practice Direction on Arbitrations supplementary to CPR Part 49. The application may be made without notice on Form 8A, referred to as the 'enforcement form'. The court will specify upon which parties it should be served and, with permission, it may be served out of the jurisdiction. Automatic directions then apply as if the matter were an arbitration application. The hearing is in private, unless the court orders otherwise. For the detailed requirements of the procedure, see para.31 of the Practice Direction.

[101G]   ***Recognition and enforcement***

Note that the party against whom recognition or enforcement of the award is invoked may object on the grounds set out in s.103.

Section 102 – Evidence to be Produced
by Party Seeking Recognition or Enforcement

102.–(1) A party seeking the recognition or enforcement of a New York Convention award must produce–

(a)   the duly authenticated original award or a duly certified copy of it, and
(b)   the original arbitration agreement or a duly certified copy of it.

(2) If the award or agreement is in a foreign language, the party must also produce a translation of it certified by an official or sworn translator or by a diplomatic or consular agent.

[102A]   Status

This section derives from the Arbitration Act 1975, s.4.

[102B]   Summary

The section sets out what evidence a person must produce when seeking recognition or enforcement of a New York Convention award. It follows closely the language of Art.IV of the Convention, which identifies the requirements.

[102C]   *The application*
As noted under s.101 above, the application procedure is set out in Part III of the Practice Direction on Arbitrations supplementary to CPR Part 49. Note that, not surprisingly, para.31.6 of the Practice Direction requires the documents mentioned in s.102 to be exhibited to the evidence in support of the application.

[102D]   *Model Law*
Art.35(2) is in similar terms.

Section 103 – Refusal of Recognition or Enforcement

103.–(1) Recognition or enforcement of a New York Convention award shall not be refused except in the following cases.

(2) Recognition or enforcement of the award may be refused if the person against whom it is invoked proves–

(a) that a party to the arbitration agreement was (under the law applicable to him) under some incapacity;

(b) that the arbitration agreement was not valid under the law to which the parties subjected it or, failing any indication thereon, under the law of the country where the award was made;

(c) that he was not given proper notice of the appointment of the arbitrator or of the arbitration proceedings or was otherwise unable to present his case;

(d) that the award deals with a difference not contemplated by or not falling within the terms of the submission to arbitration or contains decisions on matters beyond the scope of the submission to arbitration (but see subsection (4));

(e) that the composition of the arbitral tribunal or the arbitral procedure was not in accordance with the agreement of the parties or, failing such agreement, with the law of the country in which the arbitration took place;

(f) that the award has not yet become binding on the parties, or has been set aside or suspended by a competent authority of the country in which, or under the law of which, it was made.

(3) Recognition or enforcement of the award may also be refused if the award is in respect of a matter which is not capable of settlement by arbitration, or if it would be contrary to public policy to recognise or enforce the award.

(4) An award which contains decisions on matters not submitted to arbitration may be recognised or enforced to the extent that it contains decisions on matters submitted to arbitration which can be separated from those on matters not so submitted.

(5) Where an application for the setting aside or suspension of the award has been made to such a competent authority as is mentioned in subsection (2)(f), the court before which the award is sought to be relied upon may, if it considers it proper, adjourn the decision on the recognition or enforcement of the award.

It may also on the application of the party claiming recognition or enforcement of the award order the other party to give suitable security.

[103A] Status

This section derives from the Arbitration Act 1975, s.5.

[103B] Summary

The section sets out the grounds on which a person against whom recognition or enforcement of a New York Convention award is invoked may object. It follows closely the language of Arts.V and VI of the Convention.

Points

[103C]   *Mandatory recognition and enforcement*
Note that recognition or enforcement of a New York Convention award is mandatory unless one of the specified grounds of objection is made out, the onus being on the person against whom the award is invoked to prove the objection.

[103D]   *Objections to enforcement – subs.(2) and (3)*
In a recent case, the Commercial Court had to deal with a number of points arising on the question of enforcing a New York Convention award. It held that to avoid the enforcement in England of a New York Convention award, the party against whom enforcement is sought must show that the case falls within one of the exceptions in s.108(2).

While a court may take an illegality point of its own volition under s.103(3), if a party against whom enforcement is sought wishes to rely on matters within that sub-section, the burden of making good the objection rests on that party.

A party who agrees to arbitrate in a foreign jurisdiction is bound not only by the local arbitration procedure, but also by the supervisory jurisdiction of the courts of the seat of the arbitration. Any defect in the award or the conduct of the arbitration must first be sought to be remedied by that supervisory jurisdiction.

Where a party seeks to resist enforcement of a New York Convention award on the ground that enforcement would lead to substantial injustice and therefore be contrary to English public policy, the following will normally be among the relevant considerations:

(1)   the nature of the procedural injustice;
(2)   whether the objecting party has invoked the supervisory jurisdiction of the seat of the arbitration;
(3)   whether a remedy was available under that jurisdiction;
(4)   whether the courts of that jurisdiction had conclusively determined the complaint in favour of upholding the award;
(5)   if the party against whom enforcement was sought had failed to invoke that remedial jurisdiction, what reason it had for doing so and, in particular, whether it was acting unreasonably in failing to do so.

> (*Minmetals Germany GmbH* v. *Ferco Steel Ltd*
> [1999] 1 All ER (Comm.) 865

Further, where foreign arbitrators had concluded that a contract was not void for illegality, and the court having supervisory jurisdiction over the tribunal had dismissed an appeal in respect of effectively the same contention, the English court would not review that conclusion: *Westacre Investments Ltd* v. *Jugoimport-SPDR Holding Co. Ltd* [1999] 3 All ER 864 (Court of Appeal). A similar conclusion was reached in *Omnium de Traitement et de Valorisation SA* v. *Hilmarton Ltd* [1999] 2 Lloyd's Rep. 222 where the judge, following *Westacre*, held that enforcement should be allowed even if performance of a contract was contrary to public policy in the place of performance, provided it was not contrary to the domestic public policy either of the country of the proper law or of that of the curial law.

[103E] *Defending the application*

As noted under s.101 above, the application procedure is set out in Part III of the Practice Direction on Arbitrations supplementary to CPR Part 49. Such an application will often be made without notice. In such a case, by para.31.9 of the Practice Direction, within 14 days of service of the order (or such period as the court may fix if served out of jurisdiction), the respondent may apply to set aside the order. If the respondent applies to set aside the order within the period, the award cannot be enforced until the application is finally disposed of.

[103F] *Model Law*

Art.36 is in similar terms.

Section 104 – Saving for Other Bases of Recognition or Enforcement

104. Nothing in the preceding provisions of this Part affects any right to rely upon or enforce a New York Convention award at common law or under section 66.

[104A] ## Status

This section is derived from the Arbitration Act 1975, s.6.

[104B] ## Summary

The section leaves available possibilities of recognition or enforcement of a New York Convention award outside the scheme provided by ss.101 to 103.

Points

[104C] *Enforcement at common law*
One possibility is that of bringing an action founded on the arbitration award. The proceedings need not be lengthy if there is no arguable defence, since the court may then enter summary judgment on the basis of documents and sworn depositions or witness statements. Such a course may also be appropriate where the objections to the award are such as cannot be resolved without trial.

[104D] *Enforcement under s.66*
The possibility of seeking enforcement under section 66, which relates to the enforcement of domestic as well as foreign awards, is expressly referred to. Note that the methods of enforcement under s.66 are the same, and have the same result, as would be the case under s.101.

We have had some difficulty assessing how ss.66 and 99 to 104 work together. It would appear to us that a New York Convention award may be enforced under either route, but it is likely that the s.66 route presents more obstacles. In particular, recognition and enforcement of a Convention award may only be refused in the cases set out in s.103(2), and it may prove the case that enforcement under s.66, which is discretionary, provides greater opportunity for possible refusal.

Part IV

General Provisions

Section 105 – Meaning of 'The Court': Jurisdiction of High Court and County Court

105.–(1) In this Act 'the court' means the High Court or a county court, subject to the following provisions.

(2) The Lord Chancellor may by order make provision–

(a) allocating proceedings under this Act to the High Court or to county courts; or

(b) specifying proceedings under this Act which may be commenced or taken only in the High Court or in a county court.

(3) The Lord Chancellor may by order make provision requiring proceedings of any specified description under this Act in relation to which a county court has jurisdiction to be commenced or taken in one or more specified county courts.

Any jurisdiction so exercisable by a specified county court is exercisable throughout England and Wales or, as the case may be, Northern Ireland.

(4) An order under this section–

(a) may differentiate between categories of proceedings by reference to such criteria as the Lord Chancellor sees fit to specify, and

(b) may make such incidental or transitional provision as the Lord Chancellor considers necessary or expedient.

(5) An order under this section for England and Wales shall be made by statutory instrument which shall be subject to annulment in pursuance of a resolution of either House of Parliament.

(6) An order under this section for Northern Ireland shall be a statutory rule for the purposes of the Statutory Rules (Northern Ireland) Order 1979 which shall be subject to annulment in pursuance of a resolution of either House of Parliament in like manner as a statutory instrument and section 5 of the Statutory Instruments Act 1946 shall apply accordingly.

[105A] **Status**

This is a new section.

[105B] **Summary**

The section defines and explains 'the court' for the purposes of the whole Act. The Lord Chancellor is given order-making powers to allocate arbitration matters that come before the courts, as provided for in the Act, between the High Court and the county courts.

Points

[105C]   *Judge-arbitrators – Subs.(1)*
It should be borne in mind that if the tribunal is a judge-arbitrator, references to 'the court' may well have a different meaning from that which they would otherwise bear, see Sched.2 and, in particular, para.2 of that Schedule. Moreover in a number of respects a judge-arbitrator is able to exercise the power of the court himself.

[105D]   *Allocation of proceedings*
The allocation of proceedings between different courts is dealt with by the High Court and County Courts (Allocation of Arbitration Proceedings) Order 1996, SI 1996 No. 3215 as amended by SI 1999 No.1010, together with para.5 of the Practice Direction on Arbitrations under CPR Part 49.

Applications for a stay under s.9 of the Act must be made by way of application in the court in which the legal proceedings which it is sought to stay are pending. (The application is made in the legal proceedings.)

In respect of other types of arbitration application, there is an overriding provision that applications must be commenced in the Commercial Court, but they may also be commenced in a Mercantile Court or in the Central London County Court (entered into the Business List). In relation to the latter courts, the judge in charge of the list must as soon as practicable consult with the judge in charge of the commercial list and consider whether the application should be transferred to the Commercial Court or any other list. The criteria to which regard must be had in considering transfer are the financial substance of the dispute including the value of any claim or counterclaim (but disregarding considerations of interest, costs or contributory negligence), the nature of the dispute, the importance of the proceedings (and, in particular, their importance to non-parties) and whether the balance of convenience points to having the proceedings taken other than in the Commercial Court. Where the financial substance of the dispute exceeds £200,000 the proceedings must be commenced and will be retained in the Commercial Court unless they do not raise questions of general importance to non-parties. Conversely the Commercial Court may transfer arbitration applications commenced there to another list, court or Division of the High Court, adopting the same criteria. In practice, the Commercial Court seeks to exercise a supervisory jurisdiction over the Act and its development, reserving to itself applications in commercial arbitrations and those concerning significant points of arbitration law and practice. Thus, if no significant point of

arbitration law or practice is raised, rent-review arbitrations will normally be referred to the Chancery Division; building and civil engineering arbitrations will normally be referred to the Technology and
Construction Court; and ship salvage arbitrations will normally be
transferred to the Admiralty Court.

As we have observed in the Introduction, the fact that stay applications under s.9 are dealt with in the court where the relevant proceedings have been commenced, which may well not be the
Commercial Court, may unfortunately lead to differences of approach
to the application of the relevant provisions of the Act. There are
already some signs that this may be manifesting.

[105E]   *County Courts with Mercantile Lists*
There are such courts in Birmingham, Bristol, Cardiff, Leeds, Liver-
pool, Manchester and Newcastle upon Tyne.

Section 106 – Crown Application

106.-(1) Part I of this Act applies to any arbitration agreement to which Her Majesty, either in right of the Crown or of the Duchy of Lancaster or otherwise, or the Duke of Cornwall, is a party.

(2) Where Her Majesty is party to an arbitration agreement otherwise than in right of the Crown, Her Majesty shall be represented for the purposes of any arbitral proceedings–

(a) where the agreement was entered into by Her Majesty in right of the Duchy of Lancaster, by the Chancellor of the Duchy or such person as he may appoint, and

(b) in any other case, by such person as Her Majesty may appoint in writing under the Royal Sign Manual.

(3) Where the Duke of Cornwall is party to an arbitration agreement, he shall be represented for the purposes of any arbitral proceedings by such person as he may appoint.

(4) References in Part I to a party or the parties to the arbitration agreement or to arbitral proceedings shall be construed, where subsection (2) or (3) applies, as references to the person representing Her Majesty or the Duke of Cornwall.

[106A] Status

This section derives from the Arbitration Act 1950, s.30.

[106B] Summary

The section essentially preserves the law prior to this Act to the effect that the Crown is to be bound by arbitration agreements. Part I of the Act will apply to arbitral proceedings to which the Crown is a party, except so far as this section provides.

Points

[106C] *In right of the Crown*
Arbitration agreements to which the Queen is a party 'in right of the Crown' is a reference to those concerning government departments. As with other civil proceedings, the arbitral proceedings are instituted by or against the relevant department, and are conducted by authorised solicitors.

[106D] *Otherwise than in right of the Crown*
Where the Queen or the Duke of Cornwall are party to arbitration agreements in their personal capacity, then the section permits them to

avoid participating personally in the proceedings, and allows them to participate through representatives.

Depending upon the precise character of her involvement, the Queen may be represented by the Chancellor of the Duchy of Lancaster or his appointee, or by any person appointed in writing under her signature ('the Royal Sign Manual'), (subs.2).

The Duke of Cornwall may be represented by any appointee, (subs.3)

[106E]   *Definition of 'party' – Subs.(4)*

The definition of 'party' and 'parties' is extended, for the purposes of Part I, to take into account the representation of the Queen and the Duke of Cornwall in accordance with this section, see s.83.

Section 107 – Consequential Amendments and Repeals

107.–(1) The enactments specified in Schedule 3 are amended in accordance with that Schedule, the amendments being consequential on the provisions of this Act.

(2) The enactments specified in Schedule 4 are repealed to the extent specified.

[107A] Status

This is a new section.

[107B] Summary

The section brings into effect the consequential amendments and repeals specified in Schedules 3 and 4 to the Act respectively.

Points

[107C] *Scotland*

Note that s.108(3) extends the provisions of Schedules 3 and 4 to Scotland so far as they relate to enactments which extend there, with the notable exception of the Arbitration Act 1975, whose repeal is expressly confined to England and Wales and Northern Ireland, (s.108(4)). This is to enable the New York Convention regime for the recognition and enforcement of foreign arbitral awards to continue to apply to Scotland, since the re-enactment of the 1975 Act in ss.100 to 104 of this Act does not so apply.

[107D] *Appeals to Court of Appeal*

The effect of the amendment to s.18(1) of the Supreme Court Act 1981 made by S.107 and Schedule 3 is to be understood as giving effect to the exclusions (and restrictions) on the right to appeal to the Court of Appeal laid down in Part 1 of the 1996 Act and no more: *Inco Europe Ltd v. First Choice Distribution* [1999] 1 WLR 270.

Section 108 – Extent

108.–(1) The provisions of this Act extend to England and Wales and, except as mentioned below, to Northern Ireland.

(2) The following provisions of Part II do not extend to Northern Ireland–

section 92 (exclusion of Part I in relation to small claims arbitration in the county court), and
section 93 and Schedule 2 (appointment of judges as arbitrators).

(3) Sections 89, 90 and 91 (consumer arbitration agreements) extend to Scotland and the provisions of Schedules 3 and 4 (consequential amendments and repeals) extend to Scotland so far as they relate to enactments which so extend, subject as follows.

(4) The repeal of the Arbitration Act 1975 extends only to England and Wales and Northern Ireland.

[108A] Status

This section derives from the Arbitration Act 1950, s.34; the Arbitration Act 1975, s.8(4); and the Arbitration Act 1979, s.8(4).

[108B] Summary

The section determines the geographical extent of the Act.

Points

[108C] *England and Wales – Subs.(1)*
All parts of the Act extend to England and Wales without exception.

[108D] *Northern Ireland – Subss.(1), (2)*
Whereas the major provisions of the 1950 Act and the 1979 Act did not apply to Northern Ireland (which was still working with the Arbitration Act (Northern Ireland) 1937), by contrast this Act does apply to Northern Ireland. There are minor exceptions relating to the exclusion of Part I to small claims arbitration in the county court and the provisions relating to judge-arbitrators.

[108E] *Scotland – Subss.(3), (4)*
Scotland has its own arbitration law, so that in general the Act does not apply to Scotland. However, the provisions relating to consumer arbitration agreements contained in ss.89 to 91 do apply to Scotland in the interests of consistency throughout the United Kingdom.
It should be noted that the regimes for the recognition and enfor-

401

cement of foreign arbitral awards under the Geneva and New York Conventions (to which the United Kingdom is a party) continue to apply to Scotland, as they did before this Act. Part II of the Arbitration Act 1950 (enacting the Geneva Convention) applies to Scotland (see s.41 of that Act), and remains in force by virtue of s.99 of this Act. The Arbitration Act 1975 (enacting the New York Convention) is only repealed for England and Wales and Northern Ireland, (subs.(4)). It therefore remains in force so far as Scotland is concerned. The New York Convention regime is re-enacted for England and Wales and Northern Ireland by ss.100 to 104 of this Act.

Section 109 – Commencement

109.–(1) The provisions of this Act come into force on such day as the Secretary of State may appoint by order made by statutory instrument, and different days may be appointed for different purposes.

(2) An order under subsection (1) may contain such transitional provisions as appear to the Secretary of State to be appropriate.

[109A] ## Status

This is a new section.

[109B] ## Summary

The section provides for the commencement of the Act by statutory instrument.

Points

[109C] *Phased introduction*
The bulk of the Act was brought into effect on 31 January 1997. The Commencement Order is set out in the Appendices of this book, and a summary of its effect appears in paragraph 84E above.

Section 110 – Short Title

110. This Act may be cited as the Arbitration Act 1996.

[110A] **Status**

This is a new section.

[110B] **Summary**

This section sets out the title by which the Act is generally referred to and known.

Schedules

Schedule 1
Mandatory provisions of Part I

sections 9 to 11 (stay of legal proceedings);
section 12 (power of court to extend agreed time limits);
section 13 (application of Limitation Acts);
section 24 (power of court to remove arbitrator);
section 26(1) (effect of death of arbitrator);
section 28 (liability of parties for fees and expenses of arbitrators);
section 29 (immunity of arbitrator);
section 31 (objection to substantive jurisdiction of tribunal);
section 32 (determination of preliminary point of jurisdiction);
section 33 (general duty of tribunal);
section 37(2) (items to be treated as expenses of arbitrators);
section 40 (general duty of parties);
section 43 (securing the attendance of witnesses);
section 56 (power to withhold award in case of non-payment);
section 60 (effectiveness of agreement for payment of costs in any event);
section 66 (enforcement of award);
sections 67 and 68 (challenging the award: substantive jurisdiction and serious
    irregularity), and sections 70 and 71 (supplementary provisions; effect of
    order of court) so far as relating to those sections;
section 72 (saving for rights of person who takes no part in proceedings);
section 73 (loss of right to object);
section 74 (immunity of arbitral institutions, &c.);
section 75 (charge to secure payment of solicitors' costs).

Schedule 2
Modifications of Part I in relation to judge-arbitrators

Introductory
1. In this Schedule 'judge-arbitrator' means a judge of the Commercial Court or
official referee appointed as arbitrator or umpire under section 93.

General
2.–(1) Subject to the following provisions of this Schedule, references in Part I to
the court shall be construed in relation to a judge-arbitrator, or in relation to the
appointment of a judge-arbitrator, as references to the Court of Appeal.

(2) The references in sections 32(6), 45(6) and 69(8) to the Court of Appeal shall
in such a case be construed as references to the House of Lords.

Arbitrator's fees
3.–(1) The power of the court in section 28(2) to order consideration and
adjustment of the liability of a party for the fees of an arbitrator may be exercised by a judge-arbitrator.

(2) Any such exercise of the power is subject to the powers of the Court of

Appeal under sections 24(4) and 25(3)(b) (directions as to entitlement to fees or expenses in case of removal or resignation).

Exercise of court powers in support of arbitration

4.–(1) Where the arbitral tribunal consists of or includes a judge-arbitrator the powers of the court under sections 42 to 44 (enforcement of peremptory orders, summoning witnesses, and other court powers) are exercisable by the High Court and also by the judge-arbitrator himself.

(2) Anything done by a judge-arbitrator in the exercise of those powers shall be regarded as done by him in his capacity as judge of the High Court and have effect as if done by that court.

Nothing in this sub-paragraph prejudices any power vested in him as arbitrator or umpire.

Extension of time for making award

5.–(1) The power conferred by section 50 (extension of time for making award) is exercisable by the judge-arbitrator himself.

(2) Any appeal from a decision of a judge-arbitrator under that section lies to the Court of Appeal with the leave of that court.

Withholding award in case of non-payment

6.–(1) The provisions of paragraph 7 apply in place of the provisions of section 56 (power to withhold award in the case of non-payment) in relation to the withholding of an award for non-payment of the fees and expenses of a judge-arbitrator.

(2) This does not affect the application of section 56 in relation to the delivery of such an award by an arbitral or other institution or person vested by the parties with powers in relation to the delivery of the award.

7.–(1) A judge-arbitrator may refuse to deliver an award except upon payment of the fees and expenses mentioned in section 56(1).

(2) The judge-arbitrator may, on an application by a party to the arbitral proceedings, order that if he pays into the High Court the fees and expenses demanded, or such lesser amount as the judge-arbitrator may specify–

(a)   the award shall be delivered,
(b)   the amount of the fees and expenses properly payable shall be determined by such means and upon such terms as he may direct, and
(c)   out of the money paid into court there shall be paid out such fees and expenses as may be found to be properly payable and the balance of the money (if any) shall be paid out to the applicant.

(3) For this purpose the amount of fees and expenses properly payable is the amount the applicant is liable to pay under section 28 or any agreement relating to the payment of the arbitrator.

(4) No application to the judge-arbitrator under this paragraph may be made where there is any available arbitral process for appeal or review of the amount of the fees or expenses demanded.

(5) Any appeal from a decision of a judge-arbitrator under this paragraph lies to the Court of Appeal with the leave of that court.

(6) Where a party to arbitral proceedings appeals under sub-paragraph (5), an arbitrator is entitled to appear and be heard.

Correction of award or additional award

8. Subsections (4) to (6) of section 57 (correction of award or additional award: time limit for application or exercise of power) do not apply to a judge-arbitrator.

Costs

9. Where the arbitral tribunal consists of or includes a judge-arbitrator the powers of the court under section 63(4) (determination of recoverable costs) shall be exercised by the High Court.

10.–(1) The power of the court under section 64 to determine an arbitrator's reasonable fees and expenses may be exercised by a judge-arbitrator.

(2) Any such exercise of the power is subject to the powers of the Court of Appeal under sections 24(4) and 25(3)(b) (directions as to entitlement to fees or expenses in case of removal or resignation).

Enforcement of award

11. The leave of the court required by section 66 (enforcement of award) may in the case of an award of a judge-arbitrator be given by the judge-arbitrator himself.

Solicitors' costs

12. The powers of the court to make declarations and orders under the provisions applied by section 75 (power to charge property recovered in arbitral proceedings with the payment of solicitors' costs) may be exercised by the judge-arbitrator.

Powers of court in relation to service of documents

13.–(1) The power of the court under section 77(2) (powers of court in relation to service of documents) is exercisable by the judge-arbitrator.

(2) Any appeal from a decision of a judge-arbitrator under that section lies to the Court of Appeal with the leave of that court.

Powers of court to extend time limits relating to arbitral proceedings

14.–(1) The power conferred by section 79 (power of court to extend time limits relating to arbitral proceedings) is exercisable by the judge-arbitrator himself.

(2) Any appeal from a decision of a judge-arbitrator under that section lies to the Court of Appeal with the leave of that court.

Schedule 3
Consequential amendments

Merchant Shipping Act 1894 (c.60)

1. In section 496 of the Merchant Shipping Act 1894 (provisions as to deposits by owners of goods), after subsection (4) insert–

'(5) In subsection (3) the expression "legal proceedings" includes arbitral proceedings and as respects England and Wales and Northern Ireland the provisions of section 14 of the Arbitration Act 1996 apply to determine when such proceedings are commenced.'.

Stannaries Court (Abolition) Act 1896 (c.45)

2. In section 4(1) of the Stannaries Court (Abolition) Act 1896 (references of certain disputes to arbitration), for the words from 'tried before' to 'any such reference' substitute 'referred to arbitration before himself or before an arbitrator agreed on by the parties or an officer of the court'.

Tithe Act 1936 (c.43)

3. In Section 39(1) of the Tithe Act 1936 (proceedings of Tithe Redemption Commission)–

(a)   for 'the Arbitration Acts 1889 to 1934' substitute 'Part I of the Arbitration Act 1996';
(b)   for paragraph (e) substitute–
   '(e)   the making of an application to the court to determine a preliminary point of law and the bringing of an appeal to the court on a point of law;';
(c)   for 'the said Acts' substitute 'Part I of the Arbitration Act 1996'.

Education Act 1944 (c.31)

4. In section 75(2) of the Education Act 1944 (proceedings of Independent School Tribunals) for 'the Arbitration Acts 1889 to 1934' substitute 'Part I of the Arbitration Act 1996'.

Commonwealth Telegraphs Act 1949 (c.39)

5. In section 8(2) of the Commonwealth Telegraphs Act 1949 (proceedings of referees under the Act) for 'the Arbitration Acts 1889 to 1934, or the Arbitration Act (Northern Ireland) 1937,' substitute 'Part I of the Arbitration Act 1996'.

Lands Tribunal Act 1949 (c.42)

6. In section 3 of the Lands Tribunal Act 1949 (proceedings before the Lands Tribunal)–

(a)   in subsection (6)(c) (procedural rules: power to apply Arbitration Acts), and
(b)   in subsection (8) (exclusion of Arbitration Acts except as applied by rules),

for 'the Arbitration Acts 1889 to 1934' substitute 'Part I of the Arbitration Act 1996'.

Wireless Telegraphy Act 1949 (c.54)

7. In the Wireless Telegraphy Act 1949, Schedule 2 (procedure of appeals tribunal), in paragraph 3(1)–

(a)   for the words 'the Arbitration Acts 1889 to 1934' substitute 'Part I of the Arbitration Act 1996';
(b)   after the word 'Wales' insert 'or Northern Ireland'; and
(c)   for 'the said Acts' substitute 'Part I of that Act'.

Patents Act 1949 (c.87)

8. In section 67 of the Patents Act 1949 (proceedings as to infringement of pre-1978 patents referred to comptroller), for 'The Arbitration Acts 1889 to 1934' substitute 'Part I of the Arbitration Act 1996'.

Schedules

National Health Service (Amendment) Act 1949 (c.93)
9. In section 7(8) of the National Health Service (Amendment) Act 1949 (arbitration in relation to hardship arising from the National Health Service Act 1946 or the Act), for 'the Arbitration Acts 1889 to 1934' substitute 'Part I of the Arbitration Act 1996' and for 'the said Acts' substitute 'Part I of that Act'.

Arbitration Act 1950 (c.27)
10. In section 36(1) of the Arbitration Act 1950 (effect of foreign awards enforceable under Part II of that Act) for 'section 26 of this Act' substitute 'section 66 of the Arbitration Act 1996'.

Interpretation Act (Northern Ireland) 1954 (c.33 (N.I.))
11. In section 46(2) of the Interpretation Act (Northern Ireland) 1954 (miscellaneous definitions), for the definition of 'arbitrator' substitute–

'"arbitrator" has the same meaning as in Part I of the Arbitration Act 1996;'.

Agricultural Marketing Act 1958 (c.47)
12. In section 12(1) of the Agricultural Marketing Act 1958 (application of provisions of Arbitration Act 1950)–

(a) for the words from the beginning to 'shall apply' substitute 'Sections 45 and 69 of the Arbitration Act 1996 (which relate to the determination by the court of questions of law) and section 66 of that Act (enforcement of awards) apply'; and
(b) for 'an arbitration' substitute 'arbitral proceedings'.

Carriage by Air Act 1961 (c.27)
13.–(1) The Carriage by Air Act 1961 is amended as follows.

(2) In section 5(3) (time for bringing proceedings)–

(a) for 'an arbitration' in the first place where it occurs substitute 'arbitral proceedings'; and
(b) for the words from 'and subsections (3) and (4)' to the end substitute 'and the provisions of section 14 of the Arbitration Act 1996 apply to determine when such proceedings are commenced.'.

(3) In section 11(c) (application of section 5 to Scotland)–

(a) for 'subsections (3) and (4)' substitute 'the provisions of section 14 of the Arbitration Act 1996'; and
(b) for 'an arbitration' substitute 'arbitral proceedings'.

Factories Act 1961 (c.34)
14. In the Factories Act 1961, for section 171 (application of Arbitration Act 1950), substitute–

'Application of the Arbitration Act 1996. 171. Part I of the Arbitration Act 1996 does not apply to proceedings under this Act except in so far as it may be applied by regulations made under this Act.'.

Clergy Pensions Measure 1961 (No. 3)
15. In the Clergy Pensions Measure 1961, section 38(4) (determination of questions), for the words 'The Arbitration Act 1950' substitute 'Part I of the Arbitration Act 1996'.

Transport Act 1962 (c.46)

16.–(1) The Transport Act 1962 is amended as follows.

(2) In section 74(6)(f) (proceedings before referees in pension disputes), for the words 'the Arbitration Act 1950' substitute 'Part I of the Arbitration Act 1996'.

(3) In section 81(7) (proceedings before referees in compensation disputes), for the words 'the Arbitration Act 1950' substitute 'Part I of the Arbitration Act 1996'.

(4) In Schedule 7, Part IV (pensions), in paragraph 17(5) for the words 'the Arbitration Act 1950' substitute 'Part I of the Arbitration Act 1996'.

Corn Rents Act 1963 (c.14)

17. In the Corn Rents Act 1963, section 1(5) (schemes for apportioning corn rents, &c.), for the words 'the Arbitration Act 1950' substitute 'Part I of the Arbitration Act 1996'.

Plant Varieties and Seeds Act 1964 (c.14)

18. In section 10(6) of the Plant Varieties and Seeds Act 1964 (meaning of 'arbitration agreement'), for 'the meaning given by section 32 of the Arbitration Act 1950' substitute 'the same meaning as in Part I of the Arbitration Act 1996'.

Lands Tribunal and Compensation Act (Northern Ireland) 1964 (c.29 (N.I.))

19. In section 9 of the Lands Tribunal and Compensation Act (Northern Ireland) 1964 (proceedings of Lands Tribunal), in subsection (3) (where Tribunal acts as arbitrator) for 'the Arbitration Act (Northern Ireland) 1937' substitute 'Part I of the Arbitration Act 1996'.

Industrial and Provident Societies Act 1965 (c.12)

20.–(1) Section 60 of the Industrial and Provident Societies Act 1965 is amended as follows.

(2) In subsection (8) (procedure for hearing disputes between society and member, &c.)–

(a) in paragraph (a) for 'the Arbitration Act 1950' substitute 'Part I of the Arbitration Act 1996'; and

(b) in paragraph (b) omit 'by virtue of section 12 of the said Act of 1950'.

(3) For subsection (9) substitute–

'(9) The court or registrar to whom any dispute is referred under subsections (2) to (7) may at the request of either party state a case on any question of law arising in the dispute for the opinion of the High Court or, as the case may be, the Court of Session.'.

Carriage of Goods by Road Act 1965 (c.37)

21. In section 7(2) of the Carriage of Goods by Road Act 1965 (arbitrations: time at which deemed to commence), for paragraphs (a) and (b) substitute–

'(a) as respects England and Wales and Northern Ireland, the provisions of section 14(3) to (5) of the Arbitration Act 1996 (which determine the time at which an arbitration is commenced) apply;'.

Factories Act (Northern Ireland) 1965 (c.20 (N.I.))

22. In section 171 of the Factories Act (Northern Ireland) 1965 (application of Arbitration Act), for 'The Arbitration Act (Northern Ireland) 1937' substitute 'Part I of the Arbitration Act 1996'.

Commonwealth Secretariat Act 1966 (c.10)

23. In section 1(3) of the Commonwealth Secretariat Act 1966 (contracts with Commonwealth Secretariat to be deemed to contain provision for arbitration), for 'the Arbitration Act 1950 and the Arbitration Act (Northern Ireland) 1937' substitute 'Part I of the Arbitration Act 1996'.

Arbitration (International Investment Disputes) Act 1966 (c.41)

24. In the Arbitration (International Investment Disputes) Act 1966, for section 3 (application of Arbitration Act 1950 and other enactments) substitute–

'Application of provisions of Arbitration Act 1996.

3.–(1) The Lord Chancellor may by order direct that any of the provisions contained in sections 36 and 38 to 44 of the Arbitration Act 1996 (provisions concerning the conduct of arbitral proceedings, &c.) shall apply to such proceedings pursuant to the Convention as are specified in the order with or without any modifications or exceptions specified in the order.

(2) Subject to subsection (1), the Arbitration Act 1996 shall not apply to proceedings pursuant to the Convention, but this subsection shall not be taken as affecting section 9 of that Act (stay of legal proceedings in respect of matter subject to arbitration).

(3) An order made under this section–

(a) may be varied or revoked by a subsequent order so made, and

(b) shall be contained in a statutory instrument.'.

Poultry Improvement Act (Northern Ireland) 1968 (c.12 (N.I.))

25. In paragraph 10(4) of the Schedule to the Poultry Improvement Act (Northern Ireland) 1968 (reference of disputes), for 'The Arbitration Act (Northern Ireland) 1937' substitute 'Part I of the Arbitration Act 1996'.

Industrial and Provident Societies Act (Northern Ireland) 1969 (c.24 (N.I.))

26.–(1) Section 69 of the Industrial and Provident Societies Act (Northern Ireland) 1969 (decision of disputes) is amended as follows.

(2) In subsection (7) (decision of disputes)–

(a) in the opening words, omit the words from 'and without prejudice' to '1937';

(b) at the beginning of paragraph (a) insert 'without prejudice to any powers exercisable by virtue of Part I of the Arbitration Act 1996,'; and

(c) in paragraph (b) omit 'the registrar or' and 'registrar or' and for the words from 'as might have been granted by the High Court' to the end substitute 'as might be granted by the registrar'.

(3) For subsection (8) substitute–

'(8) The court or registrar to whom any dispute is referred under subsections (2) to (6) may at the request of either party state a case on any question of law arising in the dispute for the opinion of the High Court.'.

Health and Personal Social Services (Northern Ireland) Order 1972 (N.I.14)

27. In Article 105(6) of the Health and Personal Social Services (Northern Ireland) Order 1972 (arbitrations under the Order), for 'the Arbitration Act (Northern Ireland) 1937' substitute 'Part I of the Arbitration Act 1996'.

Consumer Credit Act 1974 (c.39)

28.–(1) Section 146 of the Consumer Credit Act 1974 is amended as follows.

(2) In subsection (2) (solicitor engaged in contentious business), for 'section 86(1) of the Solicitors Act 1957' substitute 'section 87(1) of the Solicitors Act 1974'.

(3) In subsection (4) (solicitor in Northern Ireland engaged in contentious business), for the words from 'business done' to 'Administration of Estates (Northern Ireland) Order 1979' substitute 'contentious business (as defined in Article 3(2) of the Solicitors (Northern Ireland) Order 1976.'.

Friendly Societies Act 1974 (c.46)

29.–(1) The Friendly Societies Act 1974 is amended as follows.

(2) For section 78(1) (statement of case) substitute–

'(1) Any arbitrator, arbiter or umpire to whom a dispute falling within section 76 above is referred under the rules of a registered society or branch may at the request of either party state a case on any question of law arising in the dispute for the opinion of the High Court or, as the case may be, the Court of Session.'.

(3) In section 83(3) (procedure on objections to amalgamations &c. of friendly societies), for 'the Arbitration Act 1950 or, in Northern Ireland, the Arbitration Act (Northern Ireland) 1937' substitute 'Part I of the Arbitration Act 1996'.

Industry Act 1975 (c.68)

30. In Schedule 3 to the Industry Act (arbitration of disputes relating to vesting and compensation orders), in paragraph 14 (application of certain provisions of Arbitration Acts)–

(a) for 'the Arbitration Act 1950 or, in Northern Ireland, the Arbitration Act (Northern Ireland) 1937' substitute 'Part I of the Arbitration Act 1996', and
(b) for 'that Act' substitute 'that Part'.

Industrial Relations (Northern Ireland) Order 1976 (N.I.16)

31. In Article 59(9) of the Industrial Relations (Northern Ireland) Order 1976 (proceedings of industrial tribunal), for 'The Arbitration Act (Northern Ireland) 1937' substitute 'Part I of the Arbitration Act 1996'.

Aircraft and Shipbuilding Industries Act 1977 (c.3)

32. In Schedule 7 to the Aircraft and Shipbuilding Industries Act 1977 (procedure of Arbitration Tribunal), in paragraph 2–

(a) for 'the Arbitration Act 1950 or, in Northern Ireland, the Arbitration Act

(Northern Ireland) 1937' substitute 'Part I of the Arbitration Act 1996', and
(b)   for 'that Act' substitute 'that Part'.

Patents Act 1977 (c.37)

33. In section 130 of the Patents Act 1977 (interpretation), in subsection (8) (exclusion of Arbitration Act) for 'The Arbitration Act 1950' substitute 'Part I of the Arbitration Act 1996'.

Judicature (Northern Ireland) Act 1978 (c.23)

34.–(1) The Judicature (Northern Ireland) Act 1978 is amended as follows.

(2) In section 35(2) (restrictions on appeals to the Court of Appeal), after paragraph (f) insert–

'(fa) except as provided by Part I of the Arbitration Act 1996, from any decision of the High Court under that Part;'.

(3) In section 55(2) (rules of court) after paragraph (c) insert–

'(cc) providing for any prescribed part of the jurisdiction of the High Court in relation to the trial of any action involving matters of account to be exercised in the prescribed manner by a person agreed by the parties and for the remuneration of any such person;'.

Health and Safety at Work (Northern Ireland) Order 1978 (N.I.9)

35. In Schedule 4 to the Health and Safety at Work (Northern Ireland) Order 1978 (licensing provisions), in paragraph 3, for 'The Arbitration Act (Northern Ireland) 1937' substitute 'Part I of the Arbitration Act 1996'.

County Courts (Northern Ireland) Order 1980 (N.I.3)

36.–(1) The County Courts (Northern Ireland) Order 1980 is amended as follows.

(2) In Article 30 (civil jurisdiction exercisable by district judge)–

(a)   for paragraph (2) substitute–

'(2) Any order, decision or determination made by a district judge under this Article (other than one made in dealing with a claim by way of arbitration under paragraph (3)) shall be embodied in a decree which for all purposes (including the right of appeal under Part VI) shall have the like effect as a decree pronounced by a county court judge.';

(b)   for paragraphs (4) and (5) substitute–

'(4) Where in any action to which paragraph (1) applies the claim is dealt with by way of arbitration under paragraph (3)–

(a)   any award made by the district judge in dealing with the claim shall be embodied in a decree which for all purposes (except the right of appeal under Part VI) shall have the like effect as a decree pronounced by a county court judge;
(b)   the district judge may, and shall if so required by the High Court, state for the determination of the High Court any question of law arising out of an award so made;
(c)   except as provided by sub-paragraph (b), any award so made shall be final; and
(d)   except as otherwise provided by county court rules, no costs shall be awarded in connection with the action.

(5) Subject to paragraph (4), county court rules may–

(a) apply any of the provisions of Part I of the Arbitration Act 1996 to arbitrations under paragraph (3) with such modifications as may be prescribed;

(b) prescribe the rules of evidence to be followed on any arbitration under paragraph (3) and, in particular, make provision with respect to the manner of taking and questioning evidence.

(5A) Except as provided by virtue of paragraph (5)(a), Part I of the Arbitration Act 1996 shall not apply to an arbitration under paragraph (3).'.

(3) After Article 61 insert–

'Appeals from decisions under Part I of Arbitration Act 1996
61A.–(1) Article 61 does not apply to a decision of a county court judge made in the exercise of the jurisdiction conferred by Part I of the Arbitration Act 1996.

(2) Any party dissatisfied with a decision of the county court made in the exercise of the jurisdiction conferred by any of the following provisions of Part I of the Arbitration Act 1996, namely–

(a) section 32 (question as to substantive jurisdiction of arbitral tribunal);
(b) section 45 (question of law arising in course of arbitral proceedings);
(c) section 67 (challenging award of arbitral tribunal: substantive jurisdiction);
(d) section 68 (challenging award of arbitral tribunal: serious irregularity);
(e) section 69 (appeal on point of law),

may, subject to the provisions of that Part, appeal from that decision to the Court of Appeal.

(3) Any party dissatisfied with any decision of a county court made in the exercise of the jurisdiction conferred by any other provision of Part I of the Arbitration Act 1996 may, subject to the provisions of that Part, appeal from that decision to the High Court.

(4) The decision of the Court of Appeal on an appeal under paragraph (2) shall be final.'.

Supreme Court Act 1981 (c.54)
37.–(1) The Supreme Court Act 1981 is amended as follows.

(2) In section 18(1) (restrictions on appeals to the Court of Appeal), for paragraph (g) substitute–

'(g) except as provided by Part I of the Arbitration Act 1996, from any decision of the High Court under that Part;'.

(3) In section 151 (interpretation, &c.), in the definition of 'arbitration agreement', for 'the Arbitration Act 1950 by virtue of section 32 of that Act;' substitute 'Part I of the Arbitration Act 1996;'.

Merchant Shipping (Liner Conferences) Act 1982 (c.37)
38. In section 7(5) of the Merchant Shipping (Liner Conferences) Act 1982 (stay of legal proceedings), for the words from 'section 4(1)' to the end substitute 'section 9 of the Arbitration Act 1996 (which also provides for the staying of legal proceedings).'.

Agricultural Marketing (Northern Ireland) Order 1982 (N.I.12)

39. In Article 14 of the Agricultural Marketing (Northern Ireland) Order 1982 (application of provisions of Arbitration Act (Northern Ireland) 1937)–

(a) for the words from the beginning to 'shall apply' substitute 'Section 45 and 69 of the Arbitration Act 1996 (which relate to the determination by the court of questions of law) and section 66 of that Act (enforcement of awards)' apply; and

(b) for 'an arbitration' substitute 'arbitral proceedings'.

Mental Health Act 1983 (c.20)

40. In section 78 of the Mental Health Act 1983 (procedure of Mental Health Review Tribunals), in subsection (9) for 'The Arbitration Act 1950' substitute 'Part I of the Arbitration Act 1996'.

Registered Homes Act 1984 (c.23)

41. In section 43 of the Registered Homes Act 1984 (procedure of Registered Homes Tribunals), in subsection (3) for 'The Arbitration Act 1950' substitute 'Part I of the Arbitration Act 1996'.

Housing Act 1985 (c.68)

42. In section 47(3) of the Housing Act 1985 (agreement as to determination of matters relating to service charges) for 'section 32 of the Arbitration Act 1950' substitute 'Part I of the Arbitration Act 1996'.

Landlord and Tenant Act 1985 (c.70)

43. In section 19(3) of the Landlord and Tenant Act 1985 (agreement as to determination of matters relating to service charges), for 'section 32 of the Arbitration Act 1950' substitute 'Part I of the Arbitration Act 1996'.

Credit Unions (Northern Ireland) Order 1985 (N.I.12)

44.–(1) Article 72 of the Credit Unions (Northern Ireland) Order 1985 (decision of disputes) is amended as follows.

(2) In paragraph (7)–

(a) in the opening words, omit the words from 'and without prejudice' to '1937';

(b) at the beginning of sub-paragraph (a) insert 'without prejudice to any powers exercisable by virtue of Part I of the Arbitration Act 1996,'; and

(c) in sub-paragraph (b) omit 'the registrar or' and 'registrar or' and for the words from 'as might have been granted by the High Court' to the end substitute 'as might be granted by the registrar'.

(3) For paragraph (8) substitute–

'(8) The court or registrar to whom any dispute is referred under paragraphs (2) to (6) may at the request of either party state a case on any question of law arising in the dispute for the opinion of the High Court.'.

Agricultural Holdings Act 1986 (c.5)

45. In section 84(1) of the Agricultural Holdings Act 1986 (provisions relating to arbitration), for 'the Arbitration Act 1950' substitute 'Part I of the Arbitration Act 1996'.

Insolvency Act 1986 (c.45)

46. In the Insolvency Act 1986, after section 349 insert–

'Arbitration agreements to which bankrupt is party.

349A.–(1) This section applies where a bankrupt had become party to a contract containing an arbitration agreement before the commencement of his bankruptcy.

(2) If the trustee in bankruptcy adopts the contract, the arbitration agreement is enforceable by or against the trustee in relation to matters arising from or connected with the contract.

(3) If the trustee in bankruptcy does not adopt the contract and a matter to which the arbitration agreement applies requires to be determined in connection with or for the purposes of the bankruptcy proceedings–

(a)　the trustee with the consent of the creditors' committee, or

(b)　any other party to the agreement,

may apply to the court which may, if it thinks fit in all the circumstances of the case, order that the matter be referred to arbitration in accordance with the arbitration agreement.

(4) In this section–

'arbitration agreement' has the same meaning as in Part I of the Arbitration Act 1996; and

'the court' means the court which has jurisdiction in the bankruptcy proceedings.'.

Building Societies Act 1986 (c.53)

47. In Part II of Schedule 14 to the Building Societies Act 1986 (settlement of disputes: arbitration), in paragraph 5(6) for 'the Arbitration Act 1950 and the Arbitration Act 1979 or, in Northern Ireland, the Arbitration Act (Northern Ireland) 1937' substitute 'Part I of the Arbitration Act 1996'.

Mental Health (Northern Ireland) Order 1986 (N.I.4)

48. In Article 83 of the Mental Health (Northern Ireland) Order 1986 (procedure of Mental Health Review Tribunal), in paragraph (8) for 'The Arbitration Act (Northern Ireland) 1937' substitute 'Part I of the Arbitration Act 1996'.

Multilateral Investment Guarantee Agency Act 1988 (c.8)

49. For section 6 of the Multilateral Investment Guarantee Agency Act 1988 (application of Arbitration Act) substitute–

'Application of Arbitration Act.

6.–(1) The Lord Chancellor may by order made by statutory instrument direct that any of the provisions of sections 36 and 38 to 44 of the Arbitration Act 1996 (provisions in relation to the conduct of the arbitral proceedings, &c.) apply, with such modifications or exceptions as are specified in the order, to such arbitration proceedings pursuant to Annex II to the Convention as are specified in the order.

(2) Except as provided by an order under subsection (1) above, no provision of Part I of the Arbitration Act 1996 other than section 9 (stay of legal proceedings) applies to any such proceedings.'.

Copyright, Designs and Patents Act 1988 (c.48)

50. In section 150 of the Copyright, Designs and Patents Act 1988 (Lord Chancellor's power to make rules for Copyright Tribunal), for subsection (2) substitute–

'(2) The rules may apply in relation to the Tribunal, as respects proceedings in England and Wales or Northern Ireland, any of the provisions of Part I of the Arbitration Act 1996.'.

[Fair Employment (Northern Ireland) Act 1989 (c.32)

51. In the Fair Employment (Northern Ireland) Act 1989, section 5(7) (procedure of Fair Employment Tribunal), for 'The Arbitration Act (Northern Ireland) 1937' substitute 'Part I of the Arbitration Act 1996'.]
(Repealed by SI 1998 No. 3162 (NI 21), Art.105, Sched.5)

Limitation (Northern Ireland) Order 1989 (N.I.11)

52. In Article 2(2) of the Limitation (Northern Ireland) Order 1989 (interpretation), in the definition of 'arbitration agreement', for 'the Arbitration Act (Northern Ireland) 1937' substitute 'Part I of the Arbitration Act 1996'.

Insolvency (Northern Ireland) Order 1989 (N.I.19)

53. In the Insolvency (Northern Ireland) Order 1989, after Article 320 insert–

'Arbitration agreements to which bankrupt is party

320A.–(1) This Article applies where a bankrupt had become party to a contract containing an arbitration agreement before the commencement of his bankruptcy.

(2) If the trustee in bankruptcy adopts the contract, the arbitration agreement is enforceable by or against the trustee in relation to matters arising from or connected with the contract.

(3) If the trustee in bankruptcy does not adopt the contract and a matter to which the arbitration agreement applies requires to be determined in connection with or for the purposes of the bankruptcy proceedings–

(a)   the trustee with the consent of the creditors' committee, or
(b)   any other party to the agreement,

may apply to the court which may, if it thinks fit in all the circumstances of the case, order that the matter be referred to arbitration in accordance with the arbitration agreement.

(4) In this Article–

'arbitration agreement' has the same meaning as in Part I of the Arbitration Act 1996; and
'the court' means the court which has jurisdiction in the bankruptcy proceedings.'.

[Social Security Administration Act 1992 (c.5)

54. In section 59 of the Social Security Administration Act 1992 (procedure for inquiries, &c.), in subsection (7), for 'The Arbitration Act 1950' substitute 'Part I of the Arbitration Act 1996'.]
(Repealed by Social Security Act 1998, c.14, s.86, Sched.8)

[Social Security Administration (Northern Ireland) Act 1992 (c.8)

55. In section 57 of the Social Security Administration (Northern Ireland) Act 1992 *(procedure for inquiries, &c.), in subsection (6) for 'the Arbitration Act (Northern Ireland) 1937' substitute 'Part I of the Arbitration Act 1996'.]*
(Repealed by SI 1998 No. 150 (NI 10), Art.78, Sched.2)

Trade Union and Labour Relations (Consolidation) Act 1992 (c.52)

56. In sections 212(5) and 263(6) of the Trade Union and Labour Relations (Consolidation) Act 1992 (application of Arbitration Act) for 'the Arbitration Act 1950' substitute 'Part I of the Arbitration Act 1996'.

Industrial Relations (Northern Ireland) Order 1992 (N.I.5)

57. In Articles 84(9) and 92(5) of the Industrial Relations (Northern Ireland) Order 1992 (application of Arbitration Act) for 'The Arbitration Act (Northern Ireland) 1937' substitute 'Part I of the Arbitration Act 1996'.

Registered Homes (Northern Ireland) Order 1992 (N.I.20)

58. In Article 33(3) of the Registered Homes (Northern Ireland) Order 1992 (procedure of Registered Homes Tribunal) for 'The Arbitration Act (Northern Ireland) 1937' substitute 'Part I of the Arbitration Act 1996'.

Education Act 1993 (c.35)

59. In section 180(4) of the Education Act 1993 (procedure of Special Educational Needs Tribunal), for 'The Arbitration Act 1950' substitute 'Part I of the Arbitration Act 1996'.

Roads (Northern Ireland) Order 1993 (N.I.15)

60.–(1) The Roads (Northern Ireland) Order 1993 is amended as follows.

(2) In Article 131 (application of Arbitration Act) for 'the Arbitration Act (Northern Ireland) 1937' substitute 'Part I of the Arbitration Act 1996'.

(3) In Schedule 4 (disputes), in paragraph 3(2) for 'the Arbitration Act (Northern Ireland) 1937' substitute 'Part I of the Arbitration Act 1996'.

Merchant Shipping Act 1995 (c.21)

61. In Part II of Schedule 6 to the Merchant Shipping Act 1995 (provisions having effect in connection with Convention Relating to the Carriage of Passengers and Their Luggage by Sea), for paragraph 7 substitute–

'7. Article 16 shall apply to arbitral proceedings as it applies to an action; and, as respects England and Wales and Northern Ireland, the provisions of section 14 of the Arbitration Act 1996 apply to determine for the purposes of that Article when an arbitration is commenced.'.

Industrial Tribunals Act 1996 (c.17)

62. In section 6(2) of the Industrial Tribunals Act 1996 (procedure of industrial tribunals), for 'The Arbitration Act 1950' substitute 'Part I of the Arbitration Act 1996'.

Schedule 4
Repeals

Chapter	Short title	Extent of repeal
1892 c. 43.	Military Lands Act 1892.	In section 21(b), the words 'under the Arbitration Act 1889'.
1922 c. 51.	Allotments Act 1922.	In section 21(3), the words 'under the Arbitration Act 1889'.
1937 c. 8 (N.I.).	Arbitration Act (Northern Ireland) 1937.	The whole Act.
1949 c. 54.	Wireless Telegraphy Act 1949.	In Schedule 2, paragraph 3(3).
1949 c. 97.	National Parks and Access to the Countryside Act 1949.	In section 18(4), the words from 'Without prejudice' to 'England or Wales'.
1950 c. 27.	Arbitration Act 1950.	Part I. Section 42(3).
1958 c. 47.	Agricultural Marketing Act 1958.	Section 53(8).
1962 c. 46.	Transport Act 1962.	In Schedule 11, Part II, paragraph 7.
1964 c. 14.	Plant Varieties and Seeds Act 1964.	In section 10(4) the words from 'or in section 9' to 'three arbitrators)'. Section 39(3)(b)(i).
1964 c. 29 (N.I.).	Lands Tribunal and Compensation Act (Northern Ireland) 1964.	In section 9(3) the words from 'so, however, that' to the end.
1965 c. 12.	Industrial and Provident Societies Act 1965	In section 60(8)(b), the words 'by virtue of section 12 of the said Act of 1950'.
1965 c. 37.	Carriage of Goods by Road Act 1965.	Section 7(2)(b).
1965 c. 13 (N.I.).	New Towns Act (Northern Ireland) 1965.	In section 27(2), the words from 'under and in accordance with' to the end.
1969 c. 24 (N.I.).	Industrial and Provident Societies Act (Northern Ireland) 1969.	In section 69(7)– (a) in the opening words, the words from 'and without prejudice' to '1937'; (b) in paragraph (b), the words 'the registrar or' and 'registrar or'.
1970 c. 31.	Administration of Justice Act 1970.	Section 4. Schedule 3.
1973 c. 41.	Fair Trading Act 1973.	Section 33(2)(d).

Chapter	Short title	Extent of repeal
1973 N.I. 1.	Drainage (Northern Ireland) Order 1973.	In Article 15(4), the words from 'under and in accordance' to the end. Article 40(4). In Schedule 7, in paragraph 9(2), the words from 'under and in accordance' to the end.
1974 c. 47.	Solicitors Act 1974.	In section 87(1), in the definition of 'contentious business', the words 'appointed under the Arbitration Act 1950'.
1975 c. 3	Arbitration Act 1975.	The whole Act.
1975 c. 74.	Petroleum and Submarine Pipe-Lines Act 1975.	In Part II of Schedule 2– (a) in model clause 40(2), the words 'in accordance with the Arbitration Act 1950'; (b) in model clause 40(2B), the words 'in accordance with the Arbitration Act (Northern Ireland) 1937'. In Part II of Schedule 3, in model clause 38(2), the words 'in accordance with the Arbitration Act 1950'.
1976 N.I. 12.	Solicitors (Northern Ireland) Order 1976.	In Article 3(2), in the entry 'contentious business', the words 'appointed under the Arbitration Act (Northern Ireland) 1937'. Article 71H(3).
1977 c. 37.	Patents Act 1977.	In section 52(4) the words 'section 21 of the Arbitration Act 1950 or, as the case may be, section 22 of the Arbitration Act (Northern Ireland) 1937 (statement of cases by arbitrators); but'. Section 131(e).
1977 c. 38.	Administration of Justice Act 1977.	Section 17(2).
1978 c. 23.	Judicature (Northern Ireland) Act 1978.	In section 35(2), paragraph (g)(v). In Schedule 5, the amendment to the Arbitration Act 1950.
1979 c. 42.	Arbitration Act 1979.	The whole Act.
1980 c. 58.	Limitation Act 1980.	Section 34.
1980 N.I. 3.	County Courts (Northern Ireland) Order 1980.	Article 31(3).

Chapter	Short title	Extent of repeal
1981 c. 54.	Supreme Court Act 1981.	Section 148.
1982 c. 27.	Civil Jurisdiction and Judgments Act 1982.	Section 25(3)(c) and (5). In section 26– (a) in subsection (1), the words 'to arbitration or'; (b) in subsection (1)(a)(i), the words 'arbitration or'; (c) in subsection (2), the words 'arbitration or'.
1982 c. 53.	Administration of Justice Act 1982.	Section 15(6). In Schedule 1, Part IV.
1984 c. 5.	Merchant Shipping Act 1984.	Section 4(8).
1984 c. 12.	Telecommunications Act 1984.	Schedule 2, paragraph 13(8).
1984 c. 16.	Foreign Limitation Periods Act 1984.	Section 5.
1984 c. 28.	County Courts Act 1984.	In Schedule 2, paragraph 70.
1985 c. 61.	Administration of Justice Act 1985.	Section 58. In Schedule 9, paragraph 15.
1985 c. 68.	Housing Act 1985.	In Schedule 18, in paragraph 6(2) the words from 'and the Arbitration Act 1950' to the end.
1985 N.I. 12.	Credit Unions (Northern Ireland) Order 1985.	In Article 72(7)– (a) in the opening words, the words from 'and without prejudice' to '1937'; (b) in sub-paragraph (b), the words 'the registrar or' and 'registrar or'.
1986 c. 45.	Insolvency Act 1986.	In Schedule 14, the entry relating to the Arbitration Act 1950.
1988 c. 8.	Multilateral Investment Guarantee Agency Act 1988.	Section 8(3).
1988 c. 21.	Consumer Arbitration Agreements Act 1988.	The whole Act.
1989 N.I. 11.	Limitation (Northern Ireland) Order 1989.	Article 72. In Schedule 3, paragraph 1.
1989 N.I. 19.	Insolvency (Northern Ireland) Order 1989.	In Part II of Schedule 9, paragraph 66.
1990 c. 41.	Courts and Legal Services Act 1990.	Sections 99 and 101 to 103.
1991 N.I. 7.	Food Safety (Northern Ireland) Order 1991.	In Articles 8(8) and 11(10), the words from 'and the provisions' to the end.
1992 c. 40.	Friendly Societies Act 1992.	In Schedule 16, paragraph 30(1).
1995 c. 8.	Agricultural Tenancies Act 1995.	Section 28(4).

Chapter	Short title	Extent of repeal
1995 c. 21.	Merchant Shipping Act 1995.	Section 96(10). Section 264(9).
1995 c. 42.	Private International Law (Miscellaneous Provisions) Act 1995.	Section 3.

Part 5

Appendices

SI 1996 No. 3146 (C.96)

ARBITRATION

The Arbitration Act 1996 (Commencement No. 1) Order 1996

Made – – – – – *16th December 1996*

The Secretary of State, in exercise of the powers conferred on him by section 109 of the Arbitration Act 1996(a), hereby makes the following Order:

1. This Order may be cited as the Arbitration Act 1996 (Commencement No. 1) Order 1996.

2. The provisions of the Arbitration Act 1996 ('the Act') listed in Schedule 1 to this Order shall come into force on the day after this Order is made.

3. The rest of the Act, except sections 85 to 87, shall come into force on 31st January 1997.

4. The transitional provisions in Schedule 2 to this Order shall have effect.

John M. Taylor,
Parliamentary Under-Secretary of State
for Corporate and Consumer Affairs,
16th December 1996   Department of Trade and Industry

Schedule 1   Article 2.

Section 91 so far as it relates to the power to make orders under the section.
Section 105.
Section 107(1) and paragraph 36 of Schedule 3, so far as relating to the provision that may be made by county court rules.
Section 107(2) and the reference in Schedule 4 to the County Courts (Northern Ireland) Order 1980(a) so far as relating to the above matter.
Section 108 to 110.

Schedule 2   Article 4.

1. In this Schedule:

   (a) 'the appointed day' means the date specified in Article 3 of this Order;
   (b) 'arbitration application' means any application relating to arbitration made by or in legal proceedings, whether or not arbitral proceedings have commenced;
   (c) 'the old law' means the enactments specified in section 107 as they stood before their amendment or repeal by the Act.

2. The old law shall continue to apply to:

   (a) arbitral proceedings commenced before the appointed day;

(b)   arbitration applications commenced or made before the appointed day;

(c)   arbitration applications commenced or made on or after the appointed day relating to arbitral proceedings commenced before the appointed day

and the provisions of the Act which would otherwise be applicable shall not apply.

3.   The provisions of this Act brought into force by this Order shall apply to any other arbitration application.

4.   In the application of paragraph (b) of subsection (1) of section 46 (provision for dispute to be decided in accordance with provisions other than law) to an arbitration agreement made before the appointed day, the agreement shall have effect in accordance with the rules of law (including any conflict of laws rules) as they stood immediately before the appointed day.

Explanatory note

(This note is not part of the Regulations)

With one exception, this Order brings into force the provisions of the Arbitration Act 1996. Those provisions necessary to enable the substantive provisions to be brought into force are commenced immediately. The substantive provisions come into force on 31st January 1997. Commencement is subject to transitional provisions designed to ensure continuity of legal proceedings and to preserve the current law on what are known as 'honourable engagement' clauses in relation to existing agreements.

Sections 85 to 87, which make special provision in relation to domestic arbitration agreements, are not commenced.

The Unfair Terms in Consumer Contracts Regulations 1999 (SI 1999 No. 2083)

Citation and commencement

1. These Regulations may be cited as the Unfair Terms in Consumer Contracts Regulations 1999 and shall come into force on 1st October 1999.

Revocation

2. The Unfair Terms in Consumer Contracts Regulations 1994 are hereby revoked.

Interpretation

3.–(1) In these Regulations–

'the Community' means the European Community;

'consumer' means any natural person who, in contracts covered by these Regulations, is acting for purposes which are outside his trade, business or profession;

'court' in relation to England and Wales and Northern Ireland means a county court or the High Court, and in relation to Scotland, the Sheriff or the Court of Session;

'Director' means the Director General of Fair Trading;

'EEA Agreement' means the Agreement on the European Economic Area signed at Oporto on 2nd May 1992 as adjusted by the protocol signed at Brussels on 17th March 1993;

'Member State' means a State which is a contracting party to the EEA Agreement;

'notified' means notified in writing;

'qualifying body' means a person specified in Schedule 1;

'seller or supplier' means any natural or legal person who, in contracts covered by these Regulations, is acting for purposes relating to his trade, business or profession, whether publicly owned or privately owned;

'unfair terms' means the contractual terms referred to in regulation 5.

(2) In the application of these Regulations to Scotland for references to an 'injunction' or an 'interim injunction' there shall be substituted references to an 'interdict' or 'interim interdict' respectively.

Terms to which these Regulations apply

4.–(1) These Regulations apply in relation to unfair terms in contracts concluded between a seller or a supplier and a consumer.

(2) These Regulations do not apply to contractual terms which reflect–

(a) mandatory statutory or regulatory provisions (including such provisions under the law of any Member State or in Community legislation having effect in the United Kingdom without further enactment);

(b) the provisions or principles of international conventions to which the Member States or the Community are party.

Unfair terms

5.–(1) A contractual term which has not been individually negotiated shall be regarded as unfair if, contrary to the requirement of good faith, it causes a significant imbalance in the parties' rights and obligations arising under the contract, to the detriment of the consumer.

(2) A term shall always be regarded as not having been individually negotiated where it has been drafted in advance and the consumer has therefore not been able to influence the substance of the term.

(3) Notwithstanding that a specific term or certain aspects of it in a contract has been individually negotiated, these Regulations shall apply to the rest of a contract if an overall assessment of it indicates that it is a pre-formulated standard contract.

(4) It shall be for any seller or supplier who claims that a term was individually negotiated to show that it was.

(5) Schedule 2 to these Regulations contains an indicative and non-exhaustive list of the terms which may be regarded as unfair.

Assessment of unfair terms

6.–(1) Without prejudice to regulation 12, the unfairness of a contractual term shall be assessed, taking into account the nature of the goods or services for which the contract was concluded and by referring, at the time of conclusion of the contract, to all the circumstances attending the conclusion of the contract and to all the other terms of the contract or of another contract on which it is dependent.

(2) In so far as it is in plain intelligible language, the assessment of fairness of a term shall not relate–

(a) to the definition of the main subject matter of the contract, or
(b) to the adequacy of the price or remuneration, as against the goods or services supplied in exchange.

Written contracts

7.–(1) A seller or supplier shall ensure that any written term of a contract is expressed in plain, intelligible language.

(2) If there is doubt about the meaning of a written term, the interpretation which is most favourable to the consumer shall prevail but this rule shall not apply in proceedings brought under regulation 12.

Effect of unfair term

8.–(1) An unfair term in a contract concluded with a consumer by a seller or supplier shall not be binding on the consumer.

(2) The contract shall continue to bind the parties if it is capable of continuing in existence without the unfair term.

Choice of law clauses

9. These Regulations shall apply notwithstanding any contract term which applies or purports to apply the law of a non-Member State, if the contract has a close connection with the territory of the Member States.

Complaints – consideration by Director

10.-(1) It shall be the duty of the Director to consider any complaint made to him that any contract term drawn up for general use is unfair, unless–

(a)   the complaint appears to the Director to be frivolous or vexatious; or
(b)   a qualifying body has notified the Director that it agrees to consider the complaint.

(2) The Director shall give reasons for his decision to apply or not to apply, as the case may be, for an injunction under regulation 12 in relation to any complaint which these Regulations require him to consider.

(3) In deciding whether or not to apply for an injunction in respect of a term which the Director considers to be unfair, he may, if he considers it appropriate to do so, have regard to any undertakings given to him by or on behalf of any person as to the continued use of such a term in contracts concluded with consumers.

Complaints – consideration by qualifying bodies

11.-(1) If a qualifying body specified in Part One of Schedule 1 notifies the Director that it agrees to consider a complaint that any contract term drawn up for general use is unfair, it shall be under a duty to consider that complaint.

(2) Regulation 10(2) and (3) shall apply to a qualifying body which is under a duty to consider a complaint as they apply to the Director.

Injunctions to prevent continued use of unfair terms

12.-(1) The Director or, subject to paragraph (2), any qualifying body may apply for an injunction (including an interim injunction) against any person appearing to the Director or that body to be using, or recommending use of, an unfair term drawn up for general use in contracts concluded with consumers.

(2) A qualifying body may apply for an injunction only where–

(a)   it has notified the Director of its intention to apply at least fourteen days before the date on which the application is made, beginning with the date on which the notification was given; or
(b)   the Director consents to the application being made within a shorter period.

(3) The court on an application under this regulation may grant an injunction on such terms as it thinks fit.

(4) An injunction may relate not only to use of a particular contract term drawn up for general use but to any similar term, or a term having like effect, used or recommended for use by any person.

Powers of the Director and qualifying bodies to obtain documents and information

13.-(1) The Director may exercise the power conferred by this regulation for the purpose of–

(a)   facilitating his consideration of a complaint that a contract term drawn up for general use is unfair; or
(b)   ascertaining whether a person has complied with an undertaking or court order as to the continued use, or recommendation for use, of a term in contracts concluded with consumers.

(2) A qualifying body specified in Part One of Schedule 1 may exercise the power conferred by this regulation for the purpose of–

(a) facilitating its consideration of a complaint that a contract term drawn up for general use is unfair; or
(b) ascertaining whether a person has complied with–

> (i) an undertaking given to it or to the court following an application by that body, or
> (ii) a court order made on an application by that body,

> as to the continued use, or recommendation for use, of a term in contracts concluded with consumers.

(3) The Director may require any person to supply to him, and a qualifying body specified in Part One of Schedule 1 may require any person to supply to it–

(a) a copy of any document which that person has used or recommended for use, at the time the notice referred to in paragraph (4) below is given, as a pre-formulated standard contract in dealings with consumers;
(b) information about the use, or recommendation for use, by that person of that document or any other such document in dealings with consumers.

(4) The power conferred by this regulation is to be exercised by a notice in writing which may–

(a) specify the way in which and the time within which it is to be complied with; and
(b) be varied or revoked by a subsequent notice.

(5) Nothing in this regulation compels a person to supply any document or information which he would be entitled to refuse to produce or give in civil proceedings before the court.

(6) If a person makes default in complying with a notice under this regulation, the court may, on the application of the Director or of the qualifying body, make such order as the court thinks fit for requiring the default to be made good, and any such order may provide that all the costs or expenses of and incidental to the application shall be borne by the person in default or by any officers of a company or other association who are responsible for its default.

Notification of undertakings and orders to Director
14. A qualifying body shall notify the Director–

(a) of any undertaking given to it by or on behalf of any person as to the continued use of a term which that body considers to be unfair in contracts concluded with consumers;
(b) of the outcome of any application made by it under regulation 12, and of the terms of any undertaking given to, or order made by, the court;
(c) of the outcome of any application made by it to enforce a previous order of the court.

Publication, information and advice
15.–(1) The Director shall arrange for the publication in such form and manner as he considers appropriate, of–

(a) details of any undertaking or order notified to him under regulation 14;
(b) details of any undertaking given to him by or on behalf of any person as to

the continued use of a term which the Director considers to be unfair in contracts concluded with consumers;

(c)   details of any application made by him under regulation 12, and of the terms of any undertaking given to, or order made by, the court;

(d)   details of any application made by the Director to enforce a previous order of the court.

(2) The Director shall inform any person on request whether a particular term to which these Regulations apply has been–

(a)   the subject of an undertaking given to the Director or notified to him by a qualifying body; or

(b)   the subject of an order of the court made upon application by him or notified to him by a qualifying body;

and shall give that person details of the undertaking or a copy of the order, as the case may be, together with a copy of any amendments which the person giving the undertaking has agreed to make to the term in question.

(3) The Director may arrange for the dissemination in such form and manner as he considers appropriate of such information and advice concerning the operation of these Regulations as may appear to him to be expedient to give to the public and to all persons likely to be affected by these Regulations.

Schedule 1   Regulation 3
Qualifying bodies
Part one

1.   The Data Protection Registrar.

2.   The Director General of Electricity Supply.

3.   The Director General of Gas Supply.

4.   The Director General of Electricity Supply for Northern Ireland.

5.   The Director General of Gas for Northern Ireland.

6.   The Director General of Telecommunications.

7.   The Director General of Water Services.

8.   The Rail Regulator.

9.   Every weights and measures authority in Great Britain.

10.   The Department of Economic Development in Northern Ireland.

Part two

11.   Consumers' Association

Schedule 2
Regulation 5(5)
Indicative and non-exhaustive list of terms which may be regarded as unfair

1. Terms which have the object or effect of–

(a) excluding or limiting the legal liability of a seller or supplier in the event of the death of a consumer or personal injury to the latter resulting from an act or omission of that seller or supplier;

(b) inappropriately excluding or limiting the legal rights of the consumer vis-à-vis the seller or supplier or another party in the event of total or partial non-performance or inadequate performance by the seller or supplier of any of the contractual obligations, including the option of offsetting a debt owed to the seller or supplier against any claim which the consumer may have against him;

(c) making an agreement binding on the consumer whereas provision of services by the seller or supplier is subject to a condition whose realisation depends on his own will alone;

(d) permitting the seller or supplier to retain sums paid by the consumer where the latter decides not to conclude or perform the contract, without providing for the consumer to receive compensation of an equivalent amount from the seller or supplier where the latter is the party cancelling the contract;

(e) requiring any consumer who fails to fulfil his obligation to pay a disproportionately high sum in compensation;

(f) authorising the seller or supplier to dissolve the contract on a discretionary basis where the same facility is not granted to the consumer, or permitting the seller or supplier to retain the sums paid for services not yet supplied by him where it is the seller or supplier himself who dissolves the contract;

(g) enabling the seller or supplier to terminate a contract of indeterminate duration without reasonable notice except where there are serious grounds for doing so;

(h) automatically extending a contract of fixed duration where the consumer does not indicate otherwise, when the deadline fixed for the consumer to express his desire not to extend the contract is unreasonably early;

(i) irrevocably binding the consumer to terms with which he had no real opportunity of becoming acquainted before the conclusion of the contract;

(j) enabling the seller or supplier to alter the terms of the contract unilaterally without a valid reason which is specified in the contract;

(k) enabling the seller or supplier to alter unilaterally without a valid reason any characteristics of the product or service to be provided;

(l) providing for the price of goods to be determined at the time of delivery or allowing a seller of goods or supplier of services to increase their price without in both cases giving the consumer the corresponding right to cancel the contract if the final price is too high in relation to the price agreed when the contract was concluded;

(m) giving the seller or supplier the right to determine whether the goods or services supplied are in conformity with the contract, or giving him the exclusive right to interpret any term of the contract;

(n) limiting the seller's or supplier's obligation to respect commitments undertaken by his agents or making his commitments subject to compliance with a particular formality;

(o) obliging the consumer to fulfil all his obligations where the seller or supplier does not perform his;

(p) giving the seller or supplier the possibility of transferring his rights and obligations under the contract, where this may serve to reduce the guarantees for the consumer, without the latter's agreement;

(q) excluding or hindering the consumer's right to take legal action or exercise any other legal remedy, particularly by requiring the consumer to take disputes exclusively to arbitration not covered by legal provisions, unduly restricting the evidence available to him or imposing on him a burden of proof which, according to the applicable law, should lie with another party to the contract.

2. Scope of paragraphs 1(g), (j) and (l)

(a) Paragraph 1(g) is without hindrance to terms by which a supplier of financial services reserves the right to terminate unilaterally a contract of indeterminate duration without notice where there is a valid reason, provided that the supplier is required to inform the other contracting party or parties thereof immediately.

(b) Paragraph 1(j) is without hindrance to terms under which a supplier of financial services reserves the right to alter the rate of interest payable by the consumer or due to the latter, or the amount of other charges for financial services without notice where there is a valid reason, provided that the supplier is required to inform the other contracting party or parties thereof at the earliest opportunity and that the latter are free to dissolve the contract immediately.

Paragraph 1(j) is also without hindrance to terms under which a seller or supplier reserves the right to alter unilaterally the conditions of a contract of indeterminate duration, provided that he is required to inform the consumer with reasonable notice and that the consumer is free to dissolve the contract.

(c) Paragraphs 1(g), (j) and (l) do not apply to:
  - transactions in transferable securities, financial instruments and other products or services where the price is linked to fluctuations in a stock exchange quotation or index or a financial market rate that the seller or supplier does not control;
  - contracts for the purchase or sale of foreign currency, traveller's cheques or international money orders denominated in foreign currency;

(d) Paragraph 1(l) is without hindrance to price indexation clauses, where lawful, provided that the method by which prices vary is explicitly described.

Explanatory note
(This note is not part of the Regulations)

These Regulations revoke and replace the Unfair Terms in Consumer Contracts Regulations 1994 (SI 1994/3159) which came into force on 1st July 1995.

Those Regulations implemented Council Directive 93/13/EEC on unfair terms in consumer contracts (O.J. No. L95, 21.4.93, p.29). Regulations 3 to 9 of these Regulations re-enact Regulations 2 to 7 of the 1994 Regulations with modifications to reflect more closely the wording of the Directive.

The Regulations apply, with certain exceptions, to unfair terms in contracts concluded between a consumer and a seller or supplier (regulation 4). The Regulations provide that an unfair term is one which has not been individually negotiated and which, contrary to the requirement of good faith, causes a significant imbalance in the parties' rights and obligations under the contract to the detriment of the consumer (regulation 5). Schedule 2 contains an indicative list of terms which may be regarded as unfair.

The assessment of unfairness will take into account all the circumstances attending the conclusion of the contract. However, the assessment is not to relate to the definition of the main subject matter of the contract or the adequacy of the price or remuneration as against the goods or services supplied in exchange as long as the terms concerned are in plain, intelligible language (regulation 6). Unfair contract terms are not binding on the consumer (regulation 8).

The Regulations maintain the obligation on the Director General of Fair Trading (contained in the 1994 Regulations) to consider any complaint made to him about the fairness of any contract term drawn up for general use. He may, if he considers it appropriate to do so, seek an injunction to prevent the continued use of that term or of a term having like effect (regulations 10 and 12).

The Regulations provide for the first time that a qualifying body named in Schedule 1 (statutory regulators, trading standards departments and Consumers' Association) may also apply for an injunction to prevent the continued use of an unfair contract term provided it has notified the Director General of its intention at least 14 days before the application is made (unless the Director General consents to a shorter period) (regulation 12). A qualifying body named in Part One of Schedule 1 (public bodies) shall be under a duty to consider a complaint if it has told the Director General that it will do so (regulation 11).

The Regulations provide a new power for the Director General and the public qualifying bodies to require traders to produce copies of their standard contracts, and give information about their use, in order to facilitate investigation of complaints and ensure compliance with undertakings or court orders (regulation 13).

Qualifying bodies must notify the Director General of undertakings given to them about the continued use of an unfair term and of the outcome of any court proceedings (regulation 14). The Director General is given the power to arrange for the publication of this information in such form and manner as he considers appropriate and to offer information and advice about the operation of these Regulations (regulation 15). In addition the Director General will supply enquirers about particular standard terms with details of any relevant undertakings and court orders.

A Regulatory Impact Assessment of the costs and benefits which will result from these Regulations has been prepared by the Department of Trade and Industry and is available from Consumer Affairs Directorate, Department of Trade and Industry, Room 407, 1 Victoria Street, London SW1H 0ET (Telephone 0171 215 0341). Copies have been placed in the libraries of both Houses of Parliament.

The Arbitration Act 1950, Part II

Part II
Enforcement of certain foreign awards

Awards to which Part II applies

35.-(1) This Part of this Act applies to any award made after the twenty-eighth day of July, nineteen hundred and twenty-four–

(a) in pursuance of an agreement for arbitration to which the protocol set out in the First Schedule to this Act applies; and
(b) between persons of whom one is subject to the jurisdiction of some one of such Powers as His Majesty, being satisfied that reciprocal provisions have been made, may by Order in Council declare to be parties to the convention set out in the Second Schedule to this Act, and of whom the other is subject to the jurisdiction of some other of the Powers aforesaid; and
(c) in one of such territories as His Majesty, being satisfied that reciprocal provisions have been made, may by Order in Council declare to be territories to which the said convention applies;

and an award to which this Part of this Act applies is in this Part of this Act referred to as 'a foreign award.'

(2) His Majesty may by a subsequent Order in Council vary or revoke any Order previously made under this section.

(3) Any Order in Council under section one of the Arbitration (Foreign Awards) Act 1930, which is in force at the commencement of this Act shall have effect as if it had been made under this section.

Effect of foreign awards

36.-(1) A foreign award shall, subject to the provisions of this Part of this Act, be enforceable in England either by action or in the same manner as the award of an arbitrator is enforceable by virtue of [section 66 of the Arbitration Act 1996].

(2) Any foreign award which would be enforceable under this Part of this Act shall be treated as binding for all purposes on the persons as between whom it was made, and may accordingly be relied on by any of those persons by way of defence, set off or otherwise in any legal proceedings in England, and any references in this Part of this Act to enforcing a foreign award shall be construed as including references to relying on an award.

Conditions for enforcement of foreign awards

37.-(1) In order that a foreign award may be enforceable under this Part of this Act it must have–

(a) been made in pursuance of an agreement for arbitration which was valid under the law by which it was governed;
(b) been made by the tribunal provided for in the agreement or constituted in manner agreed upon by the parties;
(c) been made in conformity with the law governing the arbitration procedure;

(d) become final in the country in which it was made;
(e) been in respect of a matter which may lawfully be referred to arbitration under the law of England;

and the enforcement thereof must not be contrary to the public policy or the law of England.

(2) Subject to the provisions of this subsection, a foreign award shall not be enforceable under this Part of this Act if the court dealing with the case is satisfied that–

(a) the award has been annulled in the country in which it was made; or
(b) the party against whom it is sought to enforce the award was not given notice of the arbitration proceedings in sufficient time to enable him to present his case, or was under some legal incapacity and was not properly represented; or
(c) the award does not deal with all the questions referred or contains decisions on matters beyond the scope of the agreement for arbitration:

Provided that, if the award does not deal with all the questions referred, the court may, if it thinks fit, either postpone the enforcement of the award or order its enforcement subject to the giving of such security by the person seeking to enforce it as the court may think fit.

(3) If a party seeking to resist the enforcement of a foreign award proves that there is any ground other than the non-existence of the conditions specified in paragraphs (a), (b) and (c) of subsection (1) of this section, or the existence of the conditions specified in paragraphs (b) and (c) of subsection (2) of this section, entitling him to contest the validity of the award, the court may, if it thinks fit, either refuse to enforce the award or adjourn the hearing until after the expiration of such period as appears to the court to be reasonably sufficient to enable that party to take the necessary steps to have the award annulled by the competent tribunal.

Evidence
38.–(1) The party seeking to enforce a foreign award must produce–

(a) the original award or a copy thereof duly authenticated in manner required by the law of the country in which it was made; and
(b) evidence proving that the award has become final; and
(c) such evidence as may be necessary to prove that the award is a foreign award and that the conditions mentioned in paragraphs (a), (b) and (c) of subsection (1) of the last foregoing section are satisfied.

(2) In any case where any document required to be produced under subsection (1) of this section is in a foreign language, it shall be the duty of the party seeking to enforce the award to produce a translation certified as correct by a diplomatic or consular agent of the country to which that party belongs, or certified as correct in such other manner as may be sufficient according to the law of England.

(3) Subject to the provisions of this section, rules of court may be made under section [84 of the Supreme Court Act 1981], with respect to the evidence which must be furnished by a party seeking to enforce an award under this Part of this Act.

Meaning of 'final award'
39. For the purposes of this Part of this Act, an award shall not be deemed

final if any proceedings for the purpose of contesting the validity of the award are pending in the country in which it was made.

Saving for other rights, etc.

40. Nothing in this Part of this Act shall–

(a) prejudice any rights which any person would have had of enforcing in England any award or of availing himself in England of any award if neither this Part of this Act nor Part I of the Arbitration (Foreign Awards) Act 1930, had been enacted; or

(b) apply to any award made on an arbitration agreement governed by the law of England.

Application of Part II to Scotland

41.–(1) The following provisions of this section shall have effect for the purpose of the application of this Part of this Act to Scotland.

(2) For the references to England there shall be substituted references to Scotland.

(3) For subsection (1) of section thirty-six there shall be substituted the following subsection:–

'(1) A foreign award shall, subject to the provisions of this Part of this Act, be enforceable by action, or, if the agreement for arbitration contains consent to the registration of the award in the Books of Council and Session for execution and the award is so registered, it shall, subject as aforesaid, be enforceable by summary diligence.'

(4) For subsection (3) of section thirty-eight there shall be substituted the following subsection:–

'(3) The Court of Session shall, subject to the provisions of this section, have power, exercisable by statutory instrument, to make provision by Act of Sederunt with respect the evidence which must be furnished by a party seeking to enforce in Scotland an award under this Part of this Act.'

Application of Part II to Northern Ireland

42.–(1) The following provisions of this section shall have effect for the purpose of the application of this Part of this Act to Northern Ireland.

(2) For the references to England there shall be substituted references to Northern Ireland.

(3) [Subsection repealed.]

(4) [Subsection repealed.]

'(3) Subject to the provisions of this section, rules may be made under section 7 of the Northern Ireland Act 1962, with respect to the evidence which must be furnished by a party seeking to enforce an award under this Part of this Act.'

43. [Section repealed.]

Part III
General

Short title, commencement and repeal

44.–(1) This Act may be cited as the Arbitration Act, 1950.

(2) This Act shall come into operation on the first day of September, nineteen hundred and fifty.

(3) The Arbitration Act 1889, the Arbitration Clauses (Protocol) Act 1924 and the Arbitration Act 1934 are hereby repealed except in relation to arbitrations commenced (within the meaning of subsection (2) of section twenty-nine of this Act) before the commencement of this Act, and the Arbitration (Foreign Awards) Act 1930 is hereby repealed; and any reference in any Act or other document to any enactment hereby repealed shall be construed as including a reference to the corresponding provision of this Act.

First schedule

Section 35

Protocol on Arbitration Clauses signed on behalf of His Majesty at a Meeting of the Assembly of the League of Nations held on the twenty-fourth day of September, nineteen hundred and twenty-three

The undersigned, being duly authorised, declare that they accept, on behalf of the countries which they represent, the following provisions:–

1. Each of the Contracting States recognises the validity of an agreement whether relating to existing or future differences between parties, subject respectively to the jurisdiction of different Contracting States by which the parties to a contract agree to submit to arbitration all or any differences that may arise in connection with such contract relating to commercial matters or to any other matter capable of settlement by arbitration, whether or not the arbitration is to take place in a country to whose jurisdiction none of the parties is subject.

Each Contracting State reserves the right to limit the obligation mentioned above to contracts which are considered as commercial under its national law. Any Contracting State which avails itself of this right will notify the Secretary-General of the League of Nations, in order that the other Contracting States may be so informed.

2. The arbitral procedure, including the constitution of the arbitral tribunal, shall be governed by the will of the parties and by the law of the country in whose territory the arbitration takes place.

The Contracting States agree to facilitate all steps in the procedure which require to be taken in their own territories, in accordance with the provisions of their law governing arbitral procedure applicable to existing differences.

3. Each Contracting State undertakes to ensure the execution by its authorities and in accordance with the provisions of its national laws of arbitral awards made in its own territory under the preceding articles.

4. The tribunals of the Contracting Parties, on being seised of a dispute regarding a contract made between persons to whom Article 1 applies and including an arbitration agreement whether referring to present or future differences which is valid in virtue of the said article and capable of being carried into effect, shall refer the parties on the application of either of them to the decision of the arbitrators.

Such reference shall not prejudice the competence of the judicial tribunals in case the agreement or the arbitration cannot proceed or become inoperative.

5. The present Protocol, which shall remain open for signature by all States, shall be ratified. The ratifications shall be deposited as soon as possible with the Secretary-General of the League of Nations, who shall notify such deposit to all the signatory States.

6. The present Protocol shall come into force as soon as two ratifications have been deposited. Thereafter it will take effect, in the case of each Contracting State, one month after the notification by the Secretary-General of the deposit of its ratification.

7. The present Protocol may be denounced by any Contracting State on giving one year's notice. Denunciation shall be effected by a notification addressed to the Secretary-General of the League, who will immediately transmit copies of such notification to all the other signatory States and inform them of the date on which it was received. The denunciation shall take effect one year after the date on which it was notified to the Secretary-General, and shall operate only in respect of the notifying State.

8. The Contracting States may declare that their acceptance of the present Protocol does not include any or all of the under-mentioned territories: that is to say, their colonies, overseas possessions or territories, protectorates or the territories over which they exercise a mandate.

The said States may subsequently adhere separately on behalf of any territory thus excluded. The Secretary-General of the League of Nations shall be informed as soon as possible of such adhesions. He shall notify such adhesions to all signatory States. They will take effect one month after the notification by the Secretary-General to all signatory States.

The Contracting States may also denounce the Protocol separately on behalf of any of the territories referred to above. Article 7 applies to such denunciation.

Second schedule

Section 35

Convention on the Execution of Foreign Arbitral Awards signed at Geneva on behalf of His Majesty on the twenty-sixth day of September, nineteen hundred and twenty-seven

Article 1
In the territories of any High Contracting Party to which the present Convention applies, an arbitral award made in pursuance of an agreement, whether relating to existing or future differences (hereinafter called 'a submission to arbitration') covered by the Protocol on Arbitration Clauses, opened at Geneva on September 24, 1923, shall be recognised as binding and shall be enforced in accordance with the rules of the procedure of the territory where the award is relied upon, provided that the said award has been made in a territory of one of the High Contracting Parties to which the present Convention applies and between persons who are subject to the jurisdiction of one of the High Contracting Parties.

To obtain such recognition or enforcement, it shall, further, be necessary:–

(a) That the award has been made in pursuance of a submission to arbitration which is valid under the law applicable thereto;

(b) That the subject-matter of the award is capable of settlement by arbitration under the law of the country in which the award is sought to be relied upon;

(c) That the award has been made by the Arbitral Tribunal provided for in the submission to arbitration or constituted in the manner agreed upon by the parties and in conformity with the law governing the arbitration procedure;

(d) That the award has become final in the country in which it has been made, in the sense that it will not be considered as such if it is open to *opposition*, *appel* or *pourvoi en cassation* (in the countries where such forms of procedure exist) or if it is proved that any proceedings for the purpose of contesting the validity of the award are pending;

(e) That the recognition or enforcement of the award is not contrary to the public policy or to the principles of the law of the country in which it is sought to be relied upon.

Article 2

Even if the conditions laid down in Article 1 hereof are fulfilled, recognition and enforcement of the award shall be refused if the Court is satisfied:–

(a) That the award has been annulled in the country in which it was made;

(b) That the party against whom it is sought to use the award was not given notice of the arbitration proceedings in sufficient time to enable him to present his case; or, that, being under a legal incapacity, he was not properly represented;

(c) That the award does not deal with the differences contemplated by or falling within the terms of the submission to arbitration or that it contains decisions on matters beyond the scope of the submission to arbitration.

If the award has not covered all the questions submitted to the arbitral tribunal, the competent authority of the country where recognition or enforcement of the award is sought can, if it think fit, postpone such recognition or enforcement or grant it subject to such guarantee as that authority may decide.

Article 3

If the party against whom the award has been made proves that, under the law governing the arbitration procedure, there is a ground, other than the grounds referred to in Article 1(a) and (c) and Article 2(b) and (c) entitling him to contest the validity of the award in a Court of Law, the Court may, if it thinks fit, either refuse recognition or enforcement of the award or adjourn the consideration thereof, giving such party a reasonable time within which to have the award annulled by the competent tribunal.

Article 4

The party relying upon an award or claiming its enforcement must supply, in particular:–

(1) The original award or a copy thereof duly authenticated, according to the requirements of the law of the country in which it was made;

(2) Documentary or other evidence to prove that the award has become final, in the sense defined in Article 1(d), in the country in which it was made;

(3) When necessary, documentary or other evidence to prove that the conditions laid down in Article 1, paragraph 1 and paragraph 2(a) and (c) have been fulfilled.

A translation of the award and of other documents mentioned in this Article into the official language of the country where the award is sought to be relied upon may be demanded. Such translation must be certified correct by a diplomatic or consular agent of the country to which the party who seeks to rely upon the award belongs or by a sworn translator of the country where the award is sought to be relied upon.

Article 5

The provisions of the above Articles shall not deprive any interested party of the right of availing himself of an arbitral award in the manner and to the extent allowed by the law or the treaties of the country where such award is sought to be relied upon.

Article 6

The present Convention applies only to arbitral awards made after the coming into force of the Protocol on Arbitration Clauses, opened at Geneva on September 24, 1923.

Article 7

The present Convention, which will remain open to the signature of all the signatories of the Protocol of 1923 on Arbitration Clauses, shall be ratified.

It may be ratified only on behalf of those Members of the League of Nations and non-Member States on whose behalf the Protocol of 1923 shall have been ratified.

Ratifications shall be deposited as soon as possible with the Secretary-General of the League of Nations, who will notify such deposit to all the signatories.

Article 8

The present Convention shall come into force three months after it shall have been ratified on behalf of two High Contracting Parties. Thereafter, it shall take effect, in the case of each High Contracting Party, three months after the deposit of the ratification on its behalf with the Secretary-General of the League of Nations.

Article 9

The present Convention may be denounced on behalf of any Member of the League or non-Member State. Denunciation shall be notified in writing to the Secretary-General of the League of Nations, who will immediately send a copy thereof, certified to be in conformity with the notification, to all the other Contracting Parties, at the same time informing them of the date on which he received it.

The denunciation shall come into force only in respect of the High Contracting Party which shall have notified it and one year after such notification shall have reached the Secretary-General of the League of Nations.

The denunciation of the Protocol on Arbitration Clauses shall entail ipso facto the denunciation of the present Convention.

Article 10

The present Convention does not apply to the Colonies, Protectorates or territories under suzerainty or mandate of any High Contracting Party, unless they are specially mentioned.

The application of this Convention to one or more of such Colonies, Protectorates or territories to which the Protocol on Arbitration Clauses, opened at Geneva on September 24, 1923, applies, can be effected at any time by means of a declaration addressed to the Secretary-General of the League of Nations by one of the High Contracting Parties.

Such declaration shall take effect three months after the deposit thereof.

The High Contracting Parties can at any time denounce the Convention for all or any of the Colonies, Protectorates or territories referred to above. Article 9 hereof applies to such denunciation.

Article 11

A certified copy of the present Convention shall be transmitted by the Secretary-General of the League of Nations to every Member of the League of Nations and to every non-Member State which signs the same.

Table of Cases

Index to Commentary

(N.B. A few references are to section numbers only. Most are to specific paragraph numbers. In all cases the appropriate section and related paragraphs should also be consulted as they may contain relevant material. No references to the Introduction, the Tables, the Materials, institutional Rules or the Model Law are included in this index. For the Civil Procedure Rules (CPR) and arbitration applications see our General Note in Part 3, Materials, Section H and the commentary on individual sections where relevant. Institutional Rules are dealt with where relevant, towards the end of the commentary on individual sections.)